SHARING THE HARVEST

NORTHEAST REGION SARE

Sustainable Agriculture Research and Education Program

The United States Department of Agriculture (USDA) Sustainable Agriculture Research and Education (SARE) Program works to increase knowledge about—and help farmers adopt—practices that are economically viable, that are environmentally sound, and that enhance the quality of life for farmers, rural communities, and society as a whole. Authorized in the 1985 Farm Bill, SARE began funding competitive grants in 1988. Since then, it has funded more than twelve hundred projects aimed at improving the sustainability of farming. Through newsletters, books, Web sites, and other resources, SARE helps to spread information about innovative, practical, profitable, and ecological approaches to agriculture.

This book—the result of partnerships with authors Elizabeth Henderson and Robyn Van En and publisher Chelsea Green—reflects Northeast SARE's commitment to supporting food production, processing, and marketing systems that sustain rural communities and support agricultural activities in the region's rural, urban, and suburban areas.

Farmers throughout the country are under tremendous pressures—from suburban development driving up the costs of farmland to high prices for many resources they need to low prices for their products. In addition, the food we eat is often grown far away from the table—the typical produce is said to have traveled well over one thousand miles—and less-fresh and lower-quality fruits and vegetables are often the result. The tremendous interest in Community Supported Agriculture (CSA) farms among farmers as well as the public is an indication that this approach has struck a chord with people who want high-quality fresh produce and also want to have a more direct connection with that most basic of human needs—food.

For more information, please contact SARE at Hills Building, University of Vermont, Burlington, VT 05405-0082. Phone: 802-656-0471; Fax: 802-656-4656; E-mail: nesare@zoo.uvm.edu. Web site: http://www.uvm.edu/~nesare/

SHARING THE HARVEST

A CITIZEN'S GUIDE TO
Community Supported Agriculture

Revised and Expanded Edition

ELIZABETH HENDERSON
with
ROBYN VAN EN

Foreword by
JOAN DYE GUSSOW

CHELSEA GREEN PUBLISHING COMPANY
WHITE RIVER JUNCTION, VERMONT

To Robyn, and to our sons, David and Andy, the next generation, who make it worthwhile to struggle for a peaceful and sustainable world.

Copyright 1999, 2007 by Elizabeth Henderson and Robyn Van En
Illustrations copyright 1999 by Karen Kerney
All rights reserved. No part of this book may be transmitted in any form by any means without permission in writing from the publisher.

Developmental editor: Ben Watson
Copy editor: Janine Stanley-Dunham
Proofreader: Susan Barnett
Indexer: Peggy Holloway
Designer: Peter Holm, Sterling Hill Productions

Printed in the United States of America
Revised edition, November, 2007
10 9 8 7 6 5 4 3 2 1

Library of Congress Cataloging-in-Publication Data
Henderson, Elizabeth, 1943-
  Sharing the harvest : a citizen's guide to Community Supported Agriculture / Elizabeth Henderson with Robyn Van En ; foreword by Joan Dye Gussow. --
Rev. and expanded ed.
     p. cm.
  Includes bibliographical references and index.
  ISBN 978-1-933392-10-3
  1. Collective farms--United States. 2.Community gardens--United States. 3. Agriculture, Cooperative--United States.
4. Farm produce--United States--Marketing. 5. Agriculture--Economic aspects--United States. I. Van En, Robyn. II. Title.

  HD1492.U6H46 2007
  334'.6830973--dc22

                          2007021963

Our Commitment to Green Publishing
Chelsea Green sees publishing as a tool for cultural change and ecological stewardship. We strive to align our book manufacturing practices with our editorial mission and to reduce the impact of our business enterprise in the environment. We print our books and catalogs on chlorine-free recycled paper, using soy-based inks whenever possible. This book may cost slightly more because we use recycled paper, and we hope you'll agree that it's worth it. Chelsea Green is a member of the Green Press Initiative (www.greenpressinitiative.org), a nonprofit coalition of publishers, manufacturers, and authors working to protect the world's endangered forests and conserve natural resources.
  Sharing the Harvest was printed on 60# Joy White, a 30-percent postconsumer-waste old-growth-forest-free recycled paper supplied by Thomson-Shore, Inc.

Chelsea Green Publishing Company
Post Office Box 428
White River Junction, VT 05001
(802) 295-6300
www.chelseagreen.com

CONTENTS

FOREWORD

Across this country, a movement is spreading that acknowledges a long-ignored reality: Most of what we pay for our food goes to companies that transport, process, and market what comes off the farm, not to farmers themselves. The people who actually grow food don't get paid enough to keep on doing it. And so, from Maine to California, some farmers are being supported by voluntary communities of eaters organized to pay growers directly for what they produce. Bypassing the supermarket, the middlemen, and the international transportation system, these folks are getting fresh local produce in season, at reasonable prices. This is a book about eater communities who are buying what their local farmers grow, and this system is called—appropriately—Community Supported Agriculture.

If this is the first time you've heard of Community Supported Agriculture (CSA), you should turn immediately to the first chapter. There you can discover why various farmers and eaters are joining these groups, and learn what some of them experience as they strive for wholesome food and financial security. If you have heard of CSAs but know little about them, then you need to study this recounting of the many different ways in which producers and consumers have come together to find mutually satisfactory solutions to their financial, social, and culinary needs. And finally, if you are an active member of a CSA farm, you surely need to sit down and spend some time with this book written by Liz Henderson, farmer extraordinaire, and Robyn Van En, the late, great lady of the CSA movement. From them you can learn how to solve the problems that will crop up (if they have not already done so) as you and your co-sharers try to find ways of making good food available outside the dominant marketplace, for the long-term benefit of everyone.

I was two-thirds of the way through the manuscript of this book's first edition when I went to a meeting of core group members of several New York City CSAs. (Yes, there were—and still are—people in the heart of the metropolis getting food directly from farmers.) As I listened to these pioneers seeking help from one another in resolving emerging dilemmas, I kept wishing they had this book. Under the major headings "Getting Started," "Getting Organized," "The Food," and "Many Models," they would have found ideas on everything from "How to Choose a Farmer" and "Acquiring Land" through "Farmer Earnings" and "Start-up Expenses" to "Distributing the Harvest," "Regional Networking," and "Including Low-Income Members."

In listing these topics, I realize that I risk making this book sound like a dull instruction manual or a guidebook to CSAs. It is a guide, and it is instructive, but it is far more than either of those, and it is surely not dull. It is, in fact, a delight to read, since the woman who finally brought it together, after the original author succumbed to an unacceptably early death, is a strong and gifted writer. She recognizes other good writers when she reads them and articulate speakers when she hears them, and quotes many of them when it is appropriate—including many of the participants in these enterprises.

You will learn a lot from this book. I know something about most of the subjects covered but found myself highlighting the manuscript as I went, marking important facts or wonderful quotes. You will find hundreds of ideas here for dealing with issues that arise at each step along the way, but you will not find a single set of instructions for organizing or maintaining a CSA, since uniformity is not the movement's (or the authors') goal.

But if there is no single approach laid out in this

book, there is, perhaps, a single goal: namely, to encourage more people to join in the struggle to save farming. The authors are appropriately desperate to save farmers and farmland, and want more people to understand the importance of their passion. "I have searched particularly," Liz Henderson writes about her research for this unique compilation of experiences, "for ideas on how to entice, entrap, entangle, or engage as many people as possible in community support for local farming." This edition presents new evidence of the persistent power of these ideas; even more, it describes new ways that variously defined communities have found to help their local farmers survive.

Shortly after I had visited the early CSA begun by Robyn Van En, I met another person active in the movement who made a comment I have never forgotten. "Community Supported Agriculture," he asserted, "is a kind of barter relationship. You pay this money up front in the fall to help support a local farmer, and next summer you get all this free produce." As things have turned out, not all CSAs can or want to collect money up front, and not all sharers experience the produce as "free." While generosity and community are goals, this fascinating book makes clear that it's seldom simple to achieve

them. In a society where individuality and free choice are mantras and everything has been assigned a price, self-imposed limitations do not come easily and community does not build spontaneously.

If we hope to keep eating, however, we need to keep farmers in business; and if we want to keep farmers in business, it's time for all of us, ordinary citizens and policy makers alike, to begin learning how that might be done. *Sharing the Harvest* is a great place to start.

JOAN DYE GUSSOW, PHD

Dr. Joan Dye Gussow is a serious food producer, a sometime writer, and officially a retiree from Teachers College, Columbia University, where she was the Mary Swartz Rose Professor, former chair of the Nutrition Education Program, and still teaches every fall.

Gussow lives, writes, and grows organic vegetables on the west bank of the Hudson River. Her most recent book, based on the lessons learned from 30 years of working toward growing her own food, is This Organic Life: Confessions of a Suburban Homesteader.

ACKNOWLEDGMENTS

Robyn Van En should have been the author of this book. Her untimely death from asthma took her from the work she loved on behalf of Community Supported Agriculture and from the completion of this text. I had volunteered to help out with a few chapters and to give moral and technical support. When Robyn passed, I took on what I thought would be the job of polishing her first draft for publication. Unfortunately, I discovered that there was not enough to polish, so I have rewritten the book substantially, trying to keep all of Robyn's concepts and as much of her prose as possible. For this second edition, I had the pleasure of returning to the hundreds of people I had talked to, read about, or observed for the first edition, to learn what had changed for them over these eight intervening years.

First of all, I want to thank the hundreds of CSA farmers and sharers all over the country who have contributed materials, ideas, suggestions, and enthusiasm for this book and allowed me to tell their stories. It is truly an honor to be associated with this creative, hopeful movement.

I wish to express my appreciation to c. r. lawn for reading and rereading the drafts and doing everything in his power to make it at least as lively and concrete as a Fedco Seeds catalog. His colleague at Fedco, Robin Sherman, suggested *Sharing the Harvest* as a title for the book. Thanks as well to Marianne Simmons, Portia Weiskel, Jennifer Bokaer-Smith, Brenda Reeb, and Patricia Mannix for reading parts of the manuscript and giving helpful suggestions and encouragement. And special thanks to Beth Holtzman, Communications Specialist for Northeast Sustainable Agriculture Research Education (NE SARE), for guidance, for the section on the Intervale, and for shepherding this project through the SARE process. I am also grateful to Ben Watson for his knowledgeable and tactful editing on behalf of Chelsea Green Publishing Company.

This book would not have made it into print without the generous financial and moral support of Fred Magdoff and the NE SARE Program. The NE SARE Administrative Council accepted my proposal in 1995 to provide financial support to Robyn so that she could quit her two part-time jobs and concentrate on this book, then waited patiently as the writing got under way, and agreed to transfer the support to me, as Robyn's substitute. NE SARE has also invested jointly with Chelsea Green in the publication of the book.

I also want to give thanks to the many members of the Genesee Valley Organic Community Supported Agriculture Project (GVOCSA), and especially to Alison Clarke, our fearless organizer, for joining in this great experiment in local food production with me; to my farm partners Greg Palmer, Ammie Chickering, and Katie Lavin for years of listening while hoeing, picking, and packing at my side; and to my former partner David Stern, from whom I learned a great deal about organic farming, including how risky it can be. My gratitude goes as well to Nancy Kasper for an artistic refuge in which to finish writing the first version of this book, and to my very special friend Jack Bradigan Spula for his patience with including tours of CSA farms in the itineraries of our every excursion, and for his inspiration as a writer and musician.

And finally, I would like to acknowledge the steadying presence of Harry Henderson, my father-in-law, close friend, and mentor in this writing business. I continue to have conversations with him even after his death in 2004 a few days before his eighty-ninth birthday.

Any inaccuracies or mistakes are purely my own.

INTRODUCTION

by Robyn Van En

Whatever you can do or dream you can do—begin it.
Boldness has genius, power, and magic in it. Begin it now.
—GOETHE

In 1983, my young son David and I moved to Massachusetts from the northern California redwood forests to experience New England's colored leaves of autumn and snowy winters while I finished my training as a Waldorf kindergarten teacher. We planned to be in the area for a year, to get our bearings, and then find property to buy. I would look for a place with a bit of land for me to resume growing bouquet flowers and perennial stock for landscapers, as I had done in California while I went to school. Within a month after our arrival, circumstances motivated me to find the "place," and somewhat by accident, maybe more by providence, we landed at Indian Line Farm, in the village of South Egremont.

Indian Line Farm was so named almost two hundred years ago as one of the early Dutch settlements built at an imaginary line beyond which was Indian country. Supposedly, Johnny Appleseed came along Jugend Road, planting his apple trees. Shays' Rebellion, largely a farmers' uprising against unfair taxation, which influenced the writing of the Constitution, took place right over the hill.

Indian Line is a beautiful farm with a big, old house and a huge dairy barn resting upon table-flat, fertile bottomland. Beneath the topsoil is limestone ledge, giving the soil a mellow pH. Nearby is an alkaline fen (wetland), a rare habitat protected and monitored by the Nature Conservancy. The middle of the field looks out on Jugend Mountain, formerly an Indian observation point, with the remains of

their longhouses and another old farmstead in the state park at the foot. Over the top of the mountain runs a portion of the Appalachian Trail, leading into Connecticut, just behind the first community land trust and the E. F. Schumacher Library.

I was raised in California and for years had cultivated a vegetable plot, besides the bouquet flowers and perennials, but to start managing a 60-acre, retired dairy farm in the Berkshire Hills was a huge departure for me.

Soon after I arrived in the neighborhood and joined the staple foods buying club, I had a conversation with Susan Witt from the E. F. Schumacher Society. We discussed what I would be growing at the farm the forthcoming spring, besides flowers, as I had a whole lot of available ground.

I found out that most of the buying club members had their own summer gardens but went to distant farms or the supermarket for their winter vegetables. Why shouldn't I grow those storage crops? At a co-op meeting, people said they would buy anything I grew, so I planted accordingly. By planting primarily potatoes, carrots, onions, garlic, and winter squash, I had fairly good returns, with my market ready and waiting, but I still carried all of the capitalization expenses, all of the work, and all of the risk.

I spent long periods, generally while hoeing, trying to formulate a better way to oblige both the grower and the eaters. The better way would be something cooperative, an arrangement that would allow people to draw upon their combined abilities, expertise, and resources for the mutual benefit of all concerned. It would also bring the people producing the food closer to the people who were eating the food, and the eaters closer to the land.

In the middle of my second growing season, as I pondered this agricultural conundrum, Jan Vandertuin visited the farm. He had recently returned to New England from Switzerland and was anxious to share the experience he'd had working with a couple of farmers there. These farmers had asked their regular customers to pay a share of the farm's annual production expenses in exchange for a weekly share of the produce. Shares of the vegetables, meat, and dairy products were made available to them. After talking only a few minutes, Jan and I knew that we should do the same at Indian Line Farm.

Jan Vandertuin, John Root, Jr. (co-director of Berkshire Village, a group home for handicapped adults, a stone's throw from the Schumacher Society and a mile from Indian Line), and I got busy organizing. We introduced the "share the costs to share the harvest" concept to the surrounding community by way of the Apple Project in the autumn of 1985. People paid in advance for family-sized shares of the apple harvest. Most of those folks signed up for the vegetable shares that we offered for the following spring. We were determined to make this happen, so we were very busy educating community members about the vegetable shares at the same time we were interviewing potential gardeners and farmers. No one had ever heard of being paid for vegetables in advance, before the first seed was planted, but we were finally approached by Hugh Ratcliffe, who started breaking ground that fall for the eventual spring planting. The rest of us carried on with the Apple Project and community education.

During the winter of 1985–86, we met each week to discuss and develop the logistics and procedures necessary to accomplish our goals: local food for local people at a fair price to them and a fair wage to the growers. The members' annual commitment to pay their share of the production costs and to share the risk as well as the bounty set this apart from any other agricultural initiative.

We didn't take any step of this process lightly. We discussed and debated long into the nights the necessary policies and procedures, besides the possible names for the project that would convey its full intent and purpose. We finally decided on Community Supported Agriculture, which could be transposed to Agriculture Supported Communities and say what we needed in the fewest words. CSA to ASC was the whole message. We knew it was a mouthful and doesn't fit easily into conversation or text, but to this day I can't think of a better way to name what it's all about. People have tried substituting other words ("consumer shared agriculture" is what some people call it in Canada), and I found obvious discomfort with the word *community* when I tried to explain the concept in the former Soviet Union. People have problems with *supported*, too. Please know that each word was chosen after lengthy consideration. I personally was adamant about using the word *agriculture* rather than calling the project CS Farms, because I didn't want to exclude similar initiatives from taking place on a corner lot in downtown Boston. We had to call it something fast because the project was ready to go.

To secure the land for existing and potential members, I leased my garden site (approximately five acres) to the CSA, an unincorporated association, for three years with an option to buy the land into an agricultural trust after the third year.

We offered our first shares of the vegetable harvest in the spring of 1986. Early members received a shopping bag of vegetables twice a week throughout the growing season, and twice a month from the root cellar during the winter months. This proved to be too many vegetables for most households. The next year, many found a friend or neighbor to take the second bag, and by the third year, we pared it down to one bag a week. Larger households, restaurants, or markets bought multiple shares.

We have come a long way since that first season. As the founding core group, we learned a lot and realized that there was a lot more to learn. Working with a group of people in a manner that honors and makes the best use of the collective expertise and resources, while at the same time becoming familiar

Robyn Van En. PHOTO BY CLEMENS KALISCHER.

with and adjusting to different personalities and agendas, is not easy. The logic, simplicity, and earthly need for the concept carried us through. Despite our many differences, we created a working prototype and replicable model of CSA. After four years, we separated—a rather grisly process—but, looking back, I can honestly say I wouldn't have learned nearly as much about myself, about group dynamics, about CSA's pitfalls and potential, and about the larger community, both locally and regionally, if the split had been easier.

I wrote the first edition of the CSA start-up manual, the *Basic Formula to Create Community Supported Agriculture* largely in self-defense, as I was spending hours in the field, on the phone, or in conference halls talking about our experiment at Indian Line Farm. I designed the manual to help readers answer many of their questions about CSA and to formulate questions specific to their own situations. Translating CSA to the North American landscape and mentality, which are vastly different in scale, available resources, and culture than Japan and Western Europe, was a challenge. CSA has certain fundamental logistical points that are similar no matter where or how it is practiced, but, at the same time, it is largely an evolving and highly adaptive process, as I tried to share and convey through the *Basic Formula* manual.

By the spring of 1996, close to six hundred CSA projects, engaging at least one hundred thousand people throughout the United States and Canada, were putting seeds into the earth. An armchair social study conducted by Jean-Pierre Schwartz—a founding member of the first Washington, D.C., area CSA and a founding board member of CSA of North America (CSANA)—compared the current number of CSAs, their rate of increase to date and their potential increase to the boom of bed-and-breakfast establishments springing up across the country. From this, Schwartz concluded that we might see ten thousand CSA farms and gardens by the year 2000. That could mean an average of two hundred projects in each state and Canada with nearly two million people involved, signifying an important shift in mainstream consciousness. We would be well on our way toward a "trend," with 2 percent of the overall population becoming aware of this new alternative for purchasing fresh food. People don't even need to be actual CSA members, just marginally aware. If we can achieve that much consciousness in the United States and Canada, CSA's presence certainly will have increased around the world, too. [CSAs have not multiplied as quickly as Robyn had hoped, perhaps because they are far more complex than a bed-and-breakfast. In 2007 the best estimate is that there are 1,700 CSAs in the United States with a total of perhaps 100,000 members.—EH] While the Indian Line Farm project was modeled after Jan's experience in Switzerland, I have since learned that the CSA equivalent was developed first in Japan in 1971, initiated by a group of women concerned about the use of pesticides, the increase in processed and imported foods, and the corresponding decrease in the local farm population.

The group approached a local farmer and worked

out the terms of their cooperative agreement, and the *teikei* movement was born. Literally translated, *teikei* means "partnership" or "cooperation," but according to teikei members in Japan, the more philosophical translation is "food with the farmer's face on it."

The philosophical aspects of teikei are also represented in the names of their groups, such as the Society for Reflecting the Throwaway Age, the Young Leaf Society, and the Society Protecting the Earth. This aspect is fairly common in North American CSA as well, with names like Walk Softly CSA, Twin Creek Shared Farm, Deliberate Living CSA, Gathering Together Farm, Caretaker Community Farm, and Heartbeet.

The teikei farms or garden sites in Japan tend to be quite small and intensively farmed. It is common for a group of farmers, dispersed throughout the countryside, to supply a large group of urban members by delivering their produce to one of the express train depots for transport to a central pickup point in the city. It is also common to have groups of fifteen hundred households networking with a group of fifteen farmers to get a consistent and

diverse selection of products. Typically, members are supplied with their weekly vegetables, herbs, fruit, fish, poultry, eggs, grain, and soy product needs, along with soap and candles from a community supported cottage industry. (For a more in-depth examination of teikei, see the brief history by Cayce Hill and Hiroko Kubota, chapter 22, p. 267).

In 1992 I had the great pleasure of co-hosting members from one of the teikei groups that was traveling in the United States and Canada. They had come to see organic farms in general and some CSA projects specifically. I was amazed by how similar our concerns and visions were for teikei and CSA and for the future of agriculture as the basis of all culture. It was truly a full-circle experience for me.

Equally empowering to both the community and the farmers, CSA offers solutions to common problems facing farmers and communities worldwide. Ultimately, the concept is capable of engaging and empowering people to a capacity that has been all but lost in this "modern" world.

Indian Line Farm
December 1996

PART ONE

CSA IN CONTEXT

WHAT IS COMMUNITY SUPPORTED AGRICULTURE?

You must be the change you wish to see in the world.
—MAHATMA GANDHI

Community Supported Agriculture is a connection between a nearby farmer and the people who eat the food that the farmer produces. Robyn Van En summed it up as "food producers + food consumers + annual commitment to one another = CSA and untold possibilities." The essence of the relationship is the mutual commitment: The farm feeds the people; the people support the farm and share the inherent risks and potential bounty. Doesn't sound like anything very new—for most of human history, people have been connected with the land that fed them. Growing (or hunting and gathering) food somewhere nearby is basic to human existence, as basic as breathing, drinking, and sexual reproduction. If this basic connection breaks down, there is sure to be trouble.

For the masses of people in the United States today, this connection has been broken. Most people do not know where or how their food is grown. They cannot touch the soil or talk to the farmer who tends it. Food comes from stores and restaurants and vending machines. It has been washed, processed, packaged, maybe even irradiated, and transported long distances. As trade becomes "free," the food travels even longer distances. Stores in the Northeast used to carry Florida tomatoes in the winter. Today, with the North American Free Trade Agreement (NAFTA), the tomatoes come all the way from Mexico. Under the rules of the World Trade Organization (WTO), government-supported apple juice concentrate from China undersells Washington State's juice, which for the pre-

vious decade had forced New York State orchardists to replace trees of primarily juice varieties with ones that yield fresh eating apples. Farmers alone have been shouldering the risks of this increasingly ruthless global market, which has forced millions of them from the land. CSA offers one of the most hopeful alternatives to this downward spiral, and it is the only model of farming in which customers consciously agree to share the risks and benefits with the farmers.

Robyn Van En dedicated the last ten years of her much-too-short life to spreading the word about CSA, speaking at conferences and workshops and giving advice and troubleshooting in person or over the phone. She was the founder and sole staffperson for CSA of North America (CSANA), which distributed the handbook she wrote in 1988, *Basic Formula to Create Community Supported Agriculture*. A common cliché used for people like Robyn who work selflessly for good causes is *tireless*. Well, Robyn was not tireless. She was often tired and sick. Lack of oxygen slowed her down. But it did not stop her. After each bout with asthma or bronchitis, she would assure me that, this time, she had things under control. She shared her optimistic and radiant vision of a just and sustainable society with anyone who would listen.

Just before she died, Robyn appealed to Community Supported Agriculture groups all over the country to send her copies of their brochures, letters, and newsletters. I spent three full days reading through the material from almost every state and several Canadian provinces. Twelve years ago, when I started helping Robyn with this book, we thought we would be able to interview someone from every CSA. With each growing season, the projects have proliferated to the point where no one person can do justice to them all.

Reading through the brochures was an uplifting experience—so many eloquent and thoughtful statements about the importance of growing food in harmony with nature and preserving the land for future generations. I covered the floor of my room with them, sorted first by geography, then by size of farm, then by type of project (work required or not, with core group or without, mechanized or worked by hand or horse, weekly packet delivered or picked up, subscription program or community farm), then by years of operation. So many possible categories. So many creative variations on the unifying theme of fresh, local food "with the farmer's face on it," as the Japanese organic farmers would say. So much hope, tempered by a sober awareness of the enormous scale, productivity, and destructiveness of the dominant food system. So much buoyant optimism, despite the overwhelming odds and countless hardships and obstacles.

☙ FROM THE GENESIS FARM COMMUNITY Supported Garden in New Jersey we hear:

All of us are increasingly aware that the present state of farming in the United States and around the world is in serious difficulty. Agribusiness, with its chemical-intensive, industrialized methods of crop production is replacing agriculture, with its small human-scale, and diversified character and its commitment to place and community.

Many people feel a keen sense of separation from nature and the soil resulting from our urban and suburban settlements. From childhood throughout our adult lives, we feel a deep hunger for a return to the soul of nature. The teeming life, the extraordinary beauty, and the sense of wholeness that a farm provides is a precious gift to nourish this spiritual hunger. The garden offers a way to experience the mystery of seeds and soil and to reconnect in an endless variety of possibilities for creating friendship, community, and the

strong connections that historically tied farmers and communities into harmonious relationship with the Earth and each other.

☙ FROM GORANSON FARM in Maine we hear:

Community Supported Agriculture provides a mutually beneficial arrangement between farmer and community. In exchange for financial support in the spring from the "shareholders," the farmers commit to provide healthy, locally grown food through the growing season. The goal of CSA is to reconnect people with the land that sustains them. Shareholders know how and where the food they eat is grown, and they can learn to understand the complexities of providing this food.

Joining a CSA leads to a greater awareness of our interdependence on one another and the land. It helps ensure the survival of rural life. Our farm has the ability to grow and provide food for a number of families while pursuing proper land stewardship. It is through the cooperation between farm and community that a sustainable local food supply will become a reality.

☙ FROM HARMONY VALLEY FARM in Wisconsin comes this:

Again this year, Harmony Valley Farm is making the same, high-quality produce available at a less-than-retail cost to households who wish to participate as CSA members. CSA creates a direct relationship between you and the farm. You will receive a weekly box of our fresh, in-season produce delivered to a convenient location in your neighborhood. You'll know your produce dollar goes directly to the people who plant, tend, and harvest your food. You will be supporting organic growing methods that protect soil resources and water quality and assure you of the most healthful, nutritious produce possible. For everyone, including those of us at the farm, CSA is an opportunity to connect in a meaningful way to others who care about the food we eat.

☙ FROM WINTER GREEN COMMUNITY FARM in Oregon this echoes:

Members of the Genesee Valley Organic CSA pick up their shares at the Abundance Cooperative Market in Rochester, New York. PHOTO BY CLARKE CONDÉ.

To further our vision of becoming "a farm in balance with the earth, humanity, and ourselves," we want to build the link between ourselves and the people who eat the food we produce and to help reestablish the role of agriculture in the community. We seek to provide an environment where families can strengthen their connection to the earth that sustains them. We believe that it makes more sense to grow food for the local community than for distant markets. By joining, you receive fresh, locally grown vegetables and have a direct connection to the farm and the people growing food for you. If you like, you can come and get your hands in the soil, walk the farm, or attend our farm events. In many ways, the most important reason for any of us to be involved with the Community Farm is that it is an affirmation of the kind of world we want to live in. It is a positive choice for the future.

⟶ AND EVEN FROM DISTANT PALMER, ALASKA, from Arctic Organics we hear:

Community Supported Agriculture is an answer to what is becoming a worldwide concern: the production and distribution of high-quality, carefully produced food. By joining a CSA program you are supporting your food producers directly by avoiding the middle people (i.e., the distributor and retailer). You are purchasing organically and locally grown produce, thereby avoiding the high environmental and health costs and questionable merits of agricultural chemicals and the fossil fuels and other resources necessary for shipping it long distances. In exchange, you will receive high-quality, nutrient-rich, flavorful produce on a weekly basis, freshly harvested the same day it's delivered.

CSA members all across the country appreciate the fresh vegetables but understand that their involvement with a farm means much, much more. Their reflections reveal the potential power of a food system that reconnects people with the land.

From Solas Farm in Michigan, Cathy Zarcovich exclaims:

It's not the vegetables themselves, it's what they represent to me. Community. People. Friendships. Love. Can you find love in a basket of veggies? You bet you can.[1]

HARMONY VALLEY MEMBERS IN WISCONSIN SAY:

We feel better, healthier, and we know where our food comes from.

My children . . . understand how and where we get what we eat.

AT SEVEN OAKS FARM IN VERMONT, member Martha Rosenthal writes:

Janii and Willy . . . are artists in farming and I feel privileged to eat their 'canvas.'

IN MAINE, WILLOW POND FARM member Marc Jalbert comments:

It's very comforting to know where your food is grown. Unlike a symbolic investment in stocks and bonds, we can actually go out and see the results of our investment. It's very tangible. It gives me a personal sense of security.

AT HAPPY HEART FARM IN COLORADO, member Susanne Edminster exclaims:

The farm itself is miraculous. How you grow all you do on land that seems miniscule in proportion to the bounty is awesome. And there's something about popping a radish or a leaf of lettuce into my mouth, knowing that I don't have to worry about pesticides, which is pure heaven. Sorting and processing the vegetables when I got home was not the dull task I expected. There was a harmonious feeling about it, knowing who grew them, and the love put into the effort. I enthusiastically renew my membership.

IN IOWA, A CSA MEMBER DECLARES:

We became members to support local food production. Our family received high value for our membership in all areas. Economically and health-wise, we felt that the CSA membership improved our eating habits and lowered our grocery bills.

A MEMBER OF THE GENESEE VALLEY ORGANIC CSA in New York State, Josh Tenenbaum, wrote this letter to my former partner David and me:

Growing food with the care and loving that you do can never be captured in organic or any other kind of certification. Such things are written only in our hearts. You are both visionaries, and by showing me a brief glimpse of your vision have bestowed on me a wondrous gift. It is a vision of how the world can be—not a dream world, but this world. A world of cooperation, of soil teeming with life, of noticing the natural cycles and epicycles, day alternating with night, but also planting, germination, growth, harvest, decay, and then planting again. Thank you for this gift of vision. I ask myself, how do my actions every day create or hinder such a world?

At the time of this writing, I do not know exactly how many CSA projects there are in the United States or North America. The Robyn Van En Center for CSA Resources database lists 1,208 CSAs. Localharvest.org lists 1,294, and Guillermo Paget, who manages that list, tells me that he estimates there are actually between 1,500 and 2,000 CSAs currently in existence. In "Community Farms in the 21st Century: Poised for Another Wave of Growth?" author Steven McFadden speculated that in 2004 there were as many 1,700 CSAs feeding over a hundred thousand households.[2] When I asked him where he got that number, he admitted he had extrapolated from past figures and rates of growth. I do know that CSAs vary in size from 3 shares to over 2,100. They can be found as far north as Palmer, Alaska (and even farther north in Canada), and as far south as Gainesville, Florida, or San Diego, California. The densest clusters are in the Northeast, around the Twin Cities and Madison in the Upper Midwest, and in the Bay Area of California. The number of CSAs is increasing quickly in states such as Iowa, where food activists have teamed up with the Cooperative Extension Service and the universities to provide technical assistance. Most CSAs are either organic or biodynamic in their method of production. A few are in transition to organic or to a lower use of chemicals. The CSA concept has spread from farmer to farmer and from consumer to consumer through the organic and biodynamic networks, and only recently have a few organizations and Extension agents reached out to conventional farmers. Nothing about the structure of a CSA dictates that the food be organic, but most consumers who are willing to become members do not want potentially toxic synthetic chemicals used on their fresh, local produce.

The very first CSAs in this country, Indian Line Farm in Massachusetts and the Temple-Wilton Community Farm in New Hampshire, both initiated in 1986, established the model of the "community farm," which dedicates its entire production to the members, or sharers. Indian Line divided its produce so that every sharer received an equal share or half-share. Temple-Wilton allowed sharers to take what they needed regardless of how much they paid. Only about a quarter of the farms that have adopted the CSA concept have emulated this model. Out of the forty-five CSA farms in Vermont, only one produces exclusively for sharers, while the others continue to sell to a variety of markets. The level of member participation in either growing or distributing the food varies tremendously from farm to farm. At one extreme are CSAs like the Genesee Valley Organic in New York, for which I am one of the farmers, and Fair Share Farm in Missouri, which require all sharers to do some work as part of their share payment. At the other end are what have come to be known as "subscription" CSAs, where the farm crew does all of the work and members simply receive a box or bag of produce each week. Most CSAs range somewhere in between, with members volunteering for special workdays on the farm, helping with distribution, or defraying part of their payment with "working" shares.

Jered Lawson, who initiated and then staffed CSA

> The level of member participation in either growing or distributing the food varies tremendously from farm to farm. At one extreme are CSAs that require all sharers to do some work as part of their share payment. At the other are what have come to be known as "subscription" CSAs, where the farm crew does all the work and members simply receive a box or bag of produce each week. Most CSAs range somewhere in between, with members volunteering for special workdays on the farm, helping with distribution, or defraying part of their payment with "working" shares.

West from 1994 to 1996, expected that CSAs would follow the community farm model and was at first disillusioned when large organic California farms added subscription shares to their other marketing efforts. He feared that this model would limit member involvement and the sense of connection to the farms. The larger number of members in a subscription CSA certainly reduces the intimacy of personal contact between members and farmers. Compared with smaller farms in parts of the country with more difficult weather conditions, California CSAs have also placed much less emphasis on members sharing the farmers' risks. Dru Rivers of Full Belly Farm told me that she is nervous about pushing the risk-sharing aspect of CSAs because organic food is so readily available in California and there is so much competition. Full Belly Farm confronted this issue once, when a freak snowstorm kept them from picking and delivering one week's shares. Only a few of the members objected to sharing the risk by skipping that week's box. Farms in other parts of the country reduce the risk sharing by purchasing crops from other farms to fill out their shares.

As these larger CSAs have evolved, however, Jered has observed that, far from being a marketing add-on, the community supported component is making an important contribution to the long-term viability of the farms. Somewhat grudgingly, Jered admits that the farmers have adapted CSA effectively to their particular situation:

> Even in a hybrid CSA, or a CSA that may not have the whole farm budget as its centerpiece, there's still the fundamental desire to see the farm survive. The consumer members of the farm do come to understand that they are giving economic contributions towards that survival, and the farmers

themselves help set the terms of that contribution. Therefore, the consumers can say, "We feel comfortable that we are meeting the farm's assessment of what it needs to survive," even though it is different from the original CSA concept.[3]

Through their teachings and example, Robyn Van En and Trauger Groh shaped the "original CSA concept." Robyn tells the story of founding Indian Line Farm CSA in the introduction to this book. Trauger was one of the team of three farmers at Temple-Wilton Community Farm for many years until he retired. He is also one of the authors of *Farms of Tomorrow* (1990) and *Farms of Tomorrow Revisited* (1997), which have inspired many CSA efforts. (You will find these sources and others cited throughout this book in the Bibliography.) In these two books, Trauger and his coauthor, Steven McFadden, profile a number of CSA farms, all of them adherents of biodynamics, an approach to farming based on the writings of Rudolf Steiner and his followers. I recommend that anyone interested in CSA read *Farms of Tomorrow Revisited*. Biodynamic principles and the farms that apply them have a great deal to teach the rest of us in agriculture. But you do not have to adhere to Steiner's precepts to practice community supported agriculture.

Robyn and I agreed that the purpose of this book should be to help spread community supported agriculture by collecting as many of the best examples we could find regardless of ideology, religion, or farming methods. We did not want to craft a tight definition or try to establish the criteria for identifying "the true CSA farm." Rather, we hoped to honor the diversity of this young, but quickly spreading movement. Like the farms that host them, CSAs come in many scales and sizes. The people growing the food have different ideas about how to do it, whether efficiency is a value, how much they need or want to earn, and how many helpers they organize to get the work done. I like the way Steve Gilman talks about CSA diversity in "Our Stories": "Like grapes or garlic, CSA takes on the flavor, bouquet, and integrity of where it grows,

it's all here

we set the seeds, speak
to the sky
nurture the plants, drink
the rain, give back
to the soil, curse
the cold, dance
to the sun, sing
with the wind, weep
at the passing, dream
with the moon. we open
our hands and accept another
season of hope fulfilled
or not,
balancing burdens with blessings,
rocks and eagle feathers,
carrying the harvest home.
listen.
the birds are singing the earth
awake. the spiraling cosmos
is bursting open seeds climbing
to the light. there's a crackle
of joy in our hearts, ignited
by the sun—a flower filled
with flame. listen. the plants will tell you
of sending roots deep to survive
the dry times.
the seasons will show you how nothing
is ever really gone but keeps
turning out and over
again and again and again,
just as the ancestors
smile down from the clouds
onto the faces of children
yet unborn shining up
from rain spattered stones on the path
we walk.
listen.
it's all here.

—Sherrie Mickel, 1995

becoming appropriately adapted to each unique situation." While participants agree that CSA means a connection between a specific group of eaters and a specific piece of land and the people who farm it, healthy debates roar on about how to understand the concepts of "sharing the risk," "community," "support," and even "agriculture." You will hear passionate snippets of those debates in this book.

As an organic farmer who is active in farm politics, I am used to being in rooms or fields full of opinionated and articulate people. Sometimes we all talk at once. Lately, we have been learning to set an agenda with amounts of time for each item, so that we can speak one at a time and practice the active listening skills we are also struggling to acquire. I propose to do something like that with this book—to let many different versions of CSA speak for themselves. In Robyn's place, I will give as much as I have been able to salvage of her version and chime in with mine as well.

CHAPTER 1 NOTES

1. Cathy Zarkovich, "Love in a Basket," *The Community Farm* (Summer 2006): 2.
2. Steven McFadden, "The History of Community Supported Agriculture, Part 1: Community Farms of the 21st Century: Poised for Another Wave of Growth?" *New Farm* e-magazine, www.newfarm.org. January, 2004.
3. Steven McFadden and Trauger Groh, *Farms of Tomorrow Revisited: Community Supported Farms—Farm Supported Communities* (Kimberton, PA: BD Association, 1997), 83.

CSA AND THE GLOBAL SUPERMARKET

America of the future will be all malls connected by interstates.
All because your parents no longer can their own tomatoes.
—GARRISON KEILLOR, *A Prairie Home Companion*, March 28, 1998

"There is not one grain of anything that is sold on the free market in the world.
The free market exists in the speeches of politicians."
—DWAYNE ANDREAS, CEO of Archer Daniels Midland (ADM),
as quoted by Anna Moore Lappé, Northeast Organic Farming Association
of New York (NOFA-NY) Winter Conference, 2005

I grew up in Croton-on-Hudson, New York, a bedroom suburb of New York City, raised by parents who were deeply committed to the struggle for world peace and economic justice. They were city people through and through, as were my grandparents. No one even gardened, but at the dinner table we had long, disturbing conversations about world politics, hunger, malnutrition, and inequality. A few miles up the Hudson, the government stored millions of tons of surplus grain in cargo ships anchored along the shore. Playing hooky from school, I sometimes drove along the scenic river and puzzled over those hulking gray vessels. If all that surplus was such a big expense and there were so many starving people, I reasoned in my innocence, why couldn't they just sail those boats to wherever food was scarce and feed the hungry? The irrationality of the food system seemed unfathomable.

After forty-five years and a lot of travel, reading, and thinking, I see some pieces of the picture starting to fall into place. I do not make any claim to be an expert analyst of the world's food and agriculture systems, but after two and a half decades of living and working as an organic farmer, this is how things look to me. Food is basic to human existence. We can go a day or two, even a week, without eating; most people worldwide, however, prefer at least one daily meal. "Our bodies," as Bill Duesing, an organic educator in Connecticut, explained on the *Living on Earth* radio program, "run on the solar energy collected by plants, which are nourished by the nutrients they gather from water, air, and soil." Farmers are "managers of solar conversion," to borrow a term from Holistic Resource Management. In cooperation with the forces of nature, farmers and gardeners actually create wealth. For most of human history, human beings have not taken their food supply for granted and have regarded the creation of food as a sacred act, surrounding it with rituals of blessing and expressions of gratitude to the Earth or the gods or God. Growing food is the most basic use of the natural resources of the Earth, and through food production, we make our own working landscapes, ranging from patchworks of tiny gardens with diverse plantings to vast fields of single crops. How each society or nation produces and distributes food in large measure determines its identity.

At Farm Bureau meetings, Cornell annual agricultural economy updates, and New York Agricultural Society gatherings, I have heard farm and food industry leaders say that the U.S. food system is the safest, cheapest, and best in the world. Not only do American farmers feed the growing population of this country, but they keep the rest of the world from starving as well. With barriers to trade cleared away by the international General Agreement on Tariffs and Trade (GATT) and the North American Free Trade Agreement (NAFTA), U.S. farms, we are told, will be able to outcompete all others. The prognosticators predicted that the $57 billion worth of agricultural exports in 1997 could rise to $75 billion in the near future as the standard of living rises in China, boosting purchases of U.S. grains and meat products. U.S. farmers planted 3.7 million acres of genetically modified (GM) corn, soybeans, and cotton in 1996, the first year of significant GM commercialization, increasing to 128.3 million acres in 2005, and this is only the beginning. A team of researchers at the Department of Energy's Pacific Northwest National Laboratory predicts that agrogenetics, genetic engineering combined with plant manipulation, will reduce agricultural impacts on the environment.

The library of genetic information is doubling every twelve to twenty-four months: Monsanto, the "life sciences" company, can create ten thousand new genetic combinations a year, according to Barnaby Feder in the *New York Times*.[1] Genetically engineered crops will, it is claimed, resist pests, requiring fewer applications of pesticides, and will assimilate nutrients more efficiently, reducing the need for fertilizers. By using custom-designed, genetically engineered seed varieties, global positioning computers, and precision farming technologies, as few as fifty thousand farmers will supply 75 percent of the country's agricultural production from 50 percent of the currently cultivated farmland. Ray Goldberg writes that these farmers will work as "farm technology managers" under contract to the vertically integrated distributors and processors.[2] Armed with scanning feedback devices providing them with unparalleled amounts of intelligence on consumer finances and purchasing patterns, distributors will become the market coordinators of the domestic and global food systems.

How does this upbeat presentation fit with some of the other realities of our food supply? While I am impressed by the promise of global efficiency and technological innovation our food system leaders predict, I continually encounter evidence that the system they praise fails to provide adequate nourishment for large numbers of people, does not account for many environmental costs, and concentrates decisions over food in fewer and fewer hands. The stores in the developed world are well stocked with food, and there are no obvious shortages, yet the supply of food actually on hand in Northeastern cities would last only thirteen days should some emergency occur. The Rome Declaration, adopted by the United States and 185 other countries at the 1996 World Food Summit, pledged that we would reduce the number of undernourished people by 50 percent by 2015. Yet a 2004 U.S. Department of Agriculture (USDA) report revealed that over 38 million households *in the United States* were "food insecure," an increase of 7 million since 1999. As 2005 drew to a close, a record 27,575,192 people were using food stamps.[3] The number of food banks—community warehouses distributing salvaged and donated food to emergency food providers—rose from 75 in 1980 to 225 in 1990, and the number of pounds of food distributed rose from 25 million in 1980 to 2 billion in 2005. On its Web site, Second Harvest reports that 23 million people, including 9 million children, rely on the emergency feeding programs their network serves. In 2005 the U.S. Conference of Mayors found that requests for emergency food assistance rose by 14 percent on

> Growing food is the most basic use of the natural resources of the Earth, and through food production, we make our own working landscapes. How each society or nation produces and distributes food in large measure determines its identity.

average, increasing in 96 percent of the twenty-seven cities surveyed. Yet Congress has responded by reducing funding for food stamps! Samir Amin estimates that 3 billion people worldwide have less than $2 a day to spend on food, a mind-numbing level of human misery.[4]

Within the memory of people still alive, fruits and vegetables grew on the outskirts of most major cities. Market farms across the Hudson River in New Jersey supplied the vegetable needs of New York City dwellers. Apple orchards blossomed in the Bronx. Since that time, urban, and then suburban, sprawl has paved over cropland and planted houses and highways where cabbages used to grow. Food production has shifted to where it is most "efficient," where bigger machines can maneuver over larger, flatter fields, where chemicals and technology reduce the need for horse and human power, and where crews of poorly paid migrant workers do their jobs and then move on.

The global food system plays economic hardball. A vicious speed-up has been going on in the countryside. In 1950 a farmer could support a family with a herd of twenty cows; today two hundred are needed to eke out a living. Where once a 160-acre "section" produced enough grain for a family's livelihood, today 1,600 acres with a much greater yield per acre is just barely enough to keep a farm in business. Increasingly large and specialized farms produce basic commodities, yet rural areas no longer feed themselves. The Field to Family Community Food Project states, "Iowa is a 'textbook example' of the effects of an expansive industrial food system. The state's agricultural production, focused on grain and livestock, is highly specialized for export purposes, and because there are few food processing industries based in the state, almost all of the food consumed by the state's 2.8 million citizens (including that derived from the basic commodities produced in the state) is imported. The state depends on such imports for essentially all of its vegetables and fruits."

Until the twentieth century, the United States was largely an agrarian society. As recently as 1910, one-third of the population, some 32 million people, lived on farms. The Dust Bowl, the Depression, and World War II drove 9 million people off the land, but those major upheavals pale in significance compared with the decimation of the farming population that followed the restructuring of farm price supports in the 1950s. In 1937 there were 7 million farms; by 1993 less than 2 percent of the population was left on only 2.2 million farms—so few that the U.S. Census Bureau announced it would stop counting them. Since that time the decline in farm numbers has slowed. The Census of Agriculture reported 2,215,876 farms in 1997 and 2,128,982 in 2002.

African American farmers have been squeezed off the land even faster than white farmers. Particularly in the South, the discriminatory policies of government lending agencies and county agricultural boards dominated by white farmers compounded the economic pressures on all family-scale farms. In 1920, one out of every seven farmers was black; in 1982, the national Land Loss Fund indicated that African Americans counted for only one out of sixty-seven farmers and operated only 1 percent of the farms.

Wherever there are farms within commuting distance of cities, development is eating up prime farmland at an accelerating rate.[5] According to the National Agricultural Lands Study completed in 1981, the United States was losing one million acres of prime cropland every year, or four square miles a day. Julia Freedgood of American Farmland Trust says that this rate of loss of three thousand acres a day continued through the 1990s.[6] Between 1987 and 1992, Vermont lost seventy-three acres of farmland a day. In the 1980s and 1990s, New York State lost farms at the rate of twenty a week and farmland at the rate of one hundred thousand acres a year. The country lost 8 percent of its dairy farms in the years 1996 to 1997. The 2002 Census of Agriculture found 21.5 million fewer acres in active farming than in 1997, a loss rate of 4.2 million acres a year.

Farmers have a saying: Farmers sell wholesale and buy retail. The terms of this deal get worse and worse. The index of prices farmers pay for seed, equipment, and other necessities has risen 23 percent since 1950, while the prices paid to farmers at the farm gate have fallen 60 percent. The value of the basic commodities produced by farms is sinking: Between 1978 and 1988, the wholesale price of milk fell by 11 percent, potatoes by 9 percent, fresh vegetables by 23 percent, and red meat by 37 percent. Since the passage of the 1996 Freedom to Farm Act, the domestic price paid to farmers has fallen below the crops' support prices, and the government has to make up the difference. In 1981 dairy farmers received $13.76 per hundredweight (the national average), while consumers paid $1.86 per gallon of milk. Throughout 2006, the price per hundredweight fluctuated at less than $13 a hundredweight, lower than the 1981 price even without adjusting for inflation. After deductions for payments to their milk co-op to make up for previous losses, equity in the co-op, dues, loan repayment, and local and national promotion, dairy farmer Shannon Nichols told me her family's net earnings in 2006 were $10.74 per hundredweight. Meanwhile, the price to consumers rose to $2.47 for a gallon of whole milk.[7]

The complex and confusing federal system of loans, set-asides, loan deficiency payments, and the like has not resulted in prices that cover the cost of production on the farm. Let me give an example: In 1993 the annual cash expenses for growing an acre of corn, as calculated by the Economic Research Service of the USDA, amounted to $177.89. The total gross value of selling that corn was $227.36, for an apparent profit of $49.47 an acre. However, when the government economists added in the full "ownership costs"—expenses that must be covered to keep a farm economically viable for the long run, such as capital replacement, operating capital, land, and unpaid family labor—the bottom line was minus $59.74 per acre, and this did not include money put aside for retirement. An acre of oats came in at minus $51.18, wheat minus $52.87, and milk, per hundredweight, at minus $2.02.

According to the National Agricultural Lands Study completed in 1981, the United States was losing one million acres of prime cropland every year, or four square miles a day. Julia Freedgood of American Farmland Trust says that this rate of loss of three thousand acres a day is continued in the 1990s.

Small surprise then that, farms all over the country are going out of business. In a little booklet published in 1979, *The Loss of Our Family Farms*, Mark Ritchie, founder of the Institute for Agriculture and Trade Policy and recently elected secretary of state in Minnesota, asks whether this is the inevitable course of history or the result of conscious policies.[8] He concluded that the loss of so many farms is not the unfortunate result of policies that failed but rather the result of a concerted and unrelenting drive by agribusiness, government, banking, and university forces to restructure agriculture by reducing farm price supports, manipulating the tax structure, and conducting research and development in support of large-scale agricultural enterprises. Ritchie identifies the men who made these policies as representatives of the largest corporations, banks, and universities who saw their work, in their own words, as contributing "to the preservation and strengthening of our free society."

Midsize farms, those with gross sales ranging from $10,000 to $99,999 a year, are disappearing the fastest. Remarkably, in the years between the census reports of 1997 and 2002, two size categories have actually grown. There are 33,000 more farms with ten to forty-nine acres and 544 more with more than two thousand acres. But the farms in the middle dropped by 182,236, a decline of almost 29 percent. In *Family Farming: A New Economic Vision*, Marty Strange presented convincing evidence that farms in that middle category, particularly those in the $40,000 to $250,000 income range, make better use of their resources and are more likely to practice careful land stewardship than the largest farms. For every dollar that the family-run, middle-sized farms spend, they produce more income. Production expenses on the largest farms averaged 85 percent of gross sales, while the middle farms averaged only 72 percent.[9] (On organic CSA farms such as Peacework, production expenses are even lower, in the range of

40 to 50 percent of gross sales.) Like Ritchie, Strange concluded that the obstacles to the survival of these farms come from public policy, not from poor farm management or a lack of efficiency. Since the 1950s, public policy has been pushing farms to "get bigger or get out."

In 1979 the USDA issued a report titled *A Time to Choose*, which warned that "unless present policy and programs are changed so that they counter instead of reinforce or accelerate the trends toward ever-larger farming operations, the result will be a few large farms controlling food production in only a few years."[10] Eighteen years and over three hundred thousand lost farms later, the USDA convened a National Commission on Small Farms to examine the condition of farming and its place in the food system. The commission held a series of hearings around the country where, on very short notice, hundreds of farmers and farmer advocates testified. In January 1998 the commission released its report, *A Time to Act*. This report contains 146 recommendations for policies that would protect small farmers' access to fair markets and redirect existing federal programs, which are currently skewed to serve the interests of large agribusinesses. Written by Barbara Meister, an active member of the National Campaign for Sustainable Agriculture, *A Time to Act* contains stirring language about the vital role of small farms as the embodiment of the ideals of Thomas Jefferson, and it is perfect for waving in the faces of legislators when grassroots activists lobby for the changes it suggests. Hopefully, nineteen years hence, this latest report will not be gathering dust next to old copies of *A Time to Choose*.

As overall farm numbers shrink, ever-larger farms have become more industrialized, at a staggering cost to the environment. Attempting to dominate the biological process of growing food with chemical and mechanical technologies, these industrial farms have achieved spectacular results in terms of production per acre. With chemical fertilizers, synthetic herbicides and insecticides, hybrid varieties, irrigation, and enormous, expensive machinery, yields of basic grains—corn, wheat, and rice—have doubled and tripled. However, when the Environmental Protection Agency (EPA) tested rivers, lakes, and wetlands around the country, it found that barely half could support all uses: clean water for drinking, swimming, and recreation; fish and shellfish that are safe to eat; habitat for healthy aquatic wildlife. Agriculture accounted for 72 percent of the pollution of rivers and streams with silt, runoff from fields, and excess nutrients, such as phosphorus. Forty-six different pesticides and nitrates from nitrogen fertilizers have been found in the groundwater of twenty-five states, with the largest residues in big agricultural states, such as California and Iowa. Evidence is mounting in support of the thesis Theo Colburn put forth in the book *Our Stolen Future*: that chemicals used as pesticides act as endocrine disruptors, upsetting the normal hormonal development of frogs, seagulls, polar bears, and human beings.[11] As *Consumer Reports* states, in a study of pesticide residues in food, "No one really knows what a lifetime of consuming the tiny quantities of pesticides found on foods might to do a person." While conclusive proof of this connection is still lacking, the pervasiveness of these chemicals in the environment and in our bodies should be cause for enough alarm to jar us into action.

In a fine essay on CSA in the collection *Rooted in the Land*, Massachusetts farmer Jack Kittredge sums up the loss of soil to erosion:

> Much of the incredible productivity of North American industrial agriculture has been based on using up two irreplaceable capital assets: topsoil and petroleum. When our ancestors settled this continent, they benefited from the largesse of thousands of years of natural soil creation, virtually undiminished by agriculture. The careless practices of our modern world have seen up to half of that soil washed or blown away, and each year every acre loses an average of 7.7 tons more. Only when the last of this prehistoric legacy is washed to sea and we are on a "level playing field" with the other continents will we appreciate the magnitude of our folly.[12]

Political activist Jim Hightower has a clear grasp

of the logic that's behind the industrial farming approach: "If brute force isn't working, you're probably not using enough of it."[13] Insects, diseases, and weeds were reducing crops worldwide by 34.9 percent in 1965. Pesticide applications in the United States totaled 540 million pounds of active ingredients. By 1990, the National Center for Food and Agriculture Policy reported that losses to pests rose to 42.1 percent, even though U.S. farmers poured on 886 million pounds of pesticides. Pesticide company profits continued to climb in 1997. Novartis, the largest pesticide company in the world, sold $4.2 billion worth of agrochemicals, up 21 percent from 1996, according to the Pesticide Action Network of North America.[14] In the Worldwatch Institute's *State of the World* for 1998, Lester Brown argued that the world has surpassed the environmental limits to continued increase in agricultural yields: Soil erosion, plus declining underground water supplies, plus climate change due to global warming overbalance any further benefit from additional fertilizers, pesticides, or improved varieties.[15] This is the reality of a too-full world.

The biggest farms may be getting bigger, but the farming sector of the food system is losing control to the increasingly consolidated multinational corporations such as Syngenta and Archer Daniels Midland. Many of the once-self-employed farmers have become employees at larger farms or have found employment in the farm inputs (seed, fertilizer, pesticides, etc.), processing, and marketing sectors, which are returning 18 percent on investment and grabbing ever-larger portions of the consumer food dollar from the farms. Stewart N. Smith, former commissioner of agriculture for Maine and an agricultural economist, has traced the downward trajectory of farming.[16]

According to Smith's calculations, if current trends continue, farming as such will disappear completely by the year 2020.

The establishment of the World Trade Organization (WTO) in 1994 and the passage of NAFTA in 1997 have removed what few protections were left for U.S. agricultural producers. Under WTO regulations, U.S. agricultural exports have declined while imports have gone up 32 percent. WTO rules require that member countries allow imports of 5 percent of the volume of agricultural products consumed within their borders, whether they need them or not. In 2005 this country imported $48.7 billion worth of food products, and that wasn't all coffee, chocolate, and bananas. It included many crops we produce right here—dairy products, grains, meat, fruit, and vegetables. The favorable contribution of agricultural exports to the balance of trade has all but disappeared, outweighing imports by only $2.6 billion in 2005. Net farm income for 2006, according to the Economic Research Service of the USDA, was $54.4 billion, down $19.4 billion from 2005, a drop of 26 percent, suggesting that imports are crowding out U.S. farm products.

According to a USDA study released in 1997, the economic impact of NAFTA on the balance of agricultural trade between the United States and its two neighbors has been a negative $100 million.[17] Mexican tomatoes are underselling Florida tomatoes. U.S. corn growers, however, favor NAFTA because it opened Mexican markets to them and even helped raise the price of corn. The cost has fallen on Mexico, where Steve Suppan and Karen Lehman report that six hundred thousand to one million small corn farmers, who could not compete with the lower price of the imported corn, have been uprooted from the countryside and forced into the vast impoverished shantytowns of the unemployed in the cities. Too late to save the farms of these small-scale farmers, the ethanol fever in 2006 and 2007 has almost doubled corn prices, setting off the "tortilla crisis" among the poor in Mexico.[18]

When international trade heats up, the greatest benefits go to the corporations that control the markets. Fewer than five companies control 90 percent of the export market for corn, wheat, coffee, tea, pineapple, cotton, tobacco, jute, and forest products, according to William Heffernan. Those same big traders—Cargill, Continental Grain, Bunge, Luis Dreyfus, Andre and Co., and Mitsui/Cook—also control storage, transport, and food processing. Incidentally, Daniel Amstutz, a

15

former Cargill executive, drafted the U.S. agricultural proposal for GATT under President Reagan.

The consolidation of control of the food system inside the United States is also increasing steadily. Keeping up with the mergers in the food system alone would be a full-time job. Tom Lyson has calculated that 10 cents of every food dollar spent in this country goes to Altria (the suggestively altruistic renaming of Philip Morris),[19] a conglomerate of nine tobacco brands and the owner of Miller Brewing, 7-Up, Post Cereals, Maxwell House Coffee, Sanka, Jell-O, Oscar Mayer, Log Cabin Syrup, and more, with total sales in 1995 of $36 billion. Altria owns 84 percent of Kraft Foods, which controls 40 percent of the cheese market in North America, with annual sales of $29 billion. The *E. coli* scare of 2006 revealed that only two companies control over 70 of all the bagged salad mix in the country. As Karen Lehman and Al Krebs put it:

> Between January 1 and January 31, 1995, while most Americans were still figuring out how to break their New Year's resolutions, Philip Morris merged Kraft and General Foods into Kraft Foods; Ralston Purina sold Continental Baking Company to Interstate Bakeries Corporation, the nation's largest bread maker; Perdue Farms, the nation's fourth-largest poultry producer, acquired Showell Farms, the nation's tenth-largest poultry producer; and Grand Metropolitan proposed to acquire Pet, Inc. The brand names are all that are left of the small companies that became huge conglomerates through mergers and acquisitions.[20]

The four major packers control 83.5 percent of the beef slaughter, the greatest concentration in U.S. history.[21] What does this mean for consumers and small farmers? When there are only one or two buyers, farmers have to take what prices they can get if they want to sell their crops. If they don't cooperate with the big packers, farmers can find themselves without any buyers at all. Between 1979 and 1997, the producer share of retail beef sales dropped from 64 percent to 49 percent, as the price farmers received for slaughter steer fell 50 percent. Consumers ended up paying less, but only by 15 percent, while the packing companies enjoyed unprecedented profits.[22] The six multinational corporations that control the genetically engineered crops are forming clusters with the big grain and meat processors and the largest retailers: "Crop farmers are . . . being locked into food chain clusters through 'bundling,' or linking patented seeds with contracts, chemicals and credit."[23] Antitrust legislation, the Sherman Antitrust Act of 1890, requires that when as few as four companies gain control of 60 percent of any sector, the government must take action. Passed in response to public anger over monopolistic concentration, the act assumes that dominance of an industry or market by a few firms will damage the public by raising prices, reducing quality, and slowing technological advance. Obviously, the government has chosen not to enforce this set of laws very often.

We are living through the culmination of the era of the transnationals, the megacorporations such as Altria, Cargill, Archer Daniels Midland ("Supermarket to the World"), Wal-Mart, and ConAgra. They have no allegiance to any particular nation or group of nations, and they are manipulating the rules of the WTO, established by GATT, to increase their power to intervene in the economy of any country as they see fit.[24] The power of the WTO was demonstrated when it struck down as barriers to trade the European Union's ban on U.S. hormone-treated beef and genetically modified grains. President Bush wants the fast-track authority to expand NAFTA to the Free Trade Area of the Americas (FTAA). This would represent, in the words of some, "GATT on steroids," the final link in the choke collar the multinationals have been forging behind closed doors. If implemented, the FTAA would give corporations the freedom to do business in any country in the Americas without the control of the national government. If a government objects that a multinational is violating local environmental or labor laws, the multinational will be able to take that government to a world court, where unelected trade bureaucrats will issue the binding judgments.

The widening abyss between the wealthiest and

family business

both drizzled with grey and not so slim around the middle
anymore, a woman and a dog endure the heat
side by side in a meadow shimmering life. they have
shared many miles of the Good Red road, 4 feet 2 feet
plod and dance, trot and stumble, lope and scramble
right on down to this afternoon hunkered over under
the sun so hot. the dog's head is nestled in sweet
grass at the streambank, the woman's is bent over
hairy weeds surrounding baby lettuces, and deer flies
whiz between them like agents of a curse. "sissies,"
hiss the hairy weeds; they sneer, "we will smother
you." and the woman's hands tan as the soil and just
as lined pull them out one by one and tuck them in
beside the babies.
"why is it the bad so often seem so strong
while the good get by on grace?" she asks the dog who
smiles and wags but does not raise her head. so the
woman calls out clear as crows discussing family business
across the pasture at first light, "what I mean is, how
can something that shines so true and mighty be so fragile
when shadows pass and blot out the brightness so easily
before passing away again?"
and the dog digs thoughtfully into her ear
limned with grey, cloudy brown eyes focused on something
near the horizon that the woman lifts her gaze to see
but doesn't find. she speaks once more but quiet now
as the stream murmuring to itself. "what I do understand is
you, old dog. you and the sun and these damned flies, these
hairy weeds and baby lettuces. I know what to pull and
what to save, and where to put them all. and just who will
sit out here beside me in this heat until the work is done."

—Sherrie Mickel, 1997

the rest of us is as true in agriculture as it is in other parts of the economy. In the United States, as in Europe, 70 percent of government payments go to the top 10 percent of the farms by gross income, making the rich even richer. That champion of free trade, former Federal Reserve chairman Alan Greenspan, now retired, in public testimony at the Joint Economic Committee in Congress in June 2005, pointed out that the top 20 percent of the workforce have enjoyed all of the income growth recently. At the same time, the remaining 80 percent of workers have seen little, if any growth in income and farmworkers have actually fallen farther behind. It is amazing to hear Greenspan warn that the income gap between the rich and the rest of the population has become so wide, and is growing so fast, that it might eventually threaten the stability of democratic capitalism itself. "As I've often said, this is not the type of thing which a democratic society— a capitalist democratic society—can really accept without addressing," he remarked.

From the consumer's point of view, the source of food lies hidden behind an almost impenetrable wall of plastic and petroleum. Few stores bother to label food with its point of origin. The 7 to 10 percent value of the raw food in processed products is buried by the other 90 percent—the chopping, blending, cooking, extruding, packaging, distributing, and advertising. Gary Argiropoulos, vegetable and floral manager for Hannaford Brothers in Portland, Maine, told me that *fresh* and *local* in the language of super-market produce buyers means accessible within twenty-four hours by airfreight. I once heard Michael Osterholm, the Minnesota State Epidemiologist, talking about food safety on National Public Radio. He attributed the increase in food-borne illnesses, at least in part, to eating out-of-season produce shipped from countries with low health standards; to so many of the food workers being low-paid, uneducated, and lacking in proper health care; and to the rise of drug resistance in microorganisms such as *Campylobacter* in antibiotic-fed chickens.[25] Imports alone are not to blame for our food insecurity; domestic industrialized produce and meat businesses that combine products from many farms before shipping them thousands of miles cause their share of the food poisonings as well. Judith Hoffman has summed up well the widening gulf that the global supermarket leaves between farmer and consumer: "The *inter-* of our interdependence is gone and only the dependence remains. This kind of dependence is the essence of insecurity because it permits exploitation. By the time the consumer's dollar has made it to the farmer, there are mere pennies left of it. By the time the farmer's food has made it to the consumer, it has been bled dry of flavor and passed through a gauntlet of dismayingly mysterious chemistry. It seems all that farmers and consumers have left in common as the century turns is anxiety."[26]

The Movement for Sustainable Food Systems

Every aspect of our lives is, in a sense,
a vote for the kind of world
we want to live in.
—FRANCES MOORE LAPPÉ

Ever since CSA's origins in this country back in 1985, the people who initiated Community Supported Agriculture have hoped it would provide an antidote to some of the worst aspects of the prevailing food system. Jan Vandertuin, who brought the idea from Switzerland to Robyn Van En at Indian Line Farm in Massachusetts, wrote "Vegetables for All." In this essay, Jan describes his frustrations at trying to find work producing quality vegetables: "Good-looking vegetables are often produced with the use of pesticides and herbicides. Fresh may mean storage and transportation methods that are energy-intensive—or questionable, such as irradiation. Organic produce often means overworked, underpaid agricultural workers. And reasonably priced generally means as cheap as possible without any regard for the hidden costs—government subsidies, market manipulation, exploitation of the Third World, and pollution of the environment."

Jan's search for a situation that would give recog-nition to the value of agricultural work led him from organic farms and whole-grain bakeries in the United States to the producer-consumer associations in Switzerland (see chapter 22, "CSA around the World"), which served as the inspiration for CSA in the southern Berkshires of Massachusetts. CSA has been able to spread rapidly across North America because it is part of a much larger social movement—the movement for a sustainable food system.

Populist in spirit, with strong feelings for civil rights and social justice, and an underlying spirituality, the movement for sustainable agriculture and regional food systems is not linked with any political party or religious sect. It is firmly grounded in every region in the country, encompassing organic and low-input farmers; organizations concerned with food, farming, farmworkers, community food security, social justice, domestic fair trade, and hunger; advocates for the humane treatment of animals; and environmental, consumer, and religious groups. Community Supported Agriculture is an important part of this movement.

Deciding to join such a movement is like jumping into a swiftly flowing river of icy water and swimming against the current. Yet all over the country, thousands of people with their eyes open have been doing just this. At first a trickle of isolated farmers resisted the pressures to adopt chemicals and specialize their farming on single crops. Lone consumers, hunting for food that was unadulterated and natural, joined together to create food co-ops. In the 1970s, small grassroots organizations formed: the Natural Organic Farmers Association in the Northeast, inspired by J. I. Rodale and Sir Albert Howard (renamed the Northeast Organic Farming Association in the 1990s); the Center for Rural Affairs in Walthill, Nebraska, dedicated to stopping the decline of the family farm; the Federation of Southern Co-ops, emerging from the civil rights movement to keep black farmers on their land; and many, many others.

Over the past decade, this scattering of local organizations has swelled into a significant social movement with a national network and effective policy

My farm partner Katie Lavin and I pick garlic greens with members of the Genesee Valley Organic CSA. PHOTO BY KATE LATTANZIO.

wing. In 1995 the National Campaign for Sustainable Agriculture coordinated the efforts of over five hundred member organizations and farms in lobbying for policy proposals for the Farm Bill. For the 2007 Farm Bill, an even broader range of environmental and civic organizations are negotiating their different perspectives to form a united front for change in U.S. food policy. In 1998 over 275,000 people sent in comments when the USDA proposed nationwide regulations for organic agriculture that would have made "organic" indistinguishable from conventional industrial production. Buy Local campaigns are gaining momentum in communities all over the country. The international Slow Food movement is challenging fast, cheap, but low-quality meals, and "localvores" are forming small "pods" in many areas, eating only food grown within 100, or 50, or even 25 miles of their homes. Colleges, schools, and hospitals are heeding the demand for meals made from local farm products. In a few inner cities, residents are taking over abandoned land and schoolyards to create community gardens and farm stands. Rural sociologist Tom Lyson has labeled this movement "civic agriculture," which his student Heidi Mouillesseaux-Kunzman characterizes in these terms:

> Food citizens (producers, including laborers, and consumers from rural and urban locations) who depend on food for their survival are seen as having a vested interest in and responsibility for ensuring that land is available to farm, that producers and laborers are paid well, and that the food produced is nutritious, accessible, affordable, and culturally appropriate. Likewise, nature's role in the system is not only relative to its capacity as an input provider, production determiner (weather/soil) and waste absorber, but as a factor that influences and is influenced by the other components of the system. Nature is, moreover, valued in and of itself such that proponents of a civic food system strive to ensure that

Peacework intern Andy Simmons and farmer Greg Palmer put up a hoop house with help from CSA members. PHOTO BY CRAIG DILGER.

the influence of the food system upon nature is good or healthy.[27]

Bucking the dominant system both literally and intellectually, the people involved in sustainable agriculture have had to struggle to move beyond reductionist science and "best management practices," which seek to "improve" farming one problem at a time, to thinking in terms of interdependent systems. Biodynamic and organic farmers, with their holistic approach to the farm as an integrated part of its ecosystem, as well as practitioners of permaculture design and holistic resource management, have all contributed to the "paradigm shift" in the way we look at food and farming. Although still referred to as *sustainable agriculture*, that term has come to mean food production in the context of the farm's physical, social, and economic environment. On its Web site, Sustainable Agriculture Research and Education offers this broad and interactive definition:

Sustainable agriculture does not refer to a prescribed set of practices. Instead, it challenges producers to think about the long-term implications of practices and the broad interactions and dynamics of agricultural systems. It also invites consumers to get more involved in agriculture by learning more about and becoming active participants in their food systems. A key goal is to understand agriculture from an ecological perspective—in terms of nutrient and energy dynamics, and interactions among plants, animals, insects, and other organisms in agroecosystems—then balance it with profit, community, and consumer needs.[28]

One of my daydreams is that our movement for sustainable food systems will somehow find a democratic and participatory way to create a set of holistic goals for our future. With our brothers and sisters of the land, the whole that we manage is the entire earth, the participants—all the earth's peoples, their domestic livestock and the uncountable inhabitants of the soil.

Here is a first draft set of our goals:

1. **Quality of life:** a world of peaceful, cooperative, self-reliant communities. Resources shared justly among them. No hunger; enough food that everyone is adequately nourished with food of his or her cultural preference. With adequate food recognized as a human right and food sovereignty as the right of each nation, no one is forced to leave home to seek migrant labor in a foreign land. Curiosity about other people's ways. Cultural cross-pollination on a basis of equality. Tolerance of differences. Rich spirituality.

2. **Mode of production:** many small-scale farms and gardens, run by families, tribes, or neighborhoods, clustered into cooperatives for purchasing and sales.

3. **Staple foods produced where they are eaten:** trade in surplus production at prices that cover the producers' costs while neither gouging nor undermining the economy of the buyers.

4. **Future resource base:** a world of clean air, water, and regenerated soils. Oceans and rivers teeming with fish. No pollution, no erosion, no toxic landfills or dumps. Energy from renewable sources—wind, solar, and geothermal power. Healthy farms and gardens carefully balanced with the ecology of each region.

Proliferating CSAs in every community will help make these goals a reality. The very first CSA proposal in 1985 was a sort of manifesto pronouncing similar ideals for agriculture:

Respect for
- animate Earth
- wild plants and animals
- domesticated plants and animals
- environmental limits
- cycles of nature, including the growing season and animal breeding season
- food workers' physical, social, spiritual needs

Responsibility to
- use organic/Biodynamic methods and insights, including growing own seeds/seedlings (local exchange?)
- composting kitchen wastes for garden plan with balanced variety
- quality storing/preserving for year-round supply

Be energy-conscious in
- production and distribution
- minimal machinery
- local distribution network
- human or animal power

Maintain decent working condition
- decent wages, i.e., $6–7 per hour[29] [$15 in 2007, EH]
- limited hours, i.e., 35–42 hours per week
- fully integrate and pay trainees

Emphasize the therapeutic value of agricultural work (for non-gardeners)

Support community control of the land
- balanced land-use plan
- special provisions for agriculture
- eliminate speculation

Create local social/economic forms, based on trust, which
- encourage initiative and self-reliance
- share the risks of agricultural production
- share information
- are human-scale and efficient
- charge according to needs/cost (not market)
- provide locally controlled financial services (currency, banking, insurance, etc.)

Think globally, act locally.

The Way Is as Important as the Goal[30]

As we go through the achievements of CSAs in this book, we should look back on the preceding statement of principles from time to time to see how many of these ideals are being realized.

In *Farms of Tomorrow Revisited*, Trauger Groh

reacts against the corporate miasma with an appealing utopian vision of a future in which farmland ceases to be privately owned real estate and farmers cease to be businessmen. As Trauger paints them, the "farms of tomorrow" are individualized organisms attuned to cosmic rhythms and based on three principles:

1. the highest diversity brings the highest productivity;
2. harmony needs animals for a balance of crops for market and cover crops;
3. the farm is a closed system, producing and recycling its own fertility.

The formation of community farms "liberates the farmer to work out of his spiritual intentions, not out of money considerations." The members of these community farms, in Trauger's conception, come together around shared ideals to cover all the costs and share all the risks of their farms, and the farms link together in an "associative economy" in which spirituality prevails:

> In an associative economy, we associate with our partners—active farmers among themselves, active farmers with all the member households, farm communities with other farm communities. The prevailing attitude is a striving to learn the real needs of our partners, and the ways we can best meet them. That means we do not make our self-interest the driving force of our economic behavior, but rather we take from the needs of our partners the motivation of our economic actions. We believe and trust that this will lead to the greatest welfare of all involved.[31]

In what Trauger sees as a period of diminishing life force in nature, the farm radiates life-giving energy. To CSAs he assigns a broad set of tasks: education of the young, the revival of ethics, and the renewal of human health, culture, the economy, and social life. While I find Trauger's ideas appealing, I believe we

> We may have to drop some of our favorite jargon—even the word *sustainable*—and talk, instead, about keeping farms in business for the long term, making sure that everybody gets enough nourishing food to eat, and living in a way that respects the natural limits of the world around us.

will be successful in transforming human relations only if we think in more political terms and link our community farms with the national and international networks that oppose the forces of neoliberalism and neocolonial globalization. In "Other Economies Are Possible," Ethan Miller asks the questions: "Is it possible that courageous and dedicated grassroots economic activists worldwide, forging paths that meet the basic needs of their communities while cultivating democracy and justice, are planting the seeds of another economy in our midst? Could a process of horizontal networking, linking diverse democratic alternatives and social change organizations together in webs of mutual recognition and support, generate a social movement and economic vision capable of challenging the global capitalist order?" And he responds: "To these audacious suggestions, economic activists around the world organizing under the banner of *economia solidaria*, or 'solidarity economy,' would answer a resounding 'yes!'"[32]

Wendell Berry, the revered farmer-poet-novelist and inspiration of the sustainable agriculture movement, lambastes the system of multinational corporations for its institutionalized irresponsibility. Berry is convinced that the old political parties—left, right, and center—have outlived their usefulness. To bring about a sustainable world, he suggests, we must understand that the leaders of all of those parties have sold out to the corporations and really amount to only one party—the party of the global economy, a party that is "large, though not populous, immensely powerful, self-aware, purposeful, and tightly organized." Totally unrepresented by that party are the ordinary people, both rural and urban, who share an appreciation for the value of neighborly acts; the urgency of protecting the purity of local land, water, air, and wild creatures; and the need to build cooperative links between farmers and community-minded people in nearby towns and cities.[33]

We don't seem to know it yet, but CSAs and our

During the UN Conference on Environment and Development, in 1992, a number of nongovernmental organizations (NGOs) drafted their own NGO Sustainable Agriculture Treaty, which states:

> Sustainable Agriculture is a model of social and economic organization based on an equitable and participatory vision of development, which recognizes the environment and natural resources as the foundation of economic activity. Agriculture is sustainable when it is ecologically sound, economically viable, socially just, culturally appropriate and based on a holistic scientific approach.
>
> Sustainable Agriculture preserves biodiversity, maintains soil, fertility and water purity, conserves and improves the chemical, physical and biological qualities of the soil, recycles natural resources and conserves energy.
>
> Sustainable Agriculture uses locally available renewable resources, appropriate and affordable technologies, and minimizes the use of external and purchased inputs, thereby increasing local independence and self-sufficiency and insuring a source of stable income for peasants, family farms, small farmers and rural communities, and integrates humans with their environment.
>
> Sustainable Agriculture respects the ecological principles of diversity and interdependence and uses the insights of modern science to improve rather than displace the traditional wisdom accumulated over centuries by innumerable farmers around the world.

more conservative rural neighbors belong together in the second party, what Berry calls the party of the local economy. So do the urban people who join CSAs, without any thought about the broader implications, simply to get fresh, organically grown food. I hope we will be able to spread the word about community supported agriculture in language that opens doors to people. We may have to drop some of our favorite jargon—even the word *sustainable*—and talk, instead, about keeping farms in business for the long term, making sure that everybody gets enough nourishing food to eat, and living in a way that respects the natural limits of the world around us. We need to find that place where the conservatism in organic farming meets the conservatism of small-town Republicans and Democrats and the radicalism of the Global Network of the Solidarity Socioeconomy and Via Campesina, the worldwide network of farmer's and farmworkers' organizations.

Sustainable local food production and community supported agriculture are also essential to sustainable development. Herman E. Daly emphatically distinguishes sustainable development from "sustainable growth," which he terms an oxymoron. In his words, "An economy in sustainable development adapts and improves in knowledge, organization, technical efficiency, and wisdom; it does this without assimilating or accreting an ever greater percentage of the matter-energy of the ecosystem into itself, but rather stops at a scale at which the remaining ecosystem can continue to function and renew itself year after year."[34]

Those who want to undertake sustainable development would do well to look upon CSAs as places where farming and nonfarming people are learning to create local economies that fit Daly's definition, providing for people's needs while regenerating resources. A farmer interviewed by Jack Kittredge summarized the value of CSAs:

> CSA seems like the best way to establish relationships between people who farm and consumers. It brings appreciation of the other's point of view to both sides. It helps break the wholesale/grocery store mindset, which dehumanizes farm work and makes the Earth seem like a natural resource to be used rather than cared for. It helps farmers to have a community (since the community of fellow farmers is so small). It helps provide financial stability to farmers by knowing what the year has in store for them and evens out cash flow. I like it personally, agriculturally, and financially.[35]

Sharing the Harvest will take you into the intense world of community supported agriculture. A world of heady ideals and hard physical labor. A world where people from very different backgrounds and perspectives are struggling to learn how to do practical work together to create a new food system based on these values:

- an intimate relation with our food and the land on which it is grown;

- a sense of reverence for life;
- cooperation;
- justice;
- appreciation for the beauty of the cultivated landscape; and
- a fitting humility about the place of human beings in the scheme of nature.

CHAPTER 2 NOTES

1. Barnaby J. Feder, "Getting Biotechnology Set to Hatch," *New York Times*, May 2, 1998, D1, D15.

2. Ray Goldberg, "New International Linkages Shaping the U.S. Food System," in *Food and Agricultural Markets: The Quiet Revolution*, ed. Lyle P. Schertz and Lynn M. Daft. NPA Report #270. (Washington, DC: Economic Research Service, USDA, Food and Agriculture Committee and National Planning Association, 1994).

3. Parke Wilde and Mark Nord, "Effect of Food Stamps on Food Security: A Panel Data Approach," *Review of Agricultural Economics* 27, no. 3 (Sept 2005).

4. Samir Amin, "World Poverty, Pauperization and Capital Accumulation," *Monthly Review* (October 2003).

5. See American Farmland Trust study *Farming on the Edge*, 1997. American Farmland Trust continues to update its reports on the loss of farmland on its Web site (www.farmland.org).

6. Julia Freedgood (presentation to Farmland Preservation workshop, SARE Tenth Anniversary, Austin, TX, March 5–6, 1998).

7. $2.47 was the price for a gallon of whole milk in Wegman's, the local supermarket, in Rochester, New York, in September 2006. Store prices vary, of course, and frugal shoppers can follow sales. "Milk Prices Expected to Remain Low into 2007," an article in *Grassroots*, the New York Farm Bureau monthly newspaper, quoted Robert Wellington, senior vice president of economics, communications and legislative affairs for Agri-Mark, a milk marketing co-op, as stating that farm milk prices in 2006 would likely average $2.35 per hundredweight below 2005 prices and that the current milk situation would continue into 2007.

8. "Cash Receipts and Farm Income," New York Agricultural Statistics (September 1995).

9. Marty Strange, *Family Farming: A New Economic Vision* (Lincoln: University of Nebraska Press, 1988).

10. USDA, *A Time to Choose: Summary Report on the Structure of Agriculture* (Washington, DC: USDA, 1981). USDA. *A Time to Act*. Washington, DC: USDA, January 1998.

11. Theo Colborn, Dianne Dumanoski, and John Peterson Meyers, *Our Stolen Future: Are We Threatening Our Fertility, Intelligence, and Survival?* (New York: Dutton, 1996).

12. Jack Kittredge, "Community Supported Agriculture: Rediscovering Community," in *Rooted in the Land: Essays on Community and Place*, ed. William Vitek and Wes Jackson (New Haven: Yale University Press, 1996), 26.

13. Jim Hightower, *There's Nothing in the Middle of the Road but Yellow Stripes and Dead Armadillos* (New York: Harper Collins, 1997).

14. Pesticide Action Network of North America. Sustainable Agriculture Network [SAN is a branch of SARE] List-serv, PANNA Updates Service, panna@panna.org (accessed May 1988).

15. Lester R. Brown, *State of the World 1998* (New York: W. W. Norton / Worldwatch Institute, 1998).

16. "The food and agriculture system has changed remarkably through this century under the regime of industrial agriculture, especially in shifting economic activities from the farm to the non-farm components of the system. Farmers contributed 41 percent of the system activity (and got 41 percent of the returns) in 1910, but only 9 percent in 1990. On the other hand, input suppliers increased their share from 15 percent to 24 percent, and marketers from 44 to 67 percent." Stewart N. Smith, "Farming Activities and Family Farms: Getting the Concepts Right" (presentation at the Joint Economic Committee Symposium, Washington, DC, October 21, 1992).

17. Alan Guebert, "NAFTA Is Proving to Be a Disaster," *Agri News* 9 (October 1997).

18. In addition to this displacement, there is also the grave threat to traditional corn varieties, and diversity, through the introduction—legal or otherwise—of GM corn to Mexico, which is probably its birthplace and certainly its center of diversity.

19. Tom Lyson, "The House That Tobacco Built," *Food, Farm and Consumer Forum* 6 (January 1988): 2.

20. Karen Lehman and Al Krebs, "Control of the World's Food Supply," in *The Case against the Global Economy*, ed. Jerry Mander and Edward Goldsmith. (San Francisco: Sierra Books, 1996), 124.

21. *Cattle Buyers Weekly*, "Steer and Heifer Slaughter," reported in *Feedstuffs* (6/16/03).
22. USDA Research Service, *Red Meat Yearbook* (Beltsville, MD: USDA Research Service, 1997).
23. Phil Howard, "Consolidation in Food and Agriculture," *The Natural Farmer* (Spring 2006): 18.
24. William D. Heffernan, "Domination of World Agriculture by Transnational Corporations (TNCs)," in *For All Generations: Making World Agriculture More Sustainable*, ed. J. Patrick Madden and Scott G. Chaplowe (Glendale, CA: WSAA, 1997), 173–81.
25. Michael Osterholm, Minnesota State Epidemiologist (interview on *Fresh Air*, National Public Radio, May 5, 1998).
26. Judith Hoffman, "CSA: A Two-Part Discussion." *Small Farmer's Journal* 19, no. 4 (Fall): 28–29. By mid-September 2006 the Food and Drug Administration had reported 2 deaths from eating bagged spinach and issued a nationwide advisory. Over 200 people in 26 states were sickened by the spinach. There are 73,000 cases of *E. coli* food poisoning reported annually in the United States, and 61 people die each year. In October 2006 there was a recall on Foxy brand lettuce and reports of *E. coli* in ground beef and botulism in organic carrot juice. It's worth pointing out that the huge Earthbound organic label was implicated in spinach contamination, so organic does not necessarily guarantee safety when industrial farms and food processing are at work. Michael Pollan cites an estimate from the Centers for Disease Control and Prevention that 76 million Americans suffer from food poisoning every year, putting 300,000 of them in the hospital and killing 5,000 (Pollan, "The Vegetable-Industrial Complex," *New York Times Magazine*, November 15, 2006).
27. Heidi Mouillesseaux-Kunzman, "Civic and Capitalist Food System Paradigms: A Framework for Understanding Community Supported Agriculture Impediments and Strategies for Success" (master's thesis, Cornell University, 2005).
28. Sustainable Agriculture Research and Education, www.sare.org.
29. Remember, this was 1985: $6 or $7 then would be double that today.
30. Robyn Van En, ed., *Basic Formula to Create Community Supported Agriculture* (Great Barrington, MA, 1988).
31. McFadden and Groh, *Farms of Tomorrow Revisited*, 35.
32. Ethan Miller, "Other Economics are Possible," *Dollars and Sense*, July/August, 2006, 11.
33. Berry, Wendell. "Conserving Communities." In *The Case Against the Global Economy and for a Turn Toward the Local*, ed. Jerry Mander and Edward Goldsmith. San Francisco: Sierra Books, 1996, 412.
34. Daly, Herman E. "Sustainable Growth? No Thank You." In *The Case Against the Global Economy*, ed. Jerry Mander and Edward Goldsmith. San Francisco: Sierra Books, 1996, 195.
35. Kittredge, Jack. "CSAs in the Northeast: The Farmers Speak." *The Natural Farmer* (Summer 1993): 12.

CREATING A CSA

> A world that supranational corporations and the governments and educational systems
> that serve them . . . control . . . will be . . . a postagricultural world. But as we now begin to
> see, you cannot have a postagricultural world that is not also postdemocratic, postreligious,
> and postnatural—in other words it will be posthuman, contrary to the best that we have
> meant by humanity.
>
> —WENDELL BERRY, "Conserving Communities"

Starting a Community Supported Agriculture (CSA) project is a little like having a baby—you unleash biological and social forces that may take you in directions you never expected. CSAs have come into being in many different ways: from existing farms; singly or in groups; for part or all of their production; established by institutions, such as land trusts, religious orders, or food banks; or improvised by would-be farmers or groups of consumers. There are many common elements, but each birth is unique. Here are some creation stories.

The origins of the Genesee Valley Organic CSA (GVOCSA) reach back to a box of organic vegetables I found on my doorstep in France during the summer of 1977. I had sublet an apartment from friends in the walled town of La Cadière in Provence. They had told me they were Maoists, but they had not told me that their cell was supporting two farmer comrades who delivered a box of vegetables once a week. I was already toying with the idea of market gardening, so curiosity about the box quickly led me to visit the little rented farm. My most vivid memories are of the cement-like soil and baby goats on the roof of our car. The farmer comrades were not members of the tight-knit Provencal farming community. When they moved on, their project left no trace. Twenty-five years later, the Vuillon family, whose farm is only a few miles from the plot the Maoists rented, discovered CSA and established the first Association pour le Maintien d'une Agriculture Paysanne (AMAP), setting off a movement that has spread to over three hundred farms in France (see chapter 22, "CSA around the World").

The idea stayed with me, though, and when I moved to Rose Valley in 1988 and heard about Indian Line Farm, I proposed to my partner David that we try a CSA. During the winter of 1989, David and I had a meeting with Alison Clarke, founder and staff member of the Peace and Justice Education Center (PJEC; soon to become Politics of Food, or POF, and in 2005, Rochester Roots) in Rochester, New York, and PJEC member Jim Marks, a Xerox engineer. We agreed to the broad outlines of the project and decided to put out a flyer inviting people to an organizational meeting.

By today's standards, the flyer was pretty crude, half-typed with headlines written in by hand, but when posted around the Genesee Valley Food Co-op and Ozone Brothers store in Rochester, and mailed to the members of PJEC and the local chapter of the

Human Ecology Action League (HEAL), it attracted twenty-four people. The flyer read:

> Do you want fresh-picked, organically grown vegetables? Join in Community Supported Agriculture, a cooperative experiment of Rose Valley Farm and Rochester PJEC. Support ecological agriculture by signing up for six months of farm-fresh organic vegetables. Receive a family-sized variety pack once a week, delivered to a central pickup point in Rochester. Volunteer to help make the farm work by doing a share of harvesting, packing, or weeding. Make the farm connection—don't pay the grocer!

At the initial organizational meeting, we emphasized that this was to be an experiment. To make it work, everyone would have to be flexible, willing to participate, and ready to readjust. All twenty-four households signed up and made the commitment to share in the labor. As it turned out, a few of the chemically sensitive HEAL members were too sick to do more than make occasional phone calls. The participants filled out vegetable order forms that enabled us to set the contents of the weekly packets. Vegetarian households decided to purchase two or more packets, bringing the total to thirty-one. We agreed that everyone would receive the same selection, except for members on macrobiotic diets, who could substitute turnip greens and collards for tomatoes and eggplant. We set the fee on a sliding scale of $5 to $7 a week for twenty-five weeks, a modest amount of money, but at the request of the members made the packets much smaller than those of many other CSAs. We designed a "food unit" for a two-person city household. Rather than requiring a lump sum payment at the beginning of the season, we asked for monthly payments in the hope of making CSA affordable for the lower-income members. Social justice has always been high on the agenda of the Politics of Food, a commitment shared by the people at our founding meeting.

In his 1993 survey of CSAs around the country, Tim Laird found that farmers initiated 79 percent, farmers and consumers together 6 percent, and consumers alone 5 percent of CSAs. . . . Prior to forming these CSAs, 49 percent of the farmers were not farming. Few of them had jobs even remotely connected with agriculture.

We learned a lot from our mistakes that first season. The distribution of the food was the biggest challenge. The members who worked at the farm on each harvest day transported the food the full hour's drive into the city of Rochester, where most of the members lived, and stored it in the coolers at a church. At six o'clock in the evening, two assigned members came to the church to separate the produce into shares. Even with a posted description of the process in excruciating detail—one that would have been the envy of any technical writer—the distribution did not always go smoothly. We concluded that distribution needed a trained coordinator. One person, Jamie Whitbeck, shouldered the task of scheduling all members' work for the entire season. Jamie did an incredibly conscientious job and burned himself out. (He dropped out of the group, but rejoined seventeen years later!) We learned that big jobs need to be shared by several people. Jan Cox, the bookkeeper-treasurer for the project, eventually collected all the money due but not without repeated reminders, calls, and cards to the tardy. The next year, we instituted a contract with a commitment to a definite weekly fee and a clear payment schedule. We also stopped allowing members to cancel their shares if they went away for a week or two, which had meant that every week the number of shares was different.

We held a dinner to celebrate the end of the season, the beginning of a lovely annual ritual. The consensus of the members present was that the experiment, despite a few organizational flaws, was worth repeating. All but three of the families signed up for the second year.

AN ARTICLE ON THE KIMBERTON CSA in *Mother Earth News* in 1989 inspired Jean Mills and Carol Eichelberger to turn their Alabama farm into a CSA. The two women had been partners in a nonfarm business for seven years while living on the farm

PART TWO

GETTING STARTED

where Jean grew up. They wanted an excuse to stay at home on the farm and to expand their organic garden, so they shared the article with friends and business contacts. Twenty people attended an organizing meeting and eighteen signed up. By the time the 1990 growing season began, the project had attracted twenty-nine more. Most of the members were professional people from Tuscaloosa, a nearby university town, joined by a sprinkling of old-time residents. They discussed and settled on share cost, crops, and a distribution system. To get more information about running their CSA, Jean and Carol attended the first CSA conference in Kimberton, Pennsylvania, that winter. "It's scary when I think of how little we knew when we started," Carol admits. But they survived that first season and grew to seventy members the next year, capping their membership at one hundred ever since. In the summer of 2006, they announced their decision to discontinue the CSA because of the devastating diagnosis that Carol has Huntington's disease. Loving messages from members poured in:

> "Please know I love you and look up to you both in such a huge way that it's not even conscious, you are these fundamental background heroes in my life." KB

> "I like to think of myself as your number one fan. I could never do the physical labor that you have done nor care for the earth so well as you have by your daily farming activities and your teaching and your outreach activities. Even with the ceasing of the CSA, you will remain at the top of my most admired list." SB

> "Even though your veggies are at the top of the list of things we all love about living in Tuscaloosa, and the end of the CSA is a very sad occasion, that takes a way-back seat to my heartfelt thoughts for you on this surprise turn in your journey." LK

> "You have given the community so much over the past seventeen years and will leave a gap that nothing else can or will possibly fill. I hope that in the future there will be ways that we can all begin to make at least a small contribution to your lives." E&S

With feelings as strong as these, it is hard to doubt that the members will find a way to keep the Tuscaloosa CSA going.

AFTER HEARING GREG WATSON, Massachusetts commissioner of agriculture, address the devastating effect that globalization of the food system has had on local farming, Sarah Lincoln-Harrison and her husband, Richard Harrison, wanted to do something. With five other families in Marblehead, Massachusetts, they decided to find a local organic farm to support. A suggestion from Lynda Simpkins, manager of the Natick Community Farm, led them to Edith Maxwell's one-acre farm, an hour's drive away. For the 1993 season, they contracted with Edith to pay $390 a share and to pick up the bagged produce at her farm. Edith included in the bag a list of the contents and a recipe or two. "It was easy to start," Sarah explained at a session of the Northeast CSA Conference in 1997, "without any special philosophy; we just sort of slid into it." At the end of that first season, the group wanted to expand the membership. Since Edith could not grow more, she suggested another farmer, Dick Rosenburgh, who had twelve years of experience growing organic produce for farmers' markets and who could handle the potential for growth. In the years since, the Eco-Farm CSA has gone through several transformations, for a while functioning more like a farmers' market, until settling in for the long haul as the Farm Direct Co-op, serving 345 households in three towns.

THE FOOD BANK FARM CSA began as a project to produce vegetarian chili for the Western Massachusetts Food Bank. After ten years of supplying government surplus and donated (mainly processed) foods to low-income people, the Food Bank decided to try to produce some of the food itself. Michael Docter raised $10,000 in grant money and grew the ingredients for the chili on land at Hampshire College. "It was a great idea," Michael said at a workshop at the winter 1998 Pennsylvania Association for Sustainable Agriculture (PASA) Conference, "but no one liked the chili, and it was too inefficient." Michael, having heard about CSA from Robyn Van En, visited a few farms before deciding to adopt that model and scale it up. He wrote a brochure, sent it out to the Food Bank mailing list and local environmental organizations, and plastered copies on bulletin boards. One hundred families signed up for the first season.

Meanwhile, Michael was searching for land on which to farm. The first time around, the land he eventually selected seemed like much too big a piece. In the succeeding months of the search, farmers opened his eyes to the use of tractors for growing on a larger scale. With the backing of philanthropist Ralph Taylor, the Food Bank was able to borrow $250,000 to purchase an old farm on prime Hadley loam soil and to persuade the Commonwealth of Massachusetts to sweeten the deal by acquiring the development rights. With support from CSA members, Food Bank contributors, and grants, the farm paid off the loans in only three years. In 2006, the Food Bank Farm supplied five hundred shares and an equal amount of food for the Food Bank.

FARMING MORE THAN THREE HUNDRED ACRES on the outskirts of San Diego, California, Bill Brammer has raised mainly tomatoes, squash, and cucumbers for the wholesale market. For California, Be Wise Ranch is a middle-sized operation. Bill has been watching nervously as the organic big boys—Cal Organics, Bornt Family Farm, Pavich, Earthbound—lower wholesale prices by expanding production in Mexico, where they exploit the cheaper land, water, and labor. He worries that wholesalers and distributors of organic produce are ready to source wherever prices are lowest. The CSA that Bill added to his other markets in 1993 as a community service has become a significant component of his strategy to survive the competition by switching as much as possible to direct sales. Solely by word of mouth, membership in his subscription-style CSA grew from forty to eight hundred. A single newspaper article in 1997

brought fifteen hundred calls to the farm, which had to start a waiting list. For the next eight years, Bill fought the good fight on the wholesale market until the developers closed in on his rented land. With 80 percent of his best acres gone, in 2006 direct sales through CSA and an on-farm market kept Be Wise Ranch alive and growing.

AT THE OTHER END OF THE SPECTRUM, Karen Kerney, the artist for this book, had been sharing the surplus from her half-acre garden in Jamesville, New York, with a few close female friends for several years. In 1993 they felt a mutual need to formalize their food connection as Karen's CSA. That year four women gathered weekly, starting in early spring to work beds, harvest spring greens and roots, plant seeds, and transplant. When cultivated crops were ready, they harvested together and divided everything four ways. Gradually, over the next few years, the group grew to five members, who each contributed $250 or more to cover garden expenses. Karen's goals for this project did not include making a living. In her words, "I wanted to share the sense of abundance: There is plenty of food . . . take what you need! I wanted to buck the existing system that thrives on scarcity, and to demystify the process of growing food." The members learned how to can, freeze, and dry food for the winter, and all now have gardens of their own, so Karen has reduced her role to growing corn and winter storage crops with her partner, John Sustare.

The Decision to Form a CSA

The image of the ideal CSA is tremendously alluring, the antidote to so many of the ills of the dominant global supermarket. People who have never even gardened hear of CSA and decide they want to become farmers! The dream CSA is a smoothly functioning organic or biodynamic farm dividing up all its produce among a committed group of supporters who share with the farmers the risks and benefits of farming. With a market assured and income guaranteed, the farmers concentrate on producing high-quality food and practicing careful stewardship of the land. The members get to eat the freshest, tastiest, most nutritious food they have ever experienced, as though they were master gardeners but with much less work. They and their children learn fascinating lessons about food production and, by eating seasonally, make a deep connection to a very special piece of land. They respect and honor the farmers' skills and hard work and express their appreciation through friendship, financial support, and helping on the farm. Members and farmers converge into a vital and creative community, which celebrates diversity, both social and biological, and makes food justice and security a living reality. Local, regional, and even international networks of CSAs and other sustainable food enterprises supply members year-round with ecologically produced and fairly traded foods.

That was Robyn Van En's vision and what got her out of bed in the morning. Each one of us, farmer or consumer, can decide to take a few steps toward this ideal. Although none of the existing CSAs has reached it, every single one has achieved at least some small part. Farmer or nonfarmer, we need to think through carefully when and how we might want to get involved.

The New Town Farms CSA brochure provides an excellent summary of most of the arguments in favor of CSAs (Randy Treichler has an even longer list in *The CSA Handbook*, pp. 13–18).

The Good Fruits
In the community supported farm structure, every member of the relationship benefits: the shareholders, the farmers, the farms (the Earth), and the greater community.

The Shareholders
- receive fresh, contamination-free vegetables and herbs delivered on the day of harvest
- pay close to supermarket prices for fresh, certified organic produce
- know where and how their food is grown, who grows it, and have the opportunity to partake in the miracle of growing food

- are provided with a structure through which they can support a viable local agriculture, preserve local farmland, and contribute to a healthy local economy
- have the opportunity to gain knowledge of growing food and stewardship of the Earth
- become more aware of their relation to the land, farm life, and processes that make our lives possible

The Farmers
- are given the opportunity to make a viable income by growing food in a responsible and harmonious way, directly supported by the consumer—no middleman
- have the pleasure of knowing who their product is going to and consequently feel more care, responsibility, and reward in their work
- are relieved of marketing labor and can focus more on growing food

The Farms
- are preserved from development
- are preserved from harmful farming practices
- are nurtured into a fertile, bountiful land

The Greater Community
- benefits by the preservation of open spaces, and the maintenance of an important agricultural component that is rapidly being consumed by development and industry—by preserving this diversity the community becomes a more whole and satisfying place to live
- is strengthened by the bringing together of people who share healthy concerns about our future
- gets an economic boost when food dollars remain within the community rather than supporting out-of-state corporations

To recruit members who will stay, it's also important to present an honest picture for people who may choose not to join. In "The CSA Connection,"

Career Growth
By Tom Ruggieri

I haven't always been a farmer. For the last two decades of the 20th century, I worked as an environmental engineer/health and safety professional/manager/owner.... It was a life where my every move seemed to be dictated by manmade rules. From timesheets that had to be filled out to document every minute at work, to regulations eternally haggled over by all sides. I always seemed to know that I could have a lot more fun living my life another way, but as it was the system I grew up in, the journey to the outside took several aborted attempts.

My last break took place in late 2000, when I dropped out for good. I met my mate, Rebecca Graff, when she was interning at the organic farm where my vegetables were grown. Leaving the white-collar world and finally meeting somebody going in the same direction as me was a sign I was on the right track. So here we are in 2007, in our fifth year of organic vegetable farming.

One reason I think it took me so long to get to this point is because organic farming is not promoted as a real full-time career. Well I am here to tell you that it is a profession to be taken seriously. It is a good way of life. Yet here in the US farming does not show up in TV job search ads, nor is it a vocation seriously promoted by our schools. I've filled out many a form where, for occupation, I had to check "Other".

Field of Growth

While I understand that most people have their reasons for not being able to farm, I am not sure that everyone understands the benefits and perks of organic farming. Nor do they understand how you get started in the business. The thing about organic farming is that it requires training, hard work, and capital. But so does every other business venture. And while you may read about organic farms as interest in them continues to grow, it is rarely promoted as a profession for you.

So what does it take to farm today? As in the old days, it takes existing successful farms and farmers to pass on the knowledge, expertise, and enthusiasm to those who are inclined to learn. It takes education and hands-on experience.

In the field today there are several university programs that teach organic farming, and hundreds of individual farms that do the same. When I left corporate America in 2000, I was trying to figure out the best way to invest my savings in my future. I considered going to chef's school and paying tuition, room and board — a logical scenario for beginning a new career.

But once I decided to farm, it turned out that I could save that money for the future, and get a "free" education as an intern. Most apprenticeship programs offer housing, vegetables, classes, and a stipend, in return for dedicated labor during the growing season. I found it was a chance to have no bills mailed to me (except car insurance) for most of a year. Try that going to a university.

And while organic farming is very physically demanding, there are many benefits to the lifestyle. That money you spend on going to the gym, diet aids, etc. won't be needed (though body salve and yoga help). Exercise and seven servings of fruits and vegetables a day will be built right into your life. You'll get your vitamin D directly from the sun. You'll be able to — nay, have to — eat to your heart's content.

Morning rush hour traffic can be replaced with harvesting vegetables, the sounds and smells of the highway with fresh air and birdcalls, a sore butt with other sore muscles. Economically you may also have to replace your salary with a smaller one, your house with a smaller one, and your job as an employee with one as a business owner. But if you are interested in preserving the environment, your health, and your community as a full-time job instead of a volunteer activity, farming is a career worth considering.

Make Money Caring for the Environment

I recently gave a short talk at a local university and was struck by the fact that, even though the students in the room were there because they cared about the environment and our future, few seemed to consider sustainable agriculture as a possible profession. Their beliefs in caring for the land, the community, our children, and our health were strong, but the realization that this could be their vocation seemed to be lacking.

I am dismayed at this condition in our society. When our leaders talk about entrepreneurship, or food security, or community, they don't talk about sustainable agriculture. They don't talk about the benefits to growing our food locally, investing our food dollars in our own communities, maintaining the soil and water quality of the land in and around our cities. And if they do, they don't make the leap of talking about who will do that work, what it takes, and how they can help make it happen. It is time for that to change.

The farmer is considered an American icon. Yet today the idea of becoming one is far down the list of possible careers. One reason that I have chosen to be an organic farmer is to help show that it can provide a full-time living. I hope to "prove" that the United States is as great a country as we all believe, and if you spend all day working hard growing food for your community, the community will in turn support you with a livelihood. I hope that I am right.

Vicki Dunaway gives a good summary of aspects that will scare some people away:

> CSA involves responsibility on the customer's part—to pay up front for the food, to pick up the products on an assigned day, sometimes to return baskets or other containers, and to accept some risk for failed crops. If it's a bad tomato blight year, you may not get all those ruby red tomatoes you have been dreaming of. You might find a corn earworm in the tip of your corn ear if you prefer unsprayed produce (country folks just cut off the tip and never know the difference). There might be a little soil left on your potatoes because they keep better when they are not scrubbed. Often shares are not customized, and you must have an open mind about the type of produce you will receive and be willing to try new things. You may be required to attend a meeting or to work off part of your share by spending a few hours at the farm over the season.[1]

Converting to CSA is easiest for farmers who own a piece of land and some equipment, have a few years of experience growing vegetables for market, and have established a following of customers. A farmer in this situation still needs to think through carefully whether a change in marketing strategy and increased involvement with customers is desirable. Among the many questions to consider: Will growing for a prepaid group of customers be more of a benefit or a heavy obligation? Is my farm located in an area where I can attract CSA members? How will this change my cropping systems? How many shares can I handle? Do I want to combine elements of CSA with other marketing? How much participation from members would I like on my farm? Will I enjoy going to more meetings, or is there someone else who can represent the farm? Will this help to stabilize my income or just add a lot more work servicing members? Does this fit in with my long-term goals for my farming and my life?

The obstacles multiply for farmers without land and would-be farmers without experience in growing for market. Yet around the country, many people have found ways to get started once they have made up their minds. Sarah Sheikh had a few years of experience as an apprentice and lots of community contacts in the Trumansburg, New York, area where she grew up, but no land. She was able to persuade farmer Tony Potenza to allow her to use some of his land and equipment to start a CSA. Deb Denome, Sally Howard, and their friends had access to some land, a lot of organizing skills and energy, but no farming experience. They advertised on the Internet and found Woody Wodraska, an experienced grower who was willing to take up the challenge of helping them get started on unbroken pastureland. The Acorn Community, in Mineral, Virginia, which already owned some farmland, spent three years getting its gardening, business, and organizing team in shape before launching a CSA. Hundreds of young would-be farmers are finding intern and apprentice opportunities on CSA farms, where they can learn the trade and see if they like it enough to make a lifelong commitment to farming. On more and more college campuses, from Cook College in New Jersey to the University of Montana in Missoula, groups of students are linking up with teachers of sustainable agriculture, food systems, or nutrition and using school resources to form CSA farms or gardens. (See the appendix for a profile of

Members fill out their contracts at the 2007 orientation/sign up meeting for the Genesee Valley Organic CSA at School #12 in Rochester, New York. PHOTO BY KATE KRESSMAN KEHOE.

the Hampshire College Organic Farm and the National Agriculture Library for a list of colleges with CSAs.)

For nonfarmers, the decision may be as simple as finding a local CSA. Several organizations maintain databases on CSA farms by state or region. You can visit the Web sites of the Robyn Van En Center for CSA Resources, the Biodynamic Farming and Gardening Association, or LocalHarvest to find the CSA nearest you. The Michael Fields Agricultural Institute, (414) 642-3303, rides herd on the Upper Midwest; the Southern Sustainable Agriculture Working Group, www.ssawg.org, keeps an eye on the South, and the Community Alliance with Family Farmers (CAFF) publishes its "Local Food Guide," listing the CSAs in Central Coastal California. The Kerr Center publishes a list of the CSAs in Oklahoma. The seven state Northeast Organic Farming Associations, the Maine Organic Farming and Gardening Association (MOFGA), the Carolina Farm Stewardship Association, and other state and regional farming groups also list CSAs on their Web sites. Where no CSA exists, people like the Harrisons (see page 226) or Peggy and Martin Danner, in Boise, Idaho, have reached out to other consumers and

local farmers to start new ones. The Danners founded the Boise Food Connection, a nonprofit dedicated to facilitating CSA in the Treasure Valley. In his 1993 survey of CSAs around the country, Tim Laird found that farmers initiated 79 percent, farmers and consumers together 6 percent, and consumers alone 5 percent of CSAs.[2] (Tim has graciously allowed me to use his work as a source of information for this book.) Two farms reported that organizations had started them, and one farm began because of the work of a student intern. Prior to forming these CSAs, 49 percent of the farmers were not farming. Few of them had jobs even remotely connected with agriculture. Previous jobs they listed included postal worker, agricultural consultant, plumber, laboratory technician, health club owner, teacher, truck driver, social worker, student, journalist, engineer, veterinary clinic manager, produce buyer, nurse, and underwater photographer. Five farmers went from part-time to full-time work to run CSAs.

Farmers considering the move to CSA should expect changes in their work patterns. In Tim's survey 67 percent of all the farmers felt their workload had increased, though only 55 percent of those with previous experience said this. For 30 percent the load remained the same, while only 3 percent noticed a reduction. Of the thirty-four new farmers, 85 percent said that their work hours had increased. The need to grow a greater variety of vegetables in order to make up a good share, as well as the organizational demands of running a CSA, created the extra work. Tim observed from his own experience in market gardening as compared with CSA that the members' expectations for quality and variety made production for the CSA more stressful. On the positive side, 79 percent of Tim's respondents said that their job satisfaction had increased, 17 percent said it remained the same, and only 4 percent felt a decrease. The two new farmers whose job satisfaction had diminished drew the logical conclusion and left farming, as did the one experienced farmer. CSA, clearly, is not for everyone, but for farmers such as David Inglis in Massachusetts and Jean Mills and Carol Eichelberger in Alabama, it is the only way they can imagine going on with farming. Jean told me that doing a CSA has made life in an isolated and conservative area enjoyable: "The CSA brings the world to our farm. Some members have Bush bumper stickers, others bash Bush, but the CSA magically brings out something common in us all." Dorothy Suput wrote, in the summary of her study of CSA farms in Massachusetts: "All farmers participating in a CSA organization commented on the reduction of market-related stress [leaving] extra time for training apprentices, educating members, socializing, spending time with family, organizing agricultural education programs, and giving extra care to farming."[3]

Larry Halsey told me that switching from selling at Greenmarket, the main farmers' market network in New York City, to operating a CSA with two branches, in Brooklyn and Queens, has reduced his workday from eighteen hours to six and (not surprisingly) improved his family life. In the 1996 season, the first year GVOCSA was the primary market for Rose Valley, I no longer had to spend several hours a week on the phone selling vegetables to produce managers. (There are almost no contracts in fresh market vegetables, so there is always the nagging worry that one week no one will want what you have to sell.) Our farm team did not have to pick hundreds of bunches of greens and herbs on a tight schedule. No one from the farm had to load the pickup truck, rain or shine, with heavy boxes and bags of vegetables and drive the hour or hour and a half to deliver to stores in Rochester and Ithaca. There were no emergency phone calls from apprentices asking how to change a flat tire or what to do about a truck breakdown out in the middle of nowhere. The police never called me to ask if I knew the man named Greg they held in custody.

Growing food for the GVOCSA has a more relaxed rhythm to it. On Tuesdays, I consult with Ammie, Katie, and our interns to decide what crops are ready to go in the shares for that week and we fill out our harvest sheets. Later, one of us translates that into the forms we will send in with the vegetables to instruct the distribution crew on what to put in the shares. We assemble the bulk order and make plans for the work crews. Early on the morning

before the CSA harvest, we cut, wash, and store the lettuce in our cooler to make sure that it is cooled down enough for the trip to the city. (Harvesting and cooling lettuce a day ahead actually extends its shelf life in our members' refrigerators.) Most Wednesdays and Sundays, we start picking at six or seven to be sure to get the tender greens in the cooler before the sun cooks too hot. As CSA members arrive, we slow down to a more social pace. Those mornings are an intense mixture of leisurely personal encounters and goal-oriented work. Our sense of responsibility is matched by the remarkable conscientiousness of the members.

Describing the process of growing food as "life transforming," Karen Kerney says, "The best part of the CSA was the opportunity to focus on growing food that I knew would be picked and consumed by people who cared as much as I do about how our food is grown. Relieved from the incredible amount of energy it takes to harvest, package, and market food, I was able to center my energy on the art, science, and craft of growing food."

In his poem "rural population base/distance from/stretched out possibilities," Paul Bernacky of Wayback Farm in Belmont, Waldo County, Maine, writes about how his CSA, with twenty summer and winter shares, in its eighth year in 1998, changed his work:

> *Organize with a purpose living*
> *in balance and harmony transferred*
> *stress into dynamic tension rekindles*
> *my sense of wonder.*

Steps to Forming a CSA

Whether consumer- or farmer-initiated, the steps needed to begin are similar. (See the chart on p. 39 for a summary of the process.) First, find out if enough people are interested. You can do that very informally, by polling your friends or customers, or more formally by calling a meeting. The Danners posted this engaging meeting announcement:

What's special about these vegetables? They're fresh, local, and grown for you! If you are interested in any of the following:

- A steady supply of fresh food grown locally
- Knowing the people who produce your food
- Providing direct input on what food is produced and how
- Fixed prices established before the growing season
- Trying different types of produce grown locally, along with information on how to prepare them
- Supporting local, small-scale farmers
- Participating in a grassroots effort to directly connect local growers and consumers

Then contact us for more information or attend our first planning meeting.

That first organizational meeting should set the tone for the entire project. If you want to run things and do most of the work yourself, you can present the information about your project to the participants and invite them to sign up. If you hope to make the project more participatory, however, start by giving all who attend the meeting the chance to talk about why they came, what they expect, and what they plan to contribute. Begin the agenda with some discussion of what CSA is and its history and what is going on locally. Participants need to understand why it is important to eat locally grown food and how the dominant food system endangers local farms. After you've presented some historical and philosophical perspective, you can get down to brass tacks. Is there enough energy and commitment to start a CSA? Who wants to get involved, and what specific jobs are they willing to do? If you can come out of this meeting with the beginnings of a core group, you have done a good piece of work.

At this meeting or the next, you need to agree on the values for your CSA and the policies that flow from those values. Do members want organically or biodynamically produced food, or are they satisfied to support a farm using any practices as long as it is local? Does the group want to foster ethnic, racial, economic, or age diversity among members or contribute food to the hungry? Should children be involved? Besides food, will there be an educational component? Will the project ask every member to work, or will some members be allowed to exchange labor for a lower price? Do members want to share production risks with one farm, or should provisions be made to purchase from several farms? What commodities will the CSA distribute? Once these broader questions have been answered, you can tackle the details of how the CSA will function. What farm or farms will produce the food? How much food will go in a share? What kinds of shares will you offer? How much will a share cost? Where will members pick up their food? With this information in hand, you can draw up a flyer or brochure and start recruiting members. (See sample brochure on p. 42.)

A farm that has customers already has a good base with which to start. Many CSAs have grown mostly by word of mouth, and the new members who learned about the CSA from old members are more likely to have realistic expectations. Dru Rivers of Full Belly Farm in California says that 80 percent of their new members come from word of mouth. Every year, Genesee Valley Organic CSA sends out packets of two or three brochures to current members and asks them to recruit their friends. Gradually, some entire workplaces have signed up. One member who is a nurse served our salads at hospital staff meetings and attracted a dozen or so other medical personnel. A young resident lured in the other members of his class. Caretaker Farm in Massachusetts was able to convert totally from other kinds of marketing to CSA in one season because of its established connections and excellent reputation; 125 people attended its founding meeting in September 1990. Few farms, however, could attract that many potential members all at once.

The best way to find members is to approach active groups of people who are already organized

> Converting to CSA is easiest for farmers who own a piece of land and some equipment, have a few years of experience growing vegetables for market, and have established a following of customers

Steps to Forming a CSA

(Start small and grow organically!)

1. Initiators (either farmers or group of nonfarmers) issue a call to form a CSA. You can seek members:
   a. among friends or neighbors
   b. among existing groups: day cares, environmental or consumer organizations, churches, civic groups, schools or other institutions, workplaces

2. Hold exploratory meeting of prospective sharers and farmer(s). Possible agenda:
   a. what is a CSA?
   b. why eat locally grown food?
   c. why small farms need support
   d. assess level of commitment of participants
   e. if interest is high enough, create founding core group

3. At this meeting or a subsequent meeting, come to agreement on the group's values:
   a. does the group want organic food?
   b. does the group want locally grown food?
   c. does the group want racial, ethnic, and economic diversity among members?
   d. is it important to involve children?
   e. will all members contribute work, or will some buy out by paying a higher fee?
   f. do members want to share production risks with the farm(s)?
   g. what commodities does the group want?
   h. does the group want to share mailing list with other groups?

4. Organize the core group to:
   a. decide on farmer(s)
   b. decide growing site
   c. decide how and where food will be distributed
   d. divide up member responsibilities
   e. approve the budget proposed by the farmer(s)
   f. set fee policy and payment schedule
   g. clarify expectations as to variety and quantity of food
   h. set guidelines on participation of children (if desired)
   i. decide who owns any equipment purchased

5. The core group recruits members for first season:
   a. post flyers
   b. organize recruitment meeting
   c. talk up idea with friends
   d. place notices in organizations, churches, and do flyer mailing to likely groups
   e. send out press release
   f. find friendly reporter to write story

6. Members make commitment:
   a. to pay in advance of receipt of food (whether by season, month, or other schedule) and regardless of quantity and quality of food due to weather conditions
   b. to participate in farm, distribution, and other CSA work

7. Establish the legal status of the CSA. Many groups defer decisions on legal structure for a season or two. Advice from a lawyer may be helpful. Existing options include:
   a. consumer cooperative
   b. sole proprietorship or partnership of farmer(s)
   c. corporation or limited liability company
   d. nonprofit corporation (or branch of existing one)
   e. farmer-owned co-op

8. Determine capitalization of the farm(s). Many CSAs start with minimum of rented or borrowed land and equipment. For the longer term, decisions must be made on purchase and maintenance. Options include:
   a. farmer(s) capitalize
   b. members capitalize through fees
   c. the group seeks grants
   d. the group seeks loans. Possible sources include Farm Credit, National Cooperative Bank, members, commercial banks, revolving loan funds, pre-sale of farm produce coupons

   Options for land tenure include:
   e. private holding
   f. land trust
   g. lease agreement with private owner or institution

around some related concern: schools; day cares; church groups; unions; environmental, political, or farming organizations; or workplaces. The GVOCSA started with members of PJEC and HEAL, many of whom had eaten Rose Valley vegetables sold at the Genesee Food Co-op, and expanded by recruiting members of the Sierra Club and Corpus Christi Church. The Kimberton and Brookfield Farm CSAs grew out of close relationships with Waldorf schools. Members of the Albany Catholic diocese, who had been inspired to take action in support of local farms at a New York Sustainable Agriculture Working Group meeting, approached Roxbury Farm with an offer to form a group. Joe D'Auria founded a CSA with members of the Long Island Chapter of the Northeast Organic Farming Association.

People with chemical sensitivities, cancer victims, and families with autistic or highly allergic children are attracted to CSAs by the hope of alleviating their illnesses. Helping members improve their health can be very gratifying. One member family of the GVOCSA told me that their autistic son was calmer and easier to live with when he ate our vegetables. Unfortunately, the fresh produce therapy does not always work or is started too late: our core group suffered together through the final months and eventual death from cancer of our treasurer, Joe Montesano. In a way, I guess, that is also one of the risks of farming for a community.

The workplace is another source for membership. Cass Peterson and Ward Sinclair sold shares in the Flickerville Mountain Farm and Groundhog Ranch to their former colleagues at the *Washington Post*, delivering to the newspaper office building once a week for six years. Rodale employees make up most of the sharers for George DeVault's CSA. In 1995 our neighbor Glenda Neff signed up seventeen sharers at her workplace an hour away in Syracuse. This mini-CSA lasted as long as she kept the job and did the weekly deliveries. A member of the board of Nesenkeag Farm in New Hampshire offered his corporate contacts as a way to recruit some shares. A total of thirty people in two offices signed up. Farmer Eero Ruuttila delivered the weekly shares packed four or five to a plastic foam cooler box so that the produce stayed fresh and did not take up too much space in the office buildings. He limited the size of the shares to what could be carried home easily in one plastic market bag.

Communications with members was simplified by office e-mail. In "Workplace Community Supported Agriculture: Connecting Local Farms to Local Business Employees," Denise Finney, a research associate at the Center for Environmental Farming Systems at North Carolina State University, provides a helpful guide based on the experience at the Research Triangle Institute (see chapter 18, p. 215, for details). In Massachusetts, the Community Involved in Sustaining Agriculture (CISA) "Farm to City" project persuades employers to allow CSA deliveries to their workplaces.

Dan Guenthner, who farms with his family at Common Harvest Farm near Minneapolis, Minnesota, has proposed Congregation Supported Agriculture. In *To Till It and Keep It: New Models for Congregational Involvement with the Land* (1994), Guenthner states his view that "the Bible is nothing short of an agricultural handbook!" He outlines sixteen ways in which parishes can connect with local farms, better educate their members about local food systems, and generally empower them to participate in land stewardship. His booklet, funded by the Evangelical Lutheran Church, gives several examples of church-sponsored CSAs. Support from local churches enables Common Harvest Farm CSA to distribute 15 percent of its shares to low-income families. The response will differ from parish to parish, but local churches with active social justice committees are fertile ground for would-be CSAs. (See Guenthner's list of suggestions for linking CSAs and churches on p. 41, and "Faith Community Supported Gardens" in chapter 21 on page 244) In 2006, MOFGA and the Maine Council of Churches, supported by a Sustainable Agriculture Research and Education (SARE) grant, began an alliance to connect clusters of parishioners in churches around the state with nearby farms. Russell Libby, executive director of MOFGA, hopes to double Maine's CSA membership, from 0.5 percent of the population to 1 per-

What Can We Do to Respond to the Needs of the Land?
by Dan Guenthner

- Introduce the concept of congregationally supported agriculture in your church.

- Plan to take at least six to nine months to form a core group of interested members to help organize and set up the details of a marketing relationship with a local grower.

- Find existing vegetable, dairy, fruit, and meat farmers in your area. Contact them directly to see if they would be interested in participating in a direct partnership with your church.

- If local farmers are not available, consider providing an entry-level farmer/gardener with the opportunity to get established through the benefit of an immediate market guarantee.

- Church members may have land available for a farming/gardening enterprise. Investigate and evaluate the existing resources within the church community.

- Churches can provide many initial benefits in starting a CSA—a committed community of people, access to distribution information through newsletters and bulletin boards, and weekly contact with each other. The church can help establish a fair price for food.

- Churches can also purchase shares in a congregationally supported farm for other ministries such as the local food shelf, shut-ins, and people with AIDS and other illnesses that prevent access to fresh food.

- Contact your synod or conference office to investigate if other church-owned land might be available for this type of innovative and direct form of soil stewardship.

- Approach your local or regional church camp to see if a church-supported farm can be established on tillable land at the camp. Vegetable cultivation and harvesting can become a meaningful part of the camp activities and summer program.

- Be creative! One of the most exciting things about this new form of food distribution is that each farm is unique and is determined by location, farm size, community, and climate. No matter what form the CSA takes, the expressed outcome is the same: Non-farm citizens now have access to a food supply that is socially responsible and built directly upon their helping to care for the soil.

Dan Guenthner farms at Common Harvest Farm near Minneapolis, Minnesota. Contact the Land Stewardship Project (listed in "CSA Resources") for more information on congregationally supported agriculture.

cent. The CSA Learning Center at Angelic Organics has adopted a similar approach to CSA member recruitment through church groups in the Chicago area. It will be interesting to see what response the CSA Learning Center gets to this attractive appeal:

We invite communities and congregations of all faiths to partner with the Harvest Shares Program. As communities of shared values, congregations and faith groups are in a special position to improve healthy food access and address the links among hunger and food security, loss of family farms, and environmental degradation.

Together, we explore food and farming within the context of faith and justice, and take meaningful and practical action toward a more just and sustainable local food system. The Harvest Shares Program is designed to enhance your congregation's ongoing justice, service, and hunger relief programs.

Harvest Shares Program partnership opportunities include:

- Creating food access for low-income families direct from local farmers
- Growing urban and community gardens for local food production
- Visiting and learning at—or partnering with—our urban growing sites
- Creating and providing infrastructure that supports a just and sustainable local food system
- Supporting training and direct markets for a new generation of urban and family farmers
- Facilitating food, farming, and faith workshops and discussions for your congregation
- Hosting cooking and nutrition workshops that cross boundaries of faith and culture with the universal language of food
- Exploring or holding retreats at our farm-based learning center at Angelic Organics

You should not have to pay for advertising. Brainstorm a list of likely places to display your CSA brochure: health food stores, food co-ops, and vegetarian restaurants; health clubs, chiropractic clinics, or other alternative medical offices; recycling, yoga, tai chi, or spirituality centers; and church and school bulletin boards. Local radio and television talk shows are often hungry for interesting people to interview. Find a literate reporter for a local newspaper and ask him or her to write a story about your CSA. Attracting the attention of major newspapers may be difficult, but small-town papers are usually anxious for a good story with photo opportunities. Earthcraft Farm in Indiana signed up twenty new members after an article in a local paper. And even the big rags have given CSA some attention. For years after Ward Sinclair retired from its staff to become a farmer, the *Washington Post* published his "Truckpatch" column. During the summer of 1997, the *New York Times* food editors split a share in a CSA and wrote a whole series of stories. Even though the tone of most of these stories was rather snippy and showed a remarkably low level of understanding of what CSA and eating in season is all about, they attracted new sign-ups for the CSAs mentioned.

When potential members do call in response to an ad or a flyer, have a seasoned member available to answer their questions. The office at Angelic Organics keeps a list of members who are willing to talk to new recruits. See the example of an attractive flyer from Harmony Valley Farm on page 42.

At the 1997 Northeast CSA Conference, Nancy Schauffler, an active member of Stoneledge Farm CSA in New York, listed seven recruitment techniques they tried the first year, and quantified how many members each brought in:

1. an existing parent-teacher-student environmental group—6
2. a mailing to Just Food members, who already grasped the concept—5
3. posting flyers with a tear-off phone number—20
4. talking to everyone you know—20
5. notices in organizational newsletters—5
6. an ad on public radio and an article in a newspaper—lots of members for another CSA!
7. recruitment meeting—the meeting gave the sign-up campaign a deadline to work for and solidified the group already recruited.

Nancy gave some additional pointers about the recruitment meeting. "It gives people the chance to meet the farmer, the core group, and each other. Have lots of literature available, consider showing a video, and have contracts there so people can sign up on the spot. Pass around an attendance sheet, and ask people to write areas in which they would like to volunteer, specifying newsletter, accounting, publicity, recipe collection, and the like."

Two videos and two slide shows introducing CSA are available for meetings of this kind. In 1986, Robyn Van En, Jan Vandertuin, and Downtown Productions of Great Barrington, Massachusetts, made the video *It's Not Just about Vegetables*, which is now a classic. More recently, the Center for Sustainable Living CSA project at Wilson College produced the video *CSA: Making a Difference* and two versions of a slide show, *CSA: Be Part of the Solution*. One version supplies text slides only and allows you to fill in your own farm slides, while the other includes slides of CSA farms as well. These

Farmer Rebecca Graff conducts orientation meeting for members of Fair Share Farm in Kearney, Missouri. PHOTO BY TOM RUGGIERI.

materials are well made and inexpensive. Showing John Peterson's documentary *The Real Dirt on Farmer John* gets many viewers excited about joining a CSA. (See the CSA Resources section for information on Wilson College and other CSA support materials.)

Among the more creative promotional ideas used by CSAs, John Clark of Clark Farm CSA in Connecticut and David DeWitt of First Light Organic Farm in Massachusetts give tours of their farms to schoolchildren, whose families then sometimes join. As a group, the CSA farms in Vermont offered sample shares as a benefit for supporting public radio. While this did not generate many new shares, it got CSA favorable radio coverage. CSAs in New York City attracted a lot of attention by holding a tomato tasting with 160 varieties of tomatoes at the public market. West Haven Farm at EcoVillage in Ithaca, New York, asks members who are going away for a week or so to give their shares to friends, who then often sign up themselves. The Partner Shares Program of the Madison Area CSA Coalition (MACSAC) in Wisconsin sponsors "farmathons," in which volunteers solicit financial sponsors for their work on farms and the proceeds go to purchase shares for low-income members.

Research projects such as those conducted by Gerry Cohn, Deborah Kane and Luanne Lohr, Rochelle Kelvin, Jane Kolodinsky and Leslie Pelch, and Dorothy Suput have investigated why people join CSAs. All show that members want fresh food from local farms. Some surveys show organic food as the top reason; others show fresh or local first with organic second. Support for local farming, environmental and health concerns, quality of produce, and food safety come close to the top of the lists. End-of-the-season surveys show that members stay on because of the quality of the produce, with farmwork and newsletters as close contenders. Meeting members' expectations is critical to retaining them as sharers in a CSA. (See chapter 12 for more on member retention.)

Regional CSA Support Groups

"Cooperation among cooperatives," one of the six basic principles of cooperatives around the world, is valuable for CSAs as well. Instead of competing with one another, groupings of CSAs and CSA supporters have been coming together to work for mutual benefit and CSA promotion. Like the CSAs themselves,

The Cooperative Principles from Ideology to Daily Practice
(The Rochdale Principles)

These seven principles are embraced and utilized by co-ops in all countries of the world. They represent the fundamental features of a cooperative, regardless of the type of business it is involved in. These co-op principles are an important link between the theory of working together and the application of that idea to economic institutions.

These principles were originally formulated by early co-op groups in the mid-1900s and most recently refined and adopted by the International Cooperative Alliance in 1995.

1. Open and Voluntary Membership
Co-ops do not restrict membership for any social, political, or religious reasons. They are open to all persons who can make use of their services and are willing to accept the responsibilities. Members choose to join and participate in the co-op.

2. Democratic Control: One Member/One Vote
Co-op members are each equal co-owners in the cooperative business. Each member has voting and decision-making power in the operation of the business on a one-member/one-vote basis. In this way, no one person gains control based on the amount of money invested or position held.

3. Return of Surplus to Members
Surplus or profit arising out of the operations of the co-op belongs to the members (owners). This is returned on the basis of patronage—or how much business a member did with the co-op (or based on how much the member contributed towards surplus).

4. Limited Rate of Return on Investment
Share capital (the money invested by the member/owners in the co-op as equity) can earn dividends, but it would be at a limited rate. This prevents outsiders from investing money in the co-op to make a profit purely as an investment. The rate of return is typically limited by law to 6 to 8 percent.

5. Continuous Education to Members and Public
Co-ops encourage member participation and involvement in the co-op, educating members in the principles and techniques of cooperation. Co-ops also work to educate the general public and members about their service areas: in the case of food co-ops, about nutrition, food systems, and the like.

6. Cooperation among Co-ops
To complete the cycle of cooperation, co-ops work with other local, regional, national, and international co-op organizations. Co-ops working together help each other and strengthen their individual and aggregate economic positions.

7. Concern for Community
While focusing on member needs and wishes, cooperatives work for the sustainable development of their communities.

no two of these efforts take exactly the same form. MACSAC encompasses the farms that service the city of Madison. The Midwest Organic and Sustainable Education Service (MOSES) publishes the *Upper Midwest Regional Directory*, which includes the CSAs in twelve states. A dozen or so farms in western Massachusetts and eastern New York, many of them CSAs, share resources in training apprentices through the Collaborative Regional Alliance for Farmer Training (CRAFT), and the Angelic Organics CSA Learning Center coordinates another CRAFT grouping in the Upper Midwest. Just Food in New York City recruits city dwellers, guides them in the formation of core groups, and matches them up with farms in the metropolitan area. With the support of a SARE grant, CSA farmers in Iowa formed the Iowa Network for Community Agriculture (INCA) and hired CSA farmer Jan Libbey to act as education coordinator. In the province of Quebec, Équiterre reaches out in both directions, providing technical support to farmers and energetic recruitment of members for CSAs. (More on Équiterre appears later in this chapter.) The farmers at the seven CSAs in Ester, Alaska, get together in an informal association to share ideas and organize an annual sustainable agriculture conference.

The success of MACSAC in building CSA membership to over 1,700 in a fairly small, if exceptionally progressive, region of the country proves that synergy beats competition. As Marcy Ostrom tells the story, in the fall of 1992, a group of food activists and farmers from one farm decided it was time to get more Madison consumers eating locally grown food.[4] Instead of forming one CSA, they mounted a vigorous media campaign to publicize the concept of CSA and organized an open house, where 8 farmers signed up members. Their first CSA fair drew over 300 people; Madison CSA membership went from 0 to 260 in one season. According to John Greenler, the number of CSAs rose to ten the second year and eighteen in 1997. The MACSAC Web site lists

> A farm that has customers already has a good base with which to start. Many CSAs have grown mostly by word of mouth, and the new members who learned about the CSA from old members are more likely to have realistic expectations.

twenty-four farms in 2006 and provides a map of their locations, a list of the kinds of shares and payment options each farm offers, and a guide to choosing which CSA to join. In the years since 1992, the open house has become an annual celebration, drawing up to 500 people to the city botanical gardens with speakers, farm slide shows, activities for children, and membership information on each farm. With the support of the Wisconsin Rural Development Center, MACSAC has evolved into a CSA farming association combining outreach to the public with farmer-to-farmer technical exchanges and mutual support. Similar to MOFGA or the Northeast Organic Farming Associations (NOFAs), MACSAC holds meetings on practical topics of importance to farmers—such as equipment, vegetable production, and CSA organization—and participates in a regional conference. MACSAC has not tried to define what CSA farms should be, intentionally allowing diversity to take its course as projects mature and ready to help in whatever way is needed.

Calculating that to match the Madison density of CSA membership in New York City would mean a quarter of a million member families, Just Food, the NYC Sustainable Food System Alliance, set the goal of facilitating the creation of thirty to fifty CSAs in ten years, and it met that goal ahead of schedule. New York City's fortress-like geography, transportation nightmares, and distance from market gardens made this a formidable challenge. Robyn Van En inspired and helped in the early stages of this effort, which began with informational meetings for city people in 1994. Borrowing the open house idea from MACSAC, Just Food's "CSA in NYC" program held a CSA Sign-Up Fair in the winter of 1996 attended by eight farmers and seventy potential members. The farmers gave brief presentations about their farms, consumers and farmers matched up by geographical proximity, and core groups began to form. As a result, six new CSAs started in 1996, ranging in size from fifty to over one hundred shares.

These CSAs have continued to grow in the years since, with some farms adding more pickup sites as well as additional shares. Jean-Paul Courtens of Roxbury Farm, the first farm to deliver shares to the city, has provided steady advice on the complex logistics. Sarah Milstein, a dedicated core member of the Roxbury CSA, had the job of first Just Food CSA coordinator, followed over the decade by three competent successors. At the tenth anniversary of CSA in New York City, in April 2006, Paula Lukats, the current coordinator, announced that in 2006 there are forty-two CSAs in the city with ten thousand members, supplied by sixteen core CSA farms and thirty farms that send other products. The total gross direct sales amounted to $1.7 million in 2006.[5]

Just Food anchors each CSA in a solidly formed core group of active members. A few of these cores have been resilient enough to carry on despite having to switch farms. The cores oversee site management as well as member recruitment and participation for each of the forty-two sites. They set policy on share sizes and on whether to allow flexible payment options that encourage low-income members. Many core groups are finding ways to include other food products in addition to the farms' vegetables and to provide education about how to use the food. At the 2006 Just Food CSA mini-conference, I met members of several CSA core groups, and it was obvious that tight bonds of friendship enhance the cooperative work.

Based on their experience with the first six farms, CSA in NYC has been establishing five to seven new CSAs a year. Just Food staff network with community organizations, agencies, and churches around the city to find potential core group members. Year round, the staff give monthly workshops to train CSA coordinators and core group members. At the same time, Just Food scouts for potential farms to supply the CSAs. A committee of experienced CSA farmers guides the choice of farms. In 2004, Just Food published two impressive workbooks: *The CSA in NYC Toolkit*, and the *Veggie Tip Sheet Book*. *The CSA in NYC Toolkit* is a comprehensive, practical guide to setting up a CSA, with chapters on getting started, finding a distribution site, setting up a voice mail account, outreach techniques, record keeping, running distribution, conducting a democratic core meeting, arranging for farm visits, fundraising, and evaluation. The *Toolkit* includes samples of CSA brochures, newsletters, and outreach materials; information on using food stamps; a resource directory, including a list of NYC police precincts; and a study comparing the prices of CSAs and other sources of similar foods. Just Food designed the *Veggie Tip Sheet Book* so that CSAs can easily copy pages for distribution to members.

Sarah Milstein's work with Little Seeds Gardens, a new CSA for 1998, gives a snapshot of how CSA in New York City operates. In 1996, Sarah met Claudia Kenney and Willie Denner, a young couple who were attracted to CSA farming. Sarah kept in touch with them while Claudia and Willie established Little Seeds Gardens on rented land, learned more about CSA through a year each of membership in the Hawthorne Valley and Roxbury CSAs, and got more experience growing for farmers' markets. Willie says, "I wouldn't want to enter into CSA without this previous growing experience. I am starting with trepidations even after four years of growing vegetables for market."

With her antennae always up for promising connections, Sarah found Jessica Stretton, a Just Food

---

Ever open to innovations, MACSAC partnered with Physicians Plus Insurance Company in 2005 to offer a pilot rebate of $100 for a single person and $200 for a family who joined one of two CSA farms in the Madison area. Physicians Plus had already been rewarding policyholders with rebates for a variety of fitness programs. In 2006 all twenty-four of the MACSAC CSAs participated. The response was enthusiastic—over nine hundred rebates were granted. Kathryne Auerback, director of marketing for Physicians Plus and a member of a CSA herself, told the *Capital Times,* "We wanted to find a way to bring the nutrition aspect of preventative health and wellness into our programs to encourage healthy lifestyle choice."[6]

---

member who wanted to start a CSA in her economically and ethnically diverse neighborhood, an area well located for transportation. Sarah helped Jessica devise a plan: they sent out invitations to thirty-five Just Food and Roxbury CSA members in a few zip codes. (Roxbury had a waiting list and could spare a few members.) Six people came forward to serve on a core group, and twenty expressed an interest in shares. Sarah then contacted Claudia and Willie, who said they were ready to do fifty shares in 1998 and perhaps one hundred in 1999. They met with the core group, and both sides made a two-year commitment. Sarah supplied the group with a packet of guidelines for the farmers and the core, including forms to help develop budgets. The core members found a neighborhood park to use as a distribution site, and the farmers hosted a visit to their farm for the core. Sarah kept in touch, intervening only with helpful tips, such as setting up a voice-mail box to answer queries and counseling when problems arose.

Meanwhile, Claudia and Willie were hard at work developing Little Seeds in Stuyvesant, New York. For a family with two small children, apprenticeships are awkward, so they learned farming by doing it, taking every opportunity to attend workshops and get advice from the farms in the CRAFT network. They postponed purchasing land, putting their resources for their first four years into equipment and slow, but steady growth, from one to one and a half to eight acres under cultivation. With help from family, they bought eighty-seven acres in 2000 and increased production to twenty acres. In 2005, Little Seeds passed its city CSA sites on to Hawthorne Valley and Windflower Farms and cut back to its local CSAs, fifty members who pick up at the farm and fifty in a nearby town. The long trips to the city, two and one-half hours each way, and the amount of diverse crops they needed to grow for 260 shares proved too exhausting. The Little Seeds farmers are pulling themselves up by their bootstraps, but they are doing it thoughtfully, aiming for satisfaction and community in the process.

In harmony with its mission "to increase the availability of healthful, locally grown food to the people of New York City, particularly those with little or no income," Just Food continues to explore new horizons for CSA in the big city. Hard as it is to line up middle-class members for CSAs, Just Food has been determined to discover how to make it work for low-income city dwellers as well. Just Food has encouraged all the city's CSAs to offer working shares and to donate leftover shares to food pantries or soup kitchens. Beginning in 1998, the staff focused on persuading community organizations that work with low-income people to include CSA in their missions. Explaining CSA to hard-pressed, low-income, grassroots groups that are struggling for survival amid the harsh economic and social realities of the Big Apple is a challenge. For white, middle-class organizers, building trust with other ethnic or income groups is a major hurdle to overcome. But Just Food is having success with some new approaches, which include:

- Holding community fundraisers to buy or subsidize shares for people who can't afford them.
- Establishing a revolving loan fund through which the grower could be paid up front, while the members who cannot afford a lump sum payment make installment payments over the course of the season. If these members move or become unable to pay for their shares, the farmer will not be left holding the bag. On the other hand, if the farm has a crop failure, the fund could cover the risk to members who have spent their limited food budget on the CSA by paying for food from other farms.
- Compensating with shares in a CSA the work of a group of parents in East Harlem who are enrolled in a nutrition education program.
- Through the City Farms project, creating CSAs for which the food is produced in community gardens in the city.

(For more information about how Just Food and other groups around the country are pioneering in this area, see chapter 20.)

In the province of Quebec, the remarkable nonprofit Équiterre has created a model CSA support organization that would be well worth imitating in other regions. Starting from a pilot project with one farm and twenty consumers in 1995, Équiterre has built up a network, which reached one hundred farms in 2005, supplying organic products to eighty seven hundred households. Eighty-two of the farms provide vegetable or meat shares; the other farms provide supplementary products, such as honey, apples, or cheese. Équiterre facilitates networking among the farms for mutual support and technical training and promotes CSA membership among consumers. To help with pricing, Équiterre provides a monthly comparison of organic and conventional vegetable prices. Each farm pays 2 percent of its CSA income, with a maximum payment of $350 a year, to finance the services of Équiterre. For would-be CSA farmers, Équiterre provides practical guidelines and a handbook in French, *I Cultivate, You Eat, We Share*, written by Elizabeth Hunter, the founding director of the network.[7] (For more about Équiterre, see the section "Agricultural Development" in chapter 19.)

The beginning of each new CSA is a hopeful moment. The act of selling or buying shares in a farm is full of promise and rich significance. Transforming that act into the living reality of community support, however, takes time and the willingness on all sides to change: to change how we eat, how we think about food, how we pay for it, how we manage a farm, and how farms connect with one another. The more we can share our discoveries, the faster Community Supported Agriculture will be able to grow toward agriculture supported, sustainable communities. By teaching as we learn, CSA farmers and members may not become millionaires, but we will create solid local institutions and social capital, which give us strength for the future we cannot even imagine today. Like the Mondragon cooperatives in Spain, we are building the road as we travel along it together.

CHAPTER 3 NOTES

1. Dunaway, Vicki. "The CSA Connection." *FARM: Food Alternatives with Relationship Marketing* (Summer/Fall 1996): 1–6.
2. Timothy J. Laird, "Community Supported Agriculture: A Study of an Emerging Agricultural Alternative" (master's thesis, University of Vermont, 1995), 45.
3. Ibid., 50.
4. Steven McFadden and Trauger Groh, *Farms of Tomorrow Revisited: Community Supported Farms—Farm Supported Communities* (Kimberton, PA: BD Association, 1997), 88–89.
5. In a personal communication, Paula told me that in 2007, there were 50 CSAs in New York City supplied by 18 vegetable growers. Members total more than 11,000 and income to farmers is more than $1.7 million. Thirty of the 50 CSAs use low-income payment strategies. (July 26, 2007)
6. "The Eat Healthy Rebate: A CSA Promotion from the Health Insurance Industry," *The Community Farm* (Fall 2006).
7. Elizabeth Hunter, *Je cultive, tu manges, nous partageons* (Équiterre, 2000).

CHOOSING A FARM OR A FARMER

And he gave it for his opinion . . . that whoever could make two ears of corn, or two blades
of grass, to grow upon a spot where only one grew before, would deserve better of mankind,
and do more essential service to his country, than the whole race of politicians put together.
—JONATHAN SWIFT, *Gulliver's Travels*

Not since the kings of France selected gardeners based on their ability to supply asparagus year-round have consumers had to give this much thought to the choice of their farmer. Yet choosing the person who will grow your food is surely as significant as selecting the family doctor, the pastor for your church, or the mechanic for your car. In some parts of the country, you have many CSAs to choose from; in other areas, you will have to start your own.

Choosing a CSA

With the spread of local food campaigns, there are more and more places around the country where you have a choice of CSAs. For instance, there are dense clusters around Madison, Wisconsin; Burlington, Vermont; and Berkeley, California. How is one to choose among them?

I asked the members of the Genesee Valley Organic CSA core group why they had joined. Several said they joined because they had friends who were already members. People who understand what they are getting into make good members. They know what to expect and have friends to welcome them into the CSA community. Suzanne

Wheatcraft, a core member for fifteen years, gave this thoughtful answer:

I belong to Genesee Valley Organic CSA because it's been around this area as long as I have. I feel that by being with a group of farmers who have worked for decades to learn the best crops and the whims of our weather, I am able to reap the benefits through a longer season of veggies, and a greater variety (when compared with the other CSAs). I also believe in some of the things that Genesee Valley Organic CSA (GVOCSA) does that maybe the other CSAs don't. For example, you should not be able to buy your way out of farm work. It's an important connection to what we are all about (this is the big we—humans). We eat food. Food grows in the ground. Farming makes that happen. Also, farmers farm. I sincerely appreciate that one of our farmers happens to be a great communicator, but I always feel a small amount of shame for my fellow consumers when I hear of other CSAs where the farmers are writing weekly newsletters, running all the social events, and basically holding the whole thing together. What happened to the C in CSA?

Most people join CSAs primarily for the fresh, local organic food. Some people, though, want more of an experience. When there is a choice, you can select the CSA that matches what you are looking for. Our own CSA requires work of everyone. When potential members call the core member who handles inquiries and complain when they hear about the work requirement, she sends them to the Porter Farm CSA, which limits member involvement to asking a few members to provide sites for distribution. Reciprocally, when people beg the Porters to come work on their farm, they send those people to us.

Where there are many CSAs to choose from, convenience may be the most important factor. When my son settled in Berkeley, I decided to present him with a first month's membership in a CSA. Both my son and his wife were working full-time as elementary school teachers, commuting an hour and a half each way to teach. Free time was scarce, and they did not own a car. Price, box content, and the farmer's personality were much less important than a convenient drop-off location. I went to the LocalHarvest Web site (localharvest.org), which lists CSAs by state and will sort them by zip code, and worked my way through all the CSAs that deliver in the Bay Area, until I found one that had a drop-off point in my son's neighborhood. Their close friend Gloria made a similar choice in Seattle. With no car and not much free time, she was willing to pay to have someone deliver organic food to her door. Équiterre in Quebec and Just Food in New York City provide maps of drop-off points to guide people to the most convenient food connection.

Getting to know your farmer is, of course, preferable to a choice dictated by convenience. The Madison Area CSA Coalition (MACSAC) sponsors an annual CSA festival. Each farm has a table where potential customers have the opportunity to meet their farmer, a sort of speed-dating for food. In this setting, consumers can read the brochures of each farm, chat with the farmers, compare prices and crop schedules, and make a more holistic choice. The MACSAC Web site offers this advice:

> Given the many choices, interested eaters often have difficulty choosing one of the CSA farms to be involved in. All of the farms belonging to MACSAC follow the basic CSA model, grow a wide variety of crops, and produce these crops without the use of synthetic pesticides. Beyond these similarities, all the CSA farms have characteristics which are uniquely their own. It is important to consider which farm will best meet your needs and expectations. Browse through the following list and note the factors that are important to you. As you peruse the farm list consider which farm would provide the best fit.

At Peacework Farm, farmer Ammie Chickering and apprentice Andy Simmons hoe beds in early spring. Row covers protect tender greens from deer munching and cold. PHOTO BY CRAIG DILGER.

In a much more modest way, the three farms offering CSA in the Rochester area make presentations together to groups of likely members, such as the Vegetarian Society. We distribute our brochures to the audience, and each farmer gives a brief talk about what his or her farm has to offer. We have also had good luck in getting area newspapers and magazines to report on the local CSA choices. All three CSAs benefit from this cooperation.

Finding a Farmer

If you live in an area where no CSA exists, you may want to persuade a local farmer to start one. A land-owning organization, such as a land trust, a school, a church, a state or national park, or even a town may want to hire a farmer to create a CSA.

As for any job applicant, be sure to check references. Doing a reference check before interviewing may save time. Prepare a list of the questions you want to ask before calling references and, if possible, find someone not on the candidate's list who can tell you about him or her as well.

To find appropriate candidates, place notices with state organic farming organizations, the Biodynamic Farming and Gardening Association, and regional CSA centers. The National Sustainable Agriculture Information Service (ATTRA) has addresses of organizations and networks that may be able to help (see the "CSA Resources" section on p. 284).

Unfortunately, you will probably find that few seasoned farmers are available. Most of the older farmers are settled on their own farms. Seeking Common Ground CSA made a lucky catch in Woody Wodraska, a biodynamic farmer with many years of experience under his belt, whom they found through the Internet; yet he stayed for only one full season and left in August of the second year. After a disappointing experience with another farmer, the group has reduced its numbers to twenty families who cooperate in growing the food.

Capable farmers with even a few years of experience will have a choice of positions. When the Syracuse Food Bank ran out of money for the farm

Shane LaBrake had spent a year getting started, he found sixteen farm jobs open to him. In recent years, I have seen a surprising number of advertisements for farmers to run CSAs. During the winter of 2006, Appleton Farms in Massachusetts sent out this demanding job description:

> The CSA farmer will help manage multiple aspects of the CSA operation at Appleton Farms, a property of the Trustees of Reservations [a statewide conservation organization]. Responsibilities include equipment, facilities, and infrastructure management, implementing day-to-day CSA farming tasks, and helping serve as a liaison between the CSA and the public, volunteers, and other area farmers. . . . The CSA Farmer must possess a strong appreciation for farming, the land, and the extraordinary agricultural and cultural history of Appleton Farms. The CSA Farmer must enjoy working with people. The CSA Farmer must be able to work independently with a minimum of supervision and be willing to learn new skills in order to meet the broad range of tasks that will be assigned to the position. . . . the CSA Farmer must possess excellent people skills, being around the public and able to adjust to farm methods and practices that are not normally found on a typical farm.

Institutions that hire young farmers have a high turnover rate. The problems come from both sides. Inexperienced farmers may not be able to cope with the many levels of organization required to run a CSA, while some institutions place additional burdens on staff that are not appropriate for anyone trying to farm. (Kristen Markley analyzes this problem thoughtfully in her Penn State master's thesis, "Sustainable Agriculture and Hunger."[1]) Some organizations badly underestimate the amount of knowledge and commitment it takes to run a small farm well. They offer low salaries, inadequate living conditions, and little control over the budget and then are frustrated when young farmers stay a year or two and leave. Five years ago, Phillies Bridge Farm was suffering from rapid farmer turnover, so they asked me to act as a consultant. On visiting the

Hiring a Farmer

A CSA that has land but no one to farm it might consider the following in hiring a farmer:

Previous farming experience:
- appropriate crops
- similar scale
- comparable climate and growing conditions
- apprenticeship or internship on CSA farm
- farming or gardening background
- familiarity with function, operation, and maintenance of equipment considered appropriate for given site
- formal training in organic vegetable and fruit production

Knowledge of organic or Biodynamic gardening or farming methods:
- philosophical commitment to organics
- awareness of resources available
- membership in state or regional organic farming association

Ease with people:
- evidence of organizing skills
- evidence of teaching ability
- willingness to attend meetings with CSA core group

Personality traits useful for successful CSA farming:
- well-organized
- observant
- determined and conscientious
- hard-working
- completes tasks
- self-starter
- able to share control

If a farming team:
- Are members cooperative?
- Do they divide tasks and responsibility appropriately?
- Do they have a clear decision-making process?
- Are they respectful of one another?

farm, I discovered that they provided the farmer with housing and a salary for only nine months of the year, and the farmer had to go to the board for any expenditure over $200.

A few CSAs have gotten started without a skilled farmer. The Bawa Muhaiyaddeen Fellowship, a Sufi community in Philadelphia, managed to keep the Farm Food Guild CSA going for three years with four and then two members taking the lead, and others helping out as they could. With fifty families signed up, Laura DeLind, a professor of anthropology, found herself in the role of a thirty-hour-per-week grower at the Growing in Place Community Farm in Michigan. She had expected that other members would participate more actively and that they would "learn together how our food is grown." The CSA survived the season, but Laura experienced "no joy," intense anger at those who seemed to let her down, and a lot of stress. After that hard year, the group heard and responded to Laura's frustration. As she tells it:

> The CSA worked because we all took on the responsibility of growing and learning . . . how to work with each other (harder, I think, than growing vegetables). Membership required a share fee of $275 and a minimum of ten hours of work a month, and many of the fifty-five-member households provided much more. . . . We operated for seven years. Each year, members became more aware of the realities of growing and eating good food and adjusted their lives accordingly; we were slowly becoming a community held together by our collective responsibilities to a piece of land and to the work we did there. By 2003, the CSA had become a "we" thing. On their own initiative, members wrote for grants, . . . cleaned tools and buildings, mowed common areas, built a straw bale house to store our energy cells, all the while attending to the daily tasks of keeping the garden weeded, irrigated, harvested, and pest controlled. Unfortunately, we didn't own the land we farmed (four acres) and after the business we rented it from declared bankruptcy, we lost the farm. The soil, perennial plantings, walkways, and memories now lie beneath a diesel engine repair shop and a new subdivision.

Any number of other cooperative CSA gardens are under way, such as the Genesee Community Farm in Waukesha, Wisconsin. "Learning as we grow," is their motto, and in their invitational brochure they emphasize community building and member participation. With only fifteen shares, perhaps they can avoid the strains suffered by Growing in Place and develop a more equitable division of labor and responsibility. After receiving weekly shares for three years, the five working members of Karen Kerney's CSA in Jamesville, New York, found that they were supplying day-to-day needs from their own gardens. Karen obligingly shifted to field crops of winter squash, corn, and potatoes, and the members reduced their weekly work shifts to occasional work parties at peak planting and harvest times. If they return to weekly harvests in the future, Karen says she will insist on a firm work schedule; trying to accommodate everyone's individual needs became overwhelming. It will be fascinating to see how these noncommercial gardens evolve and whether they will find a path to long-lasting cooperation.

Conceivably, a group of people could decide to form a CSA in order to give a new farmer the opportunity to get started. If everyone involved understood the situation and agreed to share the high risks, it could work. But if the farmer were not able to get up to speed within a few years, the goodwill of the members would probably evaporate. In the brochure announcing the new CSA at Maysie's Farm in Glenmoore, Pennsylvania, Sam Cantrell explains his decision to give up field biology and settle down to farm the family land:

We could "grow" houses, as most of the other farms in the area have done. Or we could again rent the fields out. . . . Or we could accept the responsibility that comes with land ownership and ecological ethic and implement a management plan that sustainably utilizes the resources available. One of us could forego the advancement of his career and, instead, invest his energy into the resurrection of the farm. That's the choice I've made . . . to have the farm become my career, to utilize this wonderful resource to fulfill my commitment to conservation and education.

Although Sam's promotional material sells shares as a contribution to the ecological preservation of land, Sam made sure he also knew how to grow high-quality vegetables by getting two years of experience marketing to local stores before venturing into CSA. After a few years in the role of farmer, Sam was able to hire a farmer to continue his work.

———————————•———————————

For the increasing numbers of bright, energetic young people who want to commit their lives to sustainable living, CSA is one of the most promising paths. Many efforts currently in motion will make this easier: new internship and hands-on training programs; land-linking to connect would-be farmers with retiring ones; the Equity Trust Fund for CSA, which helps with planning and financing; the community food security projects introducing low-income city residents to organic food production and CSAs; and much more. Hopefully, in the near future, people who need a farmer will have many more skilled practitioners from whom to choose and those who wish to farm will have many more ways to acquire the resources and skills they need.

CHAPTER 4 NOTES
1. Kristen Markley, "Sustainable Agriculture and Hunger" (master's thesis, Pennsylvania State University, 1997).

THE LAND

Whenever there are in any country uncultivated land and unemployed poor, it is clear that the laws of property have been so far extended as to violate natural right. The earth is given as common stock for men to labor and live on. . . . The small landowners are the most precious part of the State.

—THOMAS JEFFERSON, *Writings*, V. XIX

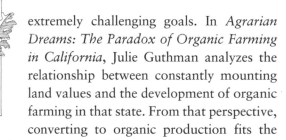

Without a piece of fertile ground, a CSA cannot exist. As Eva Cuadrado Worden succinctly put it, "The current farmland situation in the U.S. is that those who wish to farm cannot afford to own land, and those who own land cannot afford to farm."[1] CSAs can acquire the use of land in a variety of ways, but for long-term, secure tenure of that land, there are many obstacles to overcome. An existing farm can convert all or part of its production to Community Supported Agriculture (CSA). An institution, such as a church, hospital, school, state or national park, town, or food bank, can dedicate land it already holds or purchase land for CSA use. A new farmer or consumer group can rent or purchase land for a CSA. Land trusts can play an important role in any of these variations, accepting conservation easements, purchasing and then leasing land to CSAs, or sponsoring farms on land that they hold. Each of these is a special case with specific benefits and complexities.

The first CSA proposal from Indian Line Farm in 1986 called for community control of the land, balanced land-use planning, special provisions for agriculture, and the elimination of speculation. By involving members in the farms that grow their food, CSAs have begun to take real steps toward these extremely challenging goals. In *Agrarian Dreams: The Paradox of Organic Farming in California*, Julie Guthman analyzes the relationship between constantly mounting land values and the development of organic farming in that state. From that perspective, converting to organic production fits the historic pattern of California farmers seeking ever higher value crops to cover their land costs. In 1998–1999, Guthman rated 144 organic farms on a scale of 1 to 5 from least to most agro-ecological: "A 1 was given to those who took no affirmative steps but merely replaced disallowed inputs with allowable organic inputs . . . and 5, to those who managed the entire operation *by design* with minimal outside inputs and maximum attention to processes."[2] She also investigated the organic farms' labor practices to see if they used labor contractors and provided better wages and more full-time work than did conventional farms. Her findings are very significant for CSA: According to her evaluation, the farms that achieve most fully the integrated environmental and social goals of organic agriculture are the "ultradiversified farms whose cropping and direct-marketing strategies fundamentally alter the organization of production."[3] Most of these are CSA farms. She also discovered that "several of these farms have received some subsidy to land, either through inheritance, or landowner largess, or foundation and state support to buy the land." She concludes that "the reworking of nature that occurs on such farms is clearly driven by the decommodification of food and land, which opens up an economic space where social divisions can be eroded rather than accentuated."[4] In other words, CSA farms are demonstrating a real alternative to both conventional *and* organic industrialized agriculture.

In his survey of CSAs, Tim Laird found that 59 percent of the CSAs are based on farms that have converted totally or in part from other forms of marketing; 76 percent of this group purchased or inherited the land, and 17 percent received it as a gift or donation. The other 41 percent of CSAs were using land leased from private owners, land trusts, or for-profit groups. In several cases, the rental payment was a share in the produce. The more comprehensive 1999 and 2001 CSA Surveys reported respectively that 27 percent (78 farms) and 23 percent (or 69 farms) did not own any land and in 1999 that 162 reported alternative land-use arrangements for at least some of their land, renting from private landowners, nonprofits, or other institutions. That suggests that a surprisingly high percentage of CSAs do not own the land they are farming.

CSA farms tend to be small. The 1997 Census of Agriculture of all U.S. farms showed that only 30 percent farmed on fewer than forty-nine acres, in contrast with 72 percent of the CSA farms. Both the 1999 and 2001 CSA Surveys found that the median-size CSA farm is fifteen acres, with seven acres the median cropland, of which three acres is the median used for the CSA. (The survey writers stress the significance of the *median* [50 percent of farms are smaller and 50 percent larger], rather than the *mean* [the average], since a small number of much larger farms render the mean "mean"-ingless.) In 2001, 84 percent of the farms used fewer than ten acres for CSA and 65 percent used fewer than five acres.

☙ IN 1988 A GROUP OF PEOPLE from the Albany area visited the E. F. Schumacher Society and heard about the Indian Line CSA. They decided they wanted to start one in New York. They wrote an article for the food co-op newsletter inviting people to a meeting to talk about finding land and a farmer. Two landowners attended, and both offered their land. Janet Britt, who grew up on a farm in central New York, had farmed in Colorado and Missouri and had worked at the Indian Line CSA. She had never farmed on her own before, but she took the plunge and became their farmer.

Janet visited the two potential pieces of land and advised the group to select a five-acre piece on a farm in Schaghticoke. The owners were both doctors and agreed to a generously modest rent—a share in the produce. They also purchased a used tractor and leased it to Janet for $750 a year. Since the owners lived elsewhere, Janet acted as caretaker of their land and rented the farmhouse. The Hudson-Mohawk CSA existed from 1988 until 2004 with an average of one hundred shares per year. Until 1996 there was no written lease. Another farmer leased the hay fields and had a five-year lease that satisfied the local requirements to qualify for the farm rate on property taxes. When he dropped his lease, the owners offered Janet a five-year lease so they could continue to enjoy the tax break. In 2004, after a successful struggle against breast cancer, Janet passed the CSA on to another farmer, Brian Denison. Janet reflects, "As I look back, I see what I should have done. I should have delineated money for a land fund in the CSA annual budget." But Janet is pleased that another farmer continues to run the CSA and that the land remains in farming.

☙ JOHN PETERSON GREW UP on his family's farm in Caledonia, Illinois, tending dairy cows, beef cattle, chickens, and hogs and growing field crops of corn, soybeans, and grains. He took over the 186-acre farm in 1969, when his father died. Over the next decade, he expanded production to 800 acres. Then disaster struck. Like many good farmers in the early 1980s, John was overwhelmed by debt, which forced him to sell most of the land, equipment, and livestock, leaving him with only 22 acres and the farmstead. After a few years of anger, despair, wandering, writing, acting, and offering counseling services to other distressed farmers, John returned to the farm,

Tuscaloosa CSA—the beauty of the cultivated landscape. PHOTO BY JEAN MILLS.

decided to grow vegetables organically, and gave birth to Angelic Organics. With three years of experience with wholesaling vegetables and farmers' markets, John and his then farming partner, Kimberely Rector, converted most of their land to CSA. By 1998, Angelic Organics was supplying eight hundred shares in and around the city of Chicago, renting 20 acres of neighboring land for $2,500 on a year-to-year basis, and concluding the purchase of a 38-acre piece of land to their north, which allows them to fallow one-third to one-half of the land each year. Rather than incurring further debt, the farm appealed to CSA members to become investors in a limited liability company that would purchase the land at $4,500 an acre, then lease it to Angelic Organics for fifteen years with the provision that the farm could purchase the land at fair market value at any time. Simple in concept, this approach turned into a legal labyrinth from which the farm emerged with an eighty-page legal document that cost an agonizing five months and $14,000. Since then, share numbers have risen to over 1200. John's story has now become a classic that is beautifully documented in the film *The Real Dirt on Farmer John* (2005).

TWO SEASONS OF GROWING on rented land and selling through the wholesale market in Boston convinced Tom Cronin and Chris Yoder to try direct sales. In 1994 they started Vanguarden CSA, growing the food on four pieces of rented land in the town of Dover, Massachusetts. Sue Andersen joined their team for two years, from 1995 to 1997. By 1996 they had reduced wholesaling to 10 percent of production and increased shares to 193. In 2001, Tom left too, leaving Chris to run Vanguarden by himself. To make this manageable, Chris reduced the share numbers to 100 and gave up the more expensive of the rented pieces. (When I tried to reach him in March 2006, he had disconnected the CSA phone because there were so many more calls from potential members than he could handle!) Using rented land has been a limiting factor, discouraging him from investing in perennials, such as tree fruits. It is also expensive: His annual budget includes $4,000 for rent. Two of the three pieces Chris farms are on land owned by a conservation group, so he feels fairly secure that there will be no pressure to sell them out from under him. The private land he rents is pivotal to the town's open space plan. Should the owners decide to sell that, Chris believes there is a good chance it will be protected. The house he rents is on a fourth piece of land where he had constructed two greenhouses. Fortunately, the four pieces are within a mile or so of one another. Still, the reality is that Chris cannot afford to buy land for his farm.

🐚 **THE STONY BROOK–MILLSTONE WATERSHED** Association funded the establishment of a farm on a piece of its land in Pennington, New Jersey, in the early 1980s and hired a farmer. The association quickly learned that a two-acre vegetable garden could not pay the farmer's $25,000 salary. When Jim Kinsel took the job, the Watershed Association was still providing free use of the land, a two-bedroom house, equipment, and volunteer labor but expected him to earn his own salary from the farm business. Since Jim has turned the farm into Honey Brook Organic Farm, a very successful operation with 2,100 shares, the association has decided to maximize its return on the property and has required substantial lease payments for use of the farm. Starting in 1997, the farm has had three-year leases with a straight landlord/tenant relationship. The annual rent is $25 an acre, $900 a month for the apartment Jim lives in, and $3,000 for the farm buildings in 1997, rising to $7,000 in 2000, calculated as 75 percent of the market value. The Watershed board changes membership annually, and support for the farm ebbs and flows. For several years, Jim and his partner, Sherry Dudas, have nurtured the hope that they would be able to buy farmland of their own. But New Jersey land values have skyrocketed by 400 percent. In the meantime, the Watershed board has agreed to extensions of the current lease. Jim says, "I still count my blessings in renting from an organization. There's a kind of check and balance on the board; no one individual can throw me off."

🐚 **LIVING ON ZJ FARMS**, eighty acres on the outskirts of the small rural town of Solon, Iowa, Susan Jutz discovered a mimeographed copy of Robyn Van En's *Basic Guide to CSA*. She and her friend Simone Delaty studied the booklet through the winter of 1996 and decided to give CSA a try. They started

Local Harvest CSA with 18 shares in 1997, Susan supplying the vegetables from a one-acre garden and Simone providing bread, eggs, and flowers shares. When Susan's husband left, she found herself with four children to support, one of them home-schooled, and a mortgage of $1,000 a month to pay. Her teenage son pointed out that the solution was to mechanize. They purchased a Cub International, and the two brothers became the tractor drivers, enabling the farm to expand over the next two years to 140 shares without increasing the workload. Susan has never done the mechanical cultivating. An informal core group of people who have been members from the beginning help sustain Susan's elaborate local network of barter arrangements and give moral support and advice. The children, four interns, and many unpaid volunteers—fluid work shares—make up the farm's workforce. The internship experience at ZJ is total immersion; the interns live with Susan and her family, growing the vegetables and helping tend the three goats, fifty ewes, and thirty hogs.

🐚 **IN THE RAPIDLY EXPANDING SPRAWL** of San Diego, where land is priced for developers, not farmers, members played a critical role in helping Be Wise Ranch get access to land. Bill Brammer owns 20 acres and has leased as many as 300. A senior city planner and other influential members of his CSA helped him obtain a lease that ran from fifteen to twenty years on 150 acres that belong to the City of San Diego. The rest of his leases are short-term, on land in line for development. In 1998, Bill had to relinquish a piece of land he had farmed for twenty years, forcing him to dismantle and rebuild his packing shed, and in 2005 he lost what he describes as 80 percent of his best land. The next "crop" on those acres will be houses. On his much reduced acreage, Bill plans to give up wholesaling and concentrate on direct sales, the CSA, and a farm stand.

Acquiring Land

People who are thinking about starting a CSA often ask how many shares will support a farm and how

Overhead irrigation waters crops at Full Belly Farm in Guinda, CA. PHOTO BY FULL BELLY FARMERS.

much land will that require? The second part of the question is easier to answer than the first: Many farms are able to supply at least twenty shares per acre. The problem of scale is more complex, and closely linked to intensity of production. A bio-intensive grower with enough labor available can make a living on two acres. A more mechanized style of farming requires fewer helpers but more land. For a single farmer to make enough money to live on, the minimum number of shares is around one hundred if the CSA is the only market. The Gregsons of Island Meadow Farm in Washington State support themselves on the production of two acres of vegetables. They sell around thirty rather gourmet and modestly sized shares and have other markets as well. In interviewing farms for the chapter on "CSAs That Quit" (chapter 23), I found that a major reason they gave for quitting was inadequate income. Many of those CSAs were using a lot of hand labor and had thirty shares or fewer. At the other end of the spectrum, several much more mechanized farms with four hundred shares and up seem to be making a decent living and have funds to continue to capitalize their farms. But no one in his or her right mind would start a CSA with four hundred shares!

That is a scale one has to grow into. Anything over two acres requires the use of mechanical equipment; the choice of the appropriate level of technology is a critical factor in determining whether you work yourself to death for a pittance or run an efficient and at least modestly profitable farm enterprise.

New farmers or consumer groups who want to start a CSA face the challenge of finding a suitable piece of land. Fortunately, several good books tell how to find land in the country, with detailed advice on what questions to ask and how to get through the legalities of purchase with minimum legal costs. *Finding and Buying Your Dream Home in the Country* by Les Scher, despite its silly title, is a helpful book. Vern Grubinger's book, *Sustainable Vegetable Production from Start Up to Markets*, highly recommended for all new produce farmers, contains an excellent summary on finding land specifically for vegetables.

As Grubinger stresses, the quality of the soil is more important than the state of the buildings. Soil quality is crucial for organic production. While most soils can be improved by building up organic matter with compost and cover crops, contamination with heavy metals and persistent chemicals such as dioxin and PCBs cannot be eliminated. You should learn as much as you can about the history of the use of a piece of land, since testing is expensive. What was grown there? How was it produced? Was an apple orchard treated with arsenicals, DDT, or mercury? Is there a good source of water? How pure is the water supply? Do the neighbors use chemicals, air blast sprayers, or aerial spraying? Knowing specifically what to test for reduces the cost. Talking to neighboring farmers and examining the soil maps at the county extension office or the Natural Resources Conservation Service office will provide much of the information you need about the property you are considering and adjacent properties.

Residues of some pesticides are distressingly persistent. On August 15, 1996, the Connecticut State Department of Consumer Protection tested produce grown at Holcomb Farm. The CSA managers were shocked when the results showed residues of chlordane and DDE, a breakdown product of DDT. The pesticides were used on the tobacco grown at Holcomb Farm back in the early 1970s, before they were banned. While the levels of the residues were low—at or below the Environmental Protection Agency "action level"— the Hartford Food System, which oversees the Holcomb Farm, enlisted the advice and help of a wide range of experts and agencies, sharing the information collected with the CSA members. Dr. Ted Simon, a toxicologist from the Center for Disease Control (CDC), volunteered to conduct a formal risk assessment. He came to the conclusion that "the lifetime risk of cancer from consuming about eight ounces per day of Holcomb Farm vegetables throughout the twenty-week growing season for thirty years would be about one in a million." According to John Cuddy, a bioremediation specialist in Wisconsin, the organic practices used on the farm seem to be the most effective means for stimulating the microbial activity that reduces chemical residues to harmless compounds. Similar residues

were found on another organic farm in Connecticut even after thirty years of organic management.

Some aspects of land acquisition are particular to CSA farms. The location of the land is important to consider. If your plan is to have pickup at the farm, you will need to be within a half hour's drive of the population you hope to serve. CSAs that deliver to pickup sites in town can be farther away: the Decaters truck their produce three and one-half hours from Live Power Community Farm to San Francisco. Several farmers have reported CSA members are harder to recruit in rural areas, though rural Iowans, even farm families, are joining new CSAs in that state. Urban and suburban dwellers are often more likely candidates.

Land purchase within half an hour of almost any city may be prohibitively expensive. Leasing is a possible alternative. If you find land you can lease, the owners need to understand that a CSA is not the typical farming arrangement. Agricultural lease agreements are usually made for conventional production systems, where the farmer plows the ground, plants it, puts down herbicides, and then comes back a few months later for harvest. CSA production involves daily attention and potentially large groups of people coming to and going from the farm. A landowner who approached me about land he wanted to rent suddenly switched his tone from friendly to icy when he learned I had two hundred helpers. Courtesy suggests that if the landowner lives near the CSA site, a day of rest from the traffic involved should probably be made part of the agreement.

Developing land for organic management is a long-term investment. The environmental stewardship this offers may appeal to some landowners. If they have any thoughts of selling the land within a few years, however, even the lowest rent would not make your investment in soil building worthwhile, unless you are willing to take a chance, as Bill Brammer did by farming land slated for development in San Diego. The best arrangement is a clearly written lease for a term of three to five years, giving

> A careful business plan is a **basic requirement** for obtaining any commercial loan. A list of CSA members who are committed to buying your farm's products should be proof that you have a market.

The farm sign at Anchor Run Farm owned by the township of Wrightstown, Pennsylvania, and farmed by Tali Adini and Jon Thorne, who rent the land from the municipality. COURTESY OF ANCHOR RUN FARM.

the CSA the option of first refusal and applying the rent toward the eventual purchase of the land. The lease should also include consideration of any improvements the CSA makes on the property. The E. F. Schumacher Society has published a booklet titled *A New Lease on Farmland*, which suggests how such improvements can be valued so that both parties get a fair deal.

Where leasing is the best or only alternative, CSAs should consider asking for a long-term rolling lease to ensure that they have adequate notice before being forced to move. Jennifer and John Bokaer-Smith have a five-year rolling lease with the EcoVillage at Ithaca, Inc. The section of the lease describing its term reads:

1. This Lease shall be a five-year rolling Lease.
2. The Term shall commence on May 1, 1997, and shall continue in full force and effect until May 1, 2002.
3. The Lease shall be automatically renewed for one additional year, to a maximum of five years, on May 1 of each year succeeding the commencement of this Lease.

4. Termination of the Lease shall be effected only by written notice given at least six months prior to the rolling renewal date. Such termination shall have no force and effect until five years from the date of notice.

The USDA *Farmers' Bulletin #2163*, "Your Farm Lease Checklist," contains all the clauses you should think about including. The Pennsylvania State Cooperative Extension's bulletin, *Guidelines for Renting Farm Real Estate in the Northeastern United States*, has work sheets for calculating the rental fee. There are probably similar guides to agricultural rentals in other parts of the country.

While going into debt is not a good idea, borrowing money from a bank may be the only way for some farms to obtain funds to purchase farmland. Many rural communities have banks that specialize in agricultural lending. Local farmers or the cooperative extension should be able to tell you which banks to approach. A careful business plan is a basic requirement for any commercial loan. A list of CSA members who are committed to buying your farm products should be proof that you have a market.

Resources are available to help you develop a business plan. In some states, the Extension has economic specialists who will help you. The Service Corps of Retired Executives (SCORE), a national network of volunteer business executives and professionals, offers technical and managerial counseling to small businesses, such as farms. (For information on your local chapter, contact the national SCORE office at (800) 634-0245.) The SARE manual, *Building a Sustainable Farm Business* is a thorough guide to writing a business plan. In Massachusetts the New England Small Farm Institute (NESFI) offers excellent courses such as "Exploring the Small Farm Dream." An advanced course in Holistic Resource Management could help you create a plan that is both convincing to a lender and truly useful for your farm's decision-making process.

The Federal Farm Service Agency (FSA) serves as the lender of last resort, should commercial banks turn you down. The FSA is supposed to reserve some of its funds for beginning farmer loans. (A beginning farmer is someone with less than ten years' experience.) In addition, a few states, such as Iowa and Nebraska, have "aggie bond" programs to encourage loans to new farmers. Dan Looker's *Farmers for the Future* documents creative ways in which the truly determined can get into farming. *Holding Ground: A Guide to Northeast Farmland Tenure and Stewardship* by Andrea Woloschuk, Annette Higby, and Kathy Ruhf provides a wealth of information on alternatives to conventional ownership, with many case studies, decision trees for farmers and landowners, and sample long-term leases. Their introduction emphasizes the urgent need to "rethink farmland tenure."

> We need a new ethic that fosters farmland access, security, affordability and investment. . . . New approaches and tools serve to save agriculture and foster farming in our region in three ways. First, alternatives to buying land offer economic security to new and developing farmers. Eliminating substantial down-payment requirements and enormous debt can make a developing farm operation more economically viable. Second, providing alternative ways for farmers to acquire land can help preserve the working landscape and associated amenities. Third, secure tenure agreements can foster long-term stewardship of the natural resources of the farmed property.[5]

Protecting Farmland

Securing land for farming for the long term is a problem shared by farms of all scales and forms of ownership. According to American Farmland Trust, 4.3 million acres of prime and unique farmland were lost to development and suburban sprawl between 1982 and 1992, nearly 50 acres every hour. Even CSA farmers who own their land outright with no

According to American Farmland Trust, 4.3 million acres of prime and unique farmland were lost to development and suburban sprawl between 1982 and 1992, nearly 50 acres every hour.

mortgage or debt are seeking ways to ensure that the land remains in farming through future generations.

Organic and biodynamic farmers want to keep the land they are working free of chemical agriculture. The legal tools available include conservation easements, land trusts, and the purchase or transfer of development rights, but none of these offers a quick and easy formula.

Chuck Matthei, who is now deceased but who was president of Equity Trust and a pioneer in preserving lands from the inflating pressures of the real estate market, has written in "Gaining Ground: How CSAs Can Acquire, Hold and Pass On Land" (available from Equity Trust, see "CSA Resources") that the key to finding the resources to finance farmland protection "lies in distinguishing the essential personal interests in farm properties from the inherent public interests. Defining and protecting the public interests legitimizes the application of charitable and public funds to a land purchase, and thereby assures affordable access and full opportunity to the farmer."

In "This Land Shall Be Forever Stewarded," Jered Lawson delineates the private and public interests:

> The private interests in farming may include:
> 1. secure permanent tenure for the farmer;
> 2. equity in the property, which is affordable out of farming income, transferable to heirs, and recoverable through sale if necessary;
> 3. freedom and flexibility in managing the farm.
>
> The community interests in farmland may include:
> 1. an ongoing source of quality food (maintaining the land in active production);
> 2. care taken for the long-term fertility of the farm (preventing agriculturally generated pollution and environmental deterioration);
> 3. ensuring access and affordability for

successive farming, and preventing the deleterious effects of absentee ownership.[6]

Particularly in parts of the Northeast, where farming has all but disappeared, a large number of people feel strongly about the need to protect the farmland that remains. Besides the landscape value of the farms, studies focusing on cost of community services have shown conclusively that communities save money by investing in farmland and limiting development. State and local governments have established purchase of development rights (PDR) and transfer of development rights (TDR) programs. Although these programs operate differently from state to state, basically, PDR programs assess farms at their value as farmland and at their value as real estate, and pay the farmer the difference. In exchange, the state or town or a designated land trust receives an easement, which is written into the deed on the property, eliminating all future development. In most cases, the effect of PDR is to reduce the value of protected properties, so that they remain affordable as farmland. PDR does not protect the activity of farming, nor does it ensure any particular method of farming. The Massachusetts PDR program changed its easement to require continuing agricultural use. The Food Bank CSA in Massachusetts was able to acquire land at a reasonable price because the Commonwealth had purchased the development rights.

TDR is more complex: Through zoning, communities establish "sending" and "receiving" zones. To develop more densely in a receiving zone, a builder must purchase development rights from a sending zone. The cost to taxpayers is lower for TDR, but the administrative and legal complexities limit its use.

Vermont has adopted the most aggressive approach to saving farmland; the Housing and Conservation Board holds a statutory right of refusal on any farm that has received property tax consideration or other state subsidies, before it can be sold and removed from production. The Vermont board also provides financial support to local land trusts through a fund capitalized by state tax money.[7]

Land trusts provide a nongovernmental alternative. There are two kinds of land trusts: conservation land trusts and community land trusts. Usually, both are locally based, member-controlled, nonprofit corporations. The same legal and financial tools are available to both kinds of trusts. Most conservation trusts are dedicated to preserving open space, though about 10 percent of them have an active interest in farm and forest lands. Conservation trusts usually hold an easement on pieces of land, leaving the fee interest or title in the name of the farmer.

Community land trusts are largely urban and are devoted to providing the essential benefits of ownership to low-income people who have been excluded from the real estate market. A few, such as the Pioneer Valley Community Land Trust in Massachusetts, have farm holdings. Community land trusts usually retain title to the land, which they lease to residents. *The Community Land Trust Handbook*, by the Institute for Community Economics, provides an introduction to this form of land trust, with examples from around the country and guidelines for organization. The leases are often for ninety-nine years and inheritable; lessees own the buildings and other improvements. The Pioneer Valley Community Land Trust requires ecological management of the land it holds.

In finding a way to protect your farmland, the choice between a conservation trust and a community land trust may depend more on what organizations are available locally. According to Chuck Matthei, the "difference between the individual holding title and the land trust doing so may seem significant, but in fact, it may not be." More important are the details of the legal agreement, whether lease or easement.

In *The Conservation Easement Handbook*, Janet Diehl defines a conservation easement as "a legal agreement a property owner makes to restrict the type and amount of development that may take place on his or her property. . . . Each easement's restrictions are tailored to the particular property and to the interests of the individual owner. . . . The owner and the prospective easement holder identify the rights and restrictions on use that are necessary to protect the property—what can and cannot be done to it. The owner then conveys the right to enforce those restrictions to a qualified recipient, such as a public agency, a land trust, or a historic preservation organization.[8]

Diehl's handbook is a guide to understanding and writing conservation easements, providing examples as well as suggestions for monitoring and information on tax law. I have heard Chuck and others compare property rights to a bundle of sticks, which include the rights to ownership, to access to the land, to underground mining, and so forth. Development rights, governed by a conservation easement, are only one of these sticks, which you can sell while retaining all others.

Although Equity Trust is willing to hold conservation easements on CSA farms, Chuck recommended that farmers develop a relationship with a local land trust. Defining the terms of the easement on a farm and the oversight this will require from the trust may entail a long process of mutual education. Nevertheless, this will probably be easier than creating a new organization for the single purpose of protecting your farm. In beginning to explore these issues, I inquired of American Farmland Trust (AFT) whether it would be willing to hold an easement on Rose Valley Farm. My partner, David, and I owned the land outright, but we were hoping to guarantee ecological management after we are gone. Jerry Cosgrove, who ran AFT's New York office, told me AFT would accept a conservation easement but does not have the capacity to ensure that the management of the farm remains organic. He recommended I find a local organization.

〰 **THROUGH IMPRESSIVE HELP FROM** their communities, Live Power Community Farm and Fairview Gardens, both in California, Common Harvest Farm, in Wisconsin, Temple-Wilton Community Farm in New Hampshire, Indian Line Farm and Caretaker Farm in Massachusetts, and Roxbury Farm and my own Peacework Farm in New York have made the transition from leasehold to protection in perpetuity. After farming on borrowed land for over two decades, Gloria and Stephen Decater of

Live Power Community Farm faced the possibility that Richard Wilson, the owner, might need to sell the land and that they would not be able to afford to buy it at its market price. Four seasons of weekly vegetables had attracted a group of loyal farm members in Covelo and San Francisco. The Decaters turned to them for help and advice. Passions in the group ran high: There was consensus that the Decaters were the stewards everyone wanted for the land and that the Decaters deserved greater security. But the group divided over how to accomplish this; some people were in favor of private ownership, while others preferred a nonprofit arrangement.

The Decaters themselves were more concerned about the responsibilities of stewardship than private property rights. Stephen commented:

> When we contemplated private ownership, we had two areas of concern. One was that to acquire complete private ownership would entail placing ourselves in a major debt for an extended period, even if we could find a lender. We were opposed to the concept of entering into debt, and did not have a credit history, having always operated on a cash basis. The other concern was that we saw one's relationship to the land as having moral dimensions. We saw land ownership not as acquiring strictly a commodity only to be treated as we please, but also as an office bearing certain inherent responsibilities requiring wise (just) usage and stewardship. We saw land as a resource that also belonged to future generations, a resource which we had no personal right to appropriate or damage regardless of whether such a treatment was legal by current law or not. We wanted to be sure that the form of legal ownership we entered into would reflect this moral imperative.[9]

Four years of meetings and fund-raising, of encouraging advances and sudden dead ends, led to a successful combination of public and private ownership. The Decaters were able to purchase their forty acres of farmland at its value based on its potential for earning agricultural income typical of the area. This resulted in a private investment of

$69,000 ($20,000 for land and $49,000 for buildings) for the Decaters, while the CSA members raised $90,000 to purchase the development rights, all other nonagricultural use rights of the property and to cover easement costs. They could not come to terms with a local land trust, so Equity Trust in Massachusetts is holding these rights in a conservation easement on the property. The easement is a remarkably detailed document that tries to foresee every eventuality. The terms of the easement place permanent responsibility on Equity Trust (or its successor organization) for oversight of the management of the land. The "Purpose" clause of the easement reads:

> It is the purpose of this Easement and Option to assure that the Property will be forever bio-dynamically or organically farmed and, as to that portion of the Property not farmed, retained predominantly in its natural, ecological, and open condition and to prohibit any use of the Property that will impair, degrade, or interfere with the conservation values of the Property. Without limiting the general purpose of this Easement and Option, the following specific purposes are intended for this Easement and Option:
>
> (a) To conserve and protect the Property's agricultural, natural and ecological value and prevent environmental pollution and degradation due to the use of agrichemicals;
>
> (b) To prevent the conversion of agricultural land to urban and nonagricultural use; and
>
> (c) To assure that the Property will be farmed, with organic or biodynamic methods, for the production of plant and animal products for commercial or charitable purposes.

To enable Equity Trust to oversee these purposes, the easement requires that the Decaters or their successors provide reports on any intended changes at the farm and documentation of the annual organic or biodynamic certification of the farming methods. Since the easement stipulates that the landowner must make at least half of his or her income from farming, the Decaters must also submit a copy of their income tax return. Equity Trust also retains the right to inspect the property with forty-eight hours' notice. Should the Decaters or their successors violate the easement, Equity Trust has the right to demand corrections and reparation. Should the Decaters stop farming or try to sell the land for more than its agricultural value, Equity Trust has the first right to purchase the land.

The Decaters themselves participated in the creation of this easement, which ensures that at any future sale of the property the value of the land will be based on its agricultural earning potential, so that a farmer will be able to finance it from farm income. It is not some form that a state office imposes on them or any other farmer. Every farm entering into a similar arrangement has the freedom to choose the terms of a conservation easement to satisfy its own needs and desires for the future. As Stephen puts it, the easement "can be the vehicle for partnership between the private individual and a nongovernmental community-public representative." The Decaters' easement creates permanently affordable farmland, accessible on the basis of commitment to stewardship, rather than accumulated capital.

Other farmers have forged connections between community members and stewardship organizations to protect their farmland.

IN THE MINNEAPOLIS–ST. PAUL AREA, Dan Guenthner and Margaret Pennings had been running a CSA at Common Harvest Farm on rented land since 1989. To purchase land of their own, they appealed to the 170 members of their CSA for help with either loans or donations. Forty percent of the members responded with amounts ranging from $10 to several thousand dollars. Members lent $12,000 in the form of prepaid shares for the next five years. The $28,000 in donations went to the Wisconsin Farmland Conservancy to purchase the development rights to the farm. "When we were meeting with Tom Quinn, director of the Conservancy, to work out the details of the easement, we realized that this document amounts to a public declaration of what we intend to do with the land," Dan observed. "Other than getting married and having children, this is the most important commitment we have ever made. We are wedding ourselves, on all sorts of levels—spiritual, emotional, social—to this place."[10] Dan says that the conservation easement allows the farm to not be distracted by the whims of the market and allows for long-term decisions. The example of Common Harvest has inspired neighbors to protect their land from development. Within three or four miles of their farm, there are 7 landowners who have conservation easements and a land trust with fourteen hundred acres.

TO ENSURE THE FUTURE OF FAIRVIEW GARDENS in Goleta, California, Michael Ableman chose to give up any prospect of personal rights to the property and instead to form a nonprofit corporation. Ableman farmed the twelve-acre piece, first as an employee, then as an independent contractor, and finally under a variety of leases. Meanwhile, the suburbs closed in around the hundred-year-old farm, the one lingering residue of what was once a large rancho typical of the predominent land use in the fertile area near Santa Barbara. Taking the nonprofit route was a difficult decision for Ableman: "This was my life, sweat, and blood for seventeen years, my school for myself and my family. I will have no monetary ownership."

In 1992, after the death of one of the owners, Ableman realized that the time had come to take energetic measures to protect the property. The market value of the twelve acres was over $1 million, an unlikely sum to raise to save a tiny family farm. Ableman's strategy has been to stress the public values inherent in Fairview Gardens. He set out a mission that encompasses both food production and public education. The mission statement promises to:

- Preserve the agricultural heritage of this hundred-year-old farm
- Nurture the human spirit through educational programs and public activities at the farm
- Demonstrate the economic viability of sustainable agricultural methods for small farm operations
- Research and interpret the connections between

food, land stewardship, and community well-being
- Provide the local community with fresh, chemical-free fruits and vegetables

Ableman approached the Santa Barbara Land Trust to hold and monitor a conservation easement on Fairview Gardens. As the Decaters found with their local land trust, an easement that specifies the continuation of farming, and insists on organic methods to boot, presents a challenge. The Santa Barbara Land Trust hesitated about making the commitment, but timely intervention by Chuck Matthei of Equity Trust helped reassure them that overseeing a farm was within their mission. The Fairview Gardens easement presents an additional complication since it also requires that the educational work of the nonprofit be maintained in perpetuity. This is new territory for conservation land trusts.

Impressive local support, persistence, savvy organizing, and sheer hard work culminated in the October 1997 celebration of the purchase of Fairview Gardens.

FARMING IN ANOTHER HIGH-LAND-VALUE area, Temple-Wilton Community Farm, one of the two original CSAs, followed a long, winding path to a permanent home. The three founding farm families started on two pieces of rented land. After moving twice and finding himself subletting Four Corners Farm with no hope of buying it, farmer Anthony Graham came to feel there was no point in going on. He and his partner, Lincoln Geiger, were "feeling the strain." The only approach that would have worked with the owner was to bring "a suitcase full of money" to his office. Lacking those big bucks, Anthony made some smaller moves, purchasing a house near the farm and a five-acre piece across the road from his home. The adjacent piece belonged to farm members, and a member was able to buy another six-acre piece. Although the farm had an informal core, their "Thursday group," Anthony and Lincoln believed they needed a more formal entity to get traction with land acquisition. With members of their CSA, they formed the Educational Community Farm as a New Hampshire nonprofit. One board member had experience with the New Hampshire Land and Community Heritage Investment Program (LCHIP), which purchases development rights. When a forty-two-acre orchard near the farm came up for sale, the board was able to access this program to put up half the money while members raised the other half. Just as Anthony had made up his mind to build a new barn and move farm headquarters to this land, a developer did come to the owner of Four Corners Farm with a suitcase full of money. But having learned how the system works, the Educational Community board put together an even more complex packet combining state, federal, and town money to buy the development rights on Four Corners.

The developer has given them a ninety-nine-year lease on sixty acres for $99. To celebrate the farm's twentieth anniversary in 2005, Temple-Wilton finally had full use and long-term tenure on its land. Rather than proving to be a drain, this remarkable effort has brought a lot of attention to the farm, attracting a burst of new young families. In 2006 there are 104 member families and a waiting list of 95. As Anthony put it, they have succeeded in creating a protective "hedge of legal entities" so that the people doing the farmwork can be as flexible as possible.

∽ WHEN THE OWNER OF THE 18 ACRES my partners and I rent offered to sell us the entire 148-acre farm, instead of going to the bank for a mortgage, we decided to contact our local land trust. At that time, we had a five-year rolling lease with the Kraai family modeled on West Haven Farm's with EcoVillage at Ithaca. We approached the Genesee Land Trust (GLT), a conservation trust, with a proposal that they accept a conservation easement on our farm. GLT's mission is to "preserve and protect waterways, wetlands, farmland, natural and unique habitat, scenic and recreational lands." We intended to replicate what the Decaters had done at Live Power Community Farm.

We thought we could ask the members of our CSA to finance the purchase of an easement so that we could buy the farmland at its agricultural value. We knew that farmland in our area sells for $1,000 to $1,200 an acre and that the development value constitutes about half the price. To our surprise, the land trust agreed to depart from its usual practices by purchasing the farm and leasing it back to us for a very long term. In doing this, the GLT was taking the innovative step of functioning like a community land trust. Our farm business purchased the improvements on the land, a barn and a packing shed, but not the land under them.

With technical assistance from Equity Trust, the E. F. Schumacher Society, and George Parker, a Rochester lawyer, we negotiated a land lease document with the GLT. The members of the land trust board share our conviction that this land is worth preserving in perpetuity. Few of them, however, have had any experience of organic farming and none of them had ever engaged in a deal of this kind. Some of them did not understand why a long-term lease is so important to us, until we explained how long it takes to regenerate the soil and how heartbreaking it can be to do perennial plantings and then not see them mature. It also made sense to them that we could not afford to make the investments needed to upgrade the old barn unless we could be sure to use it for many years.

We had lengthy discussions of the appropriate lease fee. Under the terms of their nonprofit status, they cannot offer us a "sweetheart deal" on the rental fees we pay. They asked me to research what farmers pay to rent an acre of land in our county. We were all surprised to learn that the going rate, ranging from $35 to $50 an acre, just barely covers the land taxes. As a result, we have agreed that the farm will pay the land taxes and all insurance and other local fees but only a small administrative fee to the land trust.

Because the Kraais sold the Humbert farmhouse and kept only the land, there are no houses on the property. I was fortunate to be able to purchase a house right next to the land; Ammie and Greg commute thirty minutes from another town, and Katie rents from a neighbor. We asked the GLT if we could set aside two small corners of the farm to build houses for farmers. At first, the board

"Earth and Sky": the barns and fields at Peacework Farm in Arcadia, NY. PHOTO BY TOM RUGGIERI.

resisted: They did not like the idea of being land-lords and pictured all the problems involved with owning rural housing. A letter from Leslie Reed-Evans, director of the Williamstown Rural Lands Foundation (WRLF), the land trust that is entering into a similar arrangement with Caretaker Farm, persuaded them to see the matter differently; it said, "The WRLF needs to preserve farms. A key to pre-serving farms is to make the land and the infrastruc-ture affordable to farmers. The Caretaker Farm project gives the WRLF the opportunity to move beyond farmland preservation to farm preservation. This is an important distinction and critical to the survival of small family farms." The GLT board has realized that only by allowing the construction of homes for the farmers on the land and controlling the future sale price of those homes can they assure that farmers will be able to afford to farm there.

To raise money to pay for the 148-acre farm, the core committee of Genesee Valley Organic Community Supported Agriculture (GVOCSA) set up a special "Preserving Peacework" committee to raise

funds in coordination with the GLT. Including all of the ancillary expenses of land purchase—a survey of the property, a land stewardship fund to allow the land trust to monitor the land use on an annual basis, and so forth—the fundraising goal was $150,000. After describing the purchase and lease work in progress, the Preserving Peacework committee made this special appeal to GVOCSA members:

So, what does this mean to us? It means our CSA is going to benefit by knowing that land ownership costs and the issues around buying and selling land are not going to be issues our CSA has to deal with, nor will the farmers need to worry about a landlord who decides to sell the land out from under them. In short, in addition to reaping the benefit of knowing that Peacework Farm—"our farm"—will have a stable home, the CSA will also be a partner in the permanent preservation of high-quality organic soils, Ganargua Creek wetlands and flood-plains, and hardwood forest land with important wildlife habitat and beautiful wild flowers.

In only fourteen months, the Preserving Peacework committee raised the money to buy the farm; CSA members pledged $140,000. The GLT completed the purchase of the land in January 2006 and in March signed a twenty-five-year rolling lease with Peacework Farm. The very first contribution of $25,000 was anonymous and accompanied by this eloquent note:

> I believe that the planet is in a serious "people-created" ecological crisis motivated by greed and perpetuated by ignorance. The privilege and good fortune of eating clean local food is mine, due to the existence of the GVOCSA and Peacework Organic Farm. . . . My donation of $25,000 has caused raised eyebrows and not a few gasps. Conventional financial advice dictates "saving for a rainy day." Dear people, it is raining today, and it has been raining for a long, long time. It is rare that one has an opportunity to participate in such a fine cooperative venture. I do this with complete confidence in the ethics of the farmers, the GVOCSA, and the GLT. I participate with joy and hope so that my great-grandchildren will have safe vegetables grown on a beautiful organic farm.

Equity Trust and the E. F. Schumacher Society have helped three other CSA farms (Indian Line Farm and Caretaker Farm in Massachusetts and Roxbury Farm in New York) through similar transactions over the past few years. Together, we hope to create a new model for preserving farms and farmland. These goals are interrelated: We want to preserve farmland from development while keeping it affordable for people like ourselves who make our living as farmers. Farming on leased preserved land, we have all been able to raise money from contributions from our CSA members and other donors, establishing these farms as community property and divorcing them from the real estate market. The common threads to these stories are you need the cooperation of several entities, good legal advice is crucial, and patience will be rewarded.

Passing on the Land

Although most CSA farmers tend to be young (43.6 years is the median age, 13 years younger than the average of all U.S. farmers), the 12.5 percent of us in our late 50s and 60s are thinking about how to pass our farms on to younger people.

At Peacework, we are solving this problem by restructuring our loose partnership as a Limited Liability Company. My partners, Greg and Ammie, dance on either side of their 50th birthday, and I am up in the 60s. A few years ago, we added to our advertisement for apprentices that there would be the chance to become a partner in the farm. The universe smiled on us and sent Katie Lavin to apply. She spent the 2003 season, her 25th summer, as an apprentice with us. Her plan was to learn enough farming to go back to her grandparents' land in Fulton, New York, and start a new farm there.

That fall, I recommended Katie for the farm manager position for a farm project that was getting under way in the city of Rochester. The next spring, she came to ask our advice. Although she loved the farming work she was doing, the personnel situation was becoming unbearable. Despite her title as farm manager, her supervisor never disclosed how much money she had to spend. She found herself buying supplies out of her own pocket and using her car as the farm pickup truck. The person in charge of marketing was not doing his job and lied to her repeatedly. We suggested that Katie tell her supervisor exactly what changes she needed to remain in the job and set a deadline after which she would quit. Although the supervisor claimed she valued Katie, she was unable or unwilling to make the necessary changes, so Katie found herself out of a job. It took Ammie and me less than thirty seconds to know what to do. We invited Katie back and proposed that she consider the second year of apprenticeship a trial for staying with us permanently.

Throughout 2004 and 2005 we were working with the Genesee Land Trust on our joint project to raise the money for the GLT to buy the land we were farming and lease it back to us. Land trust leases are sometimes both long term and inheritable. Rather

69

than complicate our lease with a mechanism for succession, however, we decided to turn our farm business into a limited liability company (LLC) so that succession would be controlled by the operating agreement. Similar to a worker-owned co-op, to be a member of our LLC, you also have to be a manager, an active participant in the farming operation. We have written the agreement to require that we make major decisions, such as the choice of a new partner, by consensus. After a trial year, Katie became a member of our LLC in 2005 and worked as a full partner for two years. We were able to arrange our capital shares in the LLC in such a way that Katie was beginning to accumulate some equity in the farm. Sadly, at the end of 2006, she announced that the financial pressures of unpaid college debt were pushing her to find a better-paying job elsewhere. So, we started the search for a younger partner all over again. Our goal is to have partners spanning four decades in age so that with any luck, Peacework Farm will continue for another generation.

AFTER FOUNDING AND RUNNING Kimberton CSA Garden, one of the country's oldest CSAs, Barbara and Kerry Sullivan wanted to move closer to their families; so after some searching, in 2002, they passed the farm on to one of their apprentices. They sold the equipment, which belonged to them as sole proprietors but, like Janet Britt, did not ask any money for the farm business, since they believe it belongs to the community. Barbara told me they had always known they would not stay there indefinitely:

> We . . . had been on the lookout for our replacements for a few years. There were three apprentices who we would have been comfortable turning the CSA over to. The first two said no, but Birgit Landowne said yes. It was not that straightforward, but that's the way it turned out. In between, we had made a transfer arrangement with a young couple who had not worked with us previously. That turned out to be a big mistake, and we were relieved when they changed their minds halfway through the training year. Birgit had worked as an apprentice with us several years earlier and had

returned to work with us again. By this time she had a family. She had met her husband, Erik Landowne, when they both worked at Temple-Wilton Community Farm in New Hampshire. Both have a strong interest in biodynamics, which was important to us, and they were a good fit with the Kimberton community. Training was easy, since Birgit already knew so much about the operation. We worked with them the rest of that year, transferring all the financial management and information. And after we left, we were always available for questions. They have done a tremendous job there, and the CSA continues to be a great success.

(For the story of the Sullivans' farming at Kimberton, see *Farms of Tomorrow Revisited*.)

WHILE PASSING THEIR FARM ON to younger farmers, Sam and Elizabeth Smith at Caretaker Farm in Massachusetts also wanted to preserve the land as farmland forever, keep the farming infrastructure affordable, and assure a decent retirement for themselves—no small set of tasks. When it became clear that their children did not want to take over the farm (a daughter is farming, but far away in Colorado), the Smiths began a search for younger farmers. Initially, they invited a former apprentice, who moved to the farm with his family and spent two and a half years farming with them. By mutual consent, this match did not jell. In 2004 the Smiths put an ad in *The Natural Farmer* and other publications and began interviewing qualified couples who responded. Don Zasada and Bridget Spann were among the applicants. The Smiths were impressed with Don's farming and community skills. He had apprenticed with Dan Kaplan of Brookfield Farm, who had apprenticed with the Smiths. Over seven years at the Food Project in Boston, where he filled many different jobs, Don had become a skilled farmer. After his interview, he asked to come back and work for a day to get a fuller sense of the workings of Caretaker. Bridget researched opportunities in her field of social work with good results. Compatibility between the two families would be essential, since the Smiths continue to live on the farm, which has two decent houses. In September 2004 they decided that Don and Bridget were their first choice and sent a letter of intent. Eighteen months of working together while negotiating the complex terms of their transfer resulted in Don's taking up the reins as Caretaker farmer in 2006. Meetings with a facilitator and counselor, Sam Bittman, a former member of the farm's core group, smoothed the process.

☞ **JULIE RAWSON AND JACK KITTREDGE** at Many Hands Organic Farm in Massachusetts are planning to pass their farm on in the old-fashioned way—to their children. Their four grown children agree that the land should stay in farming and that they do not want a financial inheritance from its sale. Dan, the eldest, has been working the land with his parents for the past three years and, with his wife, Roshni, plans to build a house so that they can settle there permanently and continue farming.

☞ **AT VERMONT VALLEY COMMUNITY FARM** in Wisconsin, Barbara and David Perkins's twenty-five-year-old son, Jessie, has become a full partner in the farm. He and his dad do all of the mechanical work on the farm, while Barbara heads up harvest work crews of members. For the first time, the two younger children, twenty-three and twenty-one, also worked there voluntarily in 2006, but have not yet decided on settling down as farmers.

———————————•———————————

If our agriculture is to become a support for local communities, we will have to become much more skillful at protecting farmland. As Chuck Matthei points out, "Although we define the word 'equity' both as a financial interest in property and as a moral principle of fairness, all too often it seems that we have forgotten the necessary relationship between the two."[11] It may be that we have forgotten, and perhaps this connection is one of the many lessons we need to relearn in building the community aspect of CSA.

CHAPTER 4 NOTES

1. Eva Cuadrado Worden, "Community Supported Agriculture: Land Tenure, Social Context, Production Systems and Grower Perspectives (PhD diss., Yale University, 2000), 24.
2. Julie Guthman, *Agrarian Dreams: The Paradox of Organic Farming in California* (University of California Press, 2004), 47.
3. Ibid., 53.
4. Ibid., 185.
5. Andrea Woloschuk, Annette Higby, and Kathy Ruhf, *Holding Ground: A Guide to Northeast Farmland Tenure and Stewardship* (New England Small Farm Institute, 2004).
6. Jered Lawson, "This Land Shall be Forever Stewarded" (unpublished essay), 9.
7. Chuck Matthei, "Gaining Ground: How CSAs Can Acquire, Hold and Pass on Land" (1997 pamphlet available from Equity Trust). See "CSA Resources" section for contact information.
8. Janet Diehl, *The Conservation Easement Handbook* (Boston: Land Trust Exchange, 1988).
9. Matthei, "Gaining Ground," 7.
10. Dan Guenthner, *Growing for Market* (February 1998): 9.
11. Matthei, "Gaining Ground," 9.

PART THREE

GETTING ORGANIZED

NURTURING A SOLID CORE GROUP

Somewhere, there are people to whom we can speak
with passion without having the words catch in our throats.
Somewhere a circle of hands will open to receive us, eyes will light up as we enter,
voices will celebrate with us whenever we come into our own power.
Community means strength that joins our strength to
do the work that needs to be done.
Arms to hold us when we falter.
A circle of healings. A circle of friends.
Someplace where we can be free.

—Starhawk

For Community Supported Agriculture (CSA) to be more than just another direct marketing scheme, the growers and the members need to work together to build an institution they can share. In a consumer-initiated or organization-initiated CSA, the hired growers must feel they are more than temporary employees who serve at the will of a board that may not understand a great deal about food production. If a farm recruits people to subscribe to its produce, the growers must relinquish some of their autonomy to make decisions for their farm business. By tradition, we have given the name *core group* to the grower-member council in which we come together to run our CSAs.

To be sure, not every CSA has a core group. In fact, the 1999 CSA Survey found that 72 percent did not. Angelic Organics organized a core group only after the CSA was firmly established, when the farmers had some extra energy. A few farmers, such as David Inglis at Mahaiwe Harvest, insist that they do not need administrative or farming help from their members. David says that he and his family have the farmwork, bookkeeping, and associated efforts firmly under control. The CSA members are still, according to David, 100 percent more active in relation to the farm than to any store where they purchase food. Many other farmers would envy David's self-assurance and ability to cope.

In fact, one of the main reasons farmers have given for why they quit doing CSA has been their failure to find members who would help them.

A CSA is, in essence, a member-farmer cooperative, whoever initiates it and whatever legal form it takes. As a cooperative, a CSA is a hybrid enterprise blending worker control and customer control. No universal formula or recipe exists for creating a CSA, but organizers can tap into a rich tradition to find examples from which to learn. Its roots go back to the founding of the Rochdale Cooperative in England in 1844. Rochdale intended to offer members groceries, housing, clothing, manufacturing, jobs, and "a self-supporting home colony of united interests." Frances Moore Lappé came up with the encouraging statistic that worldwide there are more members of cooperatives than there are people who own stock in publicly traded companies.[1]

Resources for Organizing Cooperatives

If you are inexperienced, running meetings or creating a core or board may seem daunting. If you decide you want to learn, the following resources can help you improve your organizational skills.

A basic reference is Joel David Welty's *Book of Procedures for Meetings, Boards, Committees and Officers*, available for $9.95 from Midpoint Trade Books Inc.

New Society Publishers in British Columbia (www.newsociety.com) offers resource books on group process and decision making, such as Manual for Group Facilitators, Resource Manual for a Living Revolution, and Democracy in Small Groups.

In the writings of E. F. Schumacher (e.g., *Small is Beautiful*), Richard Borsodi, George Benello, Hazel Henderson, and Barbara Brandt, you can delve into the theory of small-scale economics and cooperative enterprises.

The E. F. Schumacher Society (Box 76A, RD3, Great Barrington, MA 01230) has an entire library devoted to decentrism, and the Center for Economic Democracy (P.O. Box 64, Olympia, WA 98507) is compiling a collection of basic co-op documents.

Food Not Bombs Publishing (295 Forest Ave., #314, Portland, ME 04101) carries C. T. Butler's books, *On Conflict and Consensus* and *Food Not Bombs: How to Feed the Hungry and Build Community*, as well as other books.

The National Cooperative Business Association (1401 New York Ave., NW, #510, Washington, DC 20006) provides advice and legal support.

The USDA Agricultural Cooperative Service (USDA-ACS, P.O. Box 96576, Washington, DC 20090-6576) provides support services to beginning and existing agricultural cooperatives and publishes a monthly magazine, *Farmer Cooperatives*, which is free to qualifying organizations.

The first step in building a core group is to find a few people who share a commitment to the project. Our CSA, Genesee Valley Organic (GVOCSA), was fortunate in its origins. We began as a mutually supportive effort of the Peace and Justice Education Center (which soon morphed into the Politics of Food, POF), an organization devoted to food issues in Rochester, New York, and David and me, the farmers. We quickly attracted other people with years of organizational experience in the peace movement, the Catholic Worker movement, churches, homeschooling groups, environmental groups, and businesses, as well as some newcomers to community work. This group could come up with a better solution to any organizational problem than David and I ever could on our own. This continues to be true eighteen years later when the core has grown to twenty-nine members. The second step, essential to involving committed people, is designating the group's powers and giving it real responsibility. In the GVOCSA, we agreed from the beginning that the farmers would make all production decisions about the farm: how to grow the food; what seed, fertilizers, and other materials to use; timing of operations; what to harvest when; and what equipment to purchase. The core group would not interfere in the farming in any way, beyond their initial decision to make a commitment to our farm. Our organic certification was an important factor in this choice. While we, the farmers, decide what is ready to pick each week, the members give guidance by filling out annual veggie questionnaires and making suggestions of favorite crops. Although some CSA core groups discuss the annual budget, I have not heard of a CSA in which the members actually run the farm, though I have heard of a farmer who quit because of what he considered irritating member interference in what he considered his sphere as farmer.

The GVOCSA core group is empowered to make decisions about everything that happens after the food leaves the farm. With the agreement of the other members, the core has established the basic values of the organization: an emphasis on ecologically grown food from local farms, sharing the risks of crop production with the farmers, maximum involvement of children, and a striving for social justice. From setting the price for a share to deciding what packaging to use or avoid to arranging the schedule for work at the farm and distribution sites to controlling the bank account, the core is also in charge of the day-to-day details. The core decides what other products to offer the members, such as organic strawberries, maple syrup, wine, cheese, and low-spray apples, and it sets the quality standards. From time to time, someone suggests we offer non-working shares at a higher price. The core group always votes this down.

About four years after we started the GVOCSA, a quiet coup took place within the core. For the first years of the project's existence, I acted as chair and facilitator of the monthly core meetings. After thirty-five or so years of experience with organizations, I have gotten fairly good at moving a group cheerfully through a long agenda, so people are often willing to let me do this job. In my head, the core members were doing a remarkable amount of work for my farm—but it was *my* farm. With one of those funny ironies of life, in the very same month that I brought the chapter on meeting facilitation from *The Resource Manual for a Living Revolution* to a Northeast Organic Farming Association (NOFA) Governing Council meeting, Dennis Lehmann, GVOCSA treasurer, distributed that same chapter at a core meeting! In his gentle, but firm way, he suggested we take turns facilitating the meetings, set the agenda as a group, and allocate the amount of time for each item. The core accepted his proposal in a speedy consensus. It may have been my farm, but it had become our CSA.

Besides the conscious agreement the GVOCSA core group made in dividing up areas of responsibility, we have an unspoken understanding. We do not have to agree on everything. We have no CSA party line. We do not endorse political candidates; we have no policy on abortion, gun control, or gay rights. We have taken positions in favor of recombinant bovine growth hormone (rBGH) labeling on dairy products and against food irradiation, and our newsletter provides information about NOFA, the New York Sustainable Agriculture Working Group, and the National Campaign for Sustainable Agriculture. So far, we have not had anyone infiltrate the group with a hidden agenda. Were that to

Deliberations of the core committee at Fair Share Farm in Kearney, Missouri. PHOTO BY TOM RUGGIERI.

happen, a more explicit agreement on where we agree to agree might become necessary.

As a CSA grows, you have to make decisions about how many people to employ, what jobs the employees should cover, and what work is to be done by volunteers. In 1991, after two years of operation with twenty-nine and then forty-five shares, the farm suggested that the GVOCSA grow to one hundred shares. At the end-of-season dinner meeting, we presented the members with two alternatives: The CSA could choose either to hire someone to do administration or expand our system of volunteers. The group voted overwhelmingly to do the work themselves and to distribute the jobs as widely as possible to prevent burnout. A 1991 winter letter appealed to the members to volunteer:

We can build a strong community, a responsive organization, just as farmers can. The GVOCSA will only be as good as the effort we put into it. Each one of us is the organization, is the community. If we distribute the work, then it rests happily and securely on everyone's shoulders. Please consider joining the Core Group as a scheduler, process person, outreach, distribution, or children's committee member. In this way, we can build a CSA as sustaining as the soil in which the food is planted.

Our core group for 2006 consisted of thirty members: two farm representatives, a clerk, a Webmaster, two process people, two treasurers, a scheduler for farmwork and a scheduler for distribution work, nine distribution coordinators with a person to coordinate them, two outreach people, two "inreach coordinators," a social director, two special order coordinators, and a newsletter staff of three. The process people handle telephone inquiries, incoming mail, and outgoing mailings; maintain the membership list; and generally oversee communications among the members. In 1995 we added the social director's position to develop ways of strengthening the feeling of community among the members. So far the job list includes setting up picnics for members, arranging for pickup parties once a month for each of the distribution days, and helping to organize the end-of-season dinner. The "inreach" coordinators are the most recent addition. They have two jobs: to contact all new members to make sure they understand our complicated system and to check in with any members who are more than a month late in their payments. All thirty members do not make it to all of our monthly meetings, but attendance rarely dips below twelve. Clear work responsibilities are a necessity. We have a job description for every function in our core and a procedures booklet where we have written up the jobs and the annual schedule of core responsibilities.

The GVOCSA core group makes decisions by consensus. At meetings, we discuss each new issue. If we cannot come to an agreement right away, we often table the question. By the time it comes up again (and some things never do—they either self-solve or self-destruct), agreement usually comes quickly. We keep meetings short—two hours a month, including refreshments and socializing. Although we must deal with an unending flow of details even after nineteen years, we keep the tone light, reserving our competitive drive for inventing the worst puns. While some boards recruit members based on particular skills,

we tap members who demonstrate a good sense of humor. The stability of our membership is a testament to the satisfaction of being part of a group that functions well.

Despite our efforts to spread the work evenly, for each of the nineteen years of our existence, a few people have done a disproportionate amount of the work. Remarkably, few people have burned out, since each year the extra burden has been taken up by different people. Our universal work requirement and the thirty-three miles separating our farm from most of the sharers has led to a structure that is more elaborate than farms with a closer location might require. But we are a living demonstration that distance from customers need not rule out member participation.

There is no question that organizing an effective core group takes time and energy. Susanne Long of Luna Bleu Farm in South Royalton, Vermont, wrote:

> When we started our CSA, our biggest issue—and one that remains unsolved—is how to get and maintain an active core group. We put a lot of effort into creating a core group when we began. We always tried to do it properly, but it still does not feel right. Part of the challenge of the core group is that the CSA is only a portion of our farm, and not a very big one at that. Getting a core group going and having it become its own entity that doesn't need me to kick-start it has been a problem from the beginning. The core group needs me to be its leader, but I can't spend the energy for that in the growing season, when they need me the most.[2]

At the workshop "Member Events and Involvement," Elizabeth Smith of Caretaker Farm in Williamstown, Massachusetts, expressed a sentiment I felt and have heard from a lot of farmers: "Since we had an established farm, it's been hard for us to step out of the picture and let the members take over. We almost feel guilty that we are asking all these people to take on these roles." The Caretaker Farm "Handbook for Members" outlines their complex organizational structure:

CSA Structure and Jobs

Farmers
- develop farm budget to present to core
- prepare field plans
- soil preparation
- seed selection
- planting
- cultivation
- harvesting
- machine work
- maintenance and repair of tools and buildings
- purchase farm supplies (fertilizers, pesticides, seed, fuels, packaging)
- pay farm insurance and taxes
- capitalization and financing
- bookkeeping
  - financial records
  - production records
  - certification
- educate sharers about farming
- oversee sharers' farmwork

Sharers
- decide on legal structure
- decide on commodities to purchase jointly
- agree on group values (e.g., organic, local, include
- low-income, social diversity, involve children)
- select core group
- participate in CSA
  - pay on time
  - do farmwork
  - do distribution work
  - help recruit members
  - pick up and enjoy food share
  - end-of-season survey to evaluate project

Core Group as a Whole
- selects farmers
- selects land
- selects crops desired
- sets fee policy
- sets payment schedule
- determines work policy (amount of member participation)

- sets distribution place(s)
- sets distribution procedure
- monitors progress of project

Job Assignments

Treasurers
- collect fees
- pay farmers
- keep books
- maintain bank account

Communications Coordinators
- maintain newsletter and notices
- phone, Web site, Listserv
- answer queries (calls and mail)
- post office box
- membership list
- educational materials
- annual contract

Outreach Coordinators
- recruit new members
- publicity
- links with other CSAs, farms, co-ops, other groups
- raise scholarship money

Schedulers of Member Participation
- farmwork
- distribution

Distribution Organizers
- manage distribution site
- coordinate work of other members
- organize bulk orders from other farms

Social Directors
- organize group activities (picnics, dinners, celebrations)
- oversee involvement of children
- create play area on farm
- set guidelines for parents and clear rules for children

Note: These jobs can be combined in any number of ways depending on the talents and energies of farmers and sharers.

The Steering Committee: This committee is composed of the farmers and those who wish to shape the overall direction and organization of the farm. . . . Five subgroups were created by the Steering Committee to strengthen the community and sustain it for the future.

Farming: This group is made up of Elizabeth and Sam and all the apprentices. They are responsible for the day-to-day farming decisions.

The Farm Economy: The Treasurer and the Steering Committee create the annual budget and monitor income and expenses.

Community Support: Responsibilities include recruiting distribution helpers and work support for the farm, planning our various festivals and celebrations, producing the newsletter and this Membership Handbook, running fundraisers as needed, and improving communication among the members.

Organizing: Responsibilities include putting the Annual Proposal together, keeping agendas and the minutes for the Steering Committee, and organizing the CSA's structure.

Future/Long-Range Planning: This group is working on our mission statement, exploring associative relationships with other CSAs and compatible organizations, and developing community outreach programs.

According to Sam, in just a few years the various committees ironed out the vexing issues, leaving few ongoing reasons to meet. The Steering Committee turned into an Advisory Group to be called when needed, particularly to review the annual budget. As Sam and Elizabeth approached retirement and the transfer of the farm to new ownership by a community land trust, the Long Range Planing subgroup shifted gears to head up fundraising for the Campaign for Caretaker Farm. Don and Bridget have not formed a steering committee, but the farm continues to operate much as it did under the Smiths.

When I asked Molly Bartlett of Silver Creek Farm in Ohio what was the hardest decision she ever had to make about her CSA, she answered that it was the decision to let some of the members become a working core. Her husband, Ted, agreed. Letting go of some of the responsibility for the farm was very hard, yet Silver Creek went on to have one of the most active cores of any CSA I have encountered. Two members coordinated the farmwork for each of the three weekly work sessions while Ted and Molly "hovered" nearby. A chief coordinator from among the members oversaw the work of the other core workers. The core was so well trained that when Ted had a heart attack, they were able to keep the farm rolling along. Despite all the training, the core group did not feel ready to take full responsibility for the farming when the Bartletts decided to stop doing a CSA at their farm to concentrate on educational projects.

In situations where the farm is located more than an hour's drive from the members, CSAs really depend on their city core groups to handle distribution. Roxbury Farm in Kinderhook, New York, has a separate core group for each of its four target areas. Each core determines its own budget for administrative and distribution expenses, which include the cost of free shares to members who take on site coordination. These cores organize distribution, oversee the weekly work shifts, recruit new members, and help plan farm festivals. Initially, they also handled the member database, collected fees, and kept the books. Twice a year, the cores meet together with the farm crew to discuss the annual budget and the overall direction of the farm. After a few years, the farm relieved the core groups of most of the administrative work and fee collection. Farmer Jean-Paul Courtens explains the restructuring this way, "That's very different from the way we started. It's increased our costs, and, frankly, we don't particularly enjoy doing it, but we feel there isn't much of a choice."[3]

Like Roxbury, the other CSAs that Just Food has organized in New York City have city core groups. (See the section "Regional CSA Support Groups" in chapter 3.) The Maine Organic Farming and

Gardening Association (MOFGA) has taken the core group model for its project to expand the number of CSAs in Maine in alliance with the Maine Council of Churches. The Council is organizing church groups, and MOFGA is matching them with nearby farmers.

Most of the core members for Live Power Community Farm reside in San Francisco, a three-and-a-half hour drive from the farm. This core acts as a long-range planning council and took the major responsibility for raising the $90,000 to purchase the development rights to the farmland. (See chapter 5 for more details.) A section of the core is charged with cluster organization, shaping the members into convenient neighborhood groups for food pickup, and scheduling their work. Two core members make sure that there are good communications between the members and the distant farm. Gloria Decater emphasizes the role of the core in moving the farm into "associative economics":

> We are blessed with an active volunteer core group of 10 to 20 members who help organize membership, distribution, communications, website development, member surveys, general meetings, and on-farm work days. The core group also serves as a research and advisory council for farm planning and program development, farmland acquisition and conservation, making business decisions, and fundraising. Members give input about varieties and quantities to the farmer so that the planting plan can be turned to the needs of the community. . . . Through thick and thin, our shareholders provide us with the economic and emotional backing we need to grow food and to support ourselves. . . . We believe it is essential to develop and work within this new, more conscious form of economic interrelation in order to heal our social life and our agriculture. We can achieve truly sustainable agriculture only when we have a truly sustainable economic process as a base.[4]

In the first issue of the Angelic Organics core group newsletter, "A Farm Forever," Tom Spaulding explained that the core is made up of volunteers "so

inspired by this particular community supported Farm, that we dedicate our time, energy, wisdom, and resources to help the Farm fulfill its mission. The more we do as shareholders to help with non-Farm tasks, the more our hearty and visionary farmers can concentrate on producing the delicious, colorful, nutritious, aromatic, and simply splendid veggies and fruits that we have come to depend on from June to November." Like the Live Power core, the Angelic Organics group took on strategic planning and financial development, facilitating connections with the farm and assisting the farm crew with city distribution.

After a year of enthusiastic support work, Tom moved to the farm and established the CSA Learning Center, a nonprofit organization linked to Angelic Organics. Their mission statement reads: "The CSA Learning Center empowers people to create sustainable communities of soils, plants, animals and people through education, creative and experiential programs offered in partnership with Angelic Organics, a vibrant Biodynamic community supported farm." The center offers shares at half price for low-income people, raising the money to pay the other half; organizes visits and tours of the farm; sells farm products, such as dried chili peppers and T-shirts; coordinates the Midwest Collaborative Regional Alliance for Farmer Training (CRAFT); and gives workshops for youths, adults, and families at the farm and in Chicago.[5]

Michelle Lutz of Maple Creek Farm in Yale, Michigan, was very discouraged about what she felt was a lonely struggle for her and her husband, Danny, to keep their farm going. Then, at the 2004 CSA Conference in Michigan, she heard the stories of how farms like mine and John Peterson's have activated our members. Near desperation, she decided to give it a try. That winter she and Danny wrote a long letter to their members, laying out in graphic detail the history of their struggles, the risks they had taken, and their steady improvements despite many, many setbacks. They appealed to their members to volunteer time and donate resources. The response was immediate and strong. A member purchased a chisel plow for the farm. Members came to a meeting to discuss what they could do: They volunteered to join a core group, to take over the organizing of community events, to write for the newsletter, and to help the farm with distribution and farmers' market sales. A volunteer coordinated member work on social events. A year later, Michelle looks back on their best year ever. A good growing season and skilled hired workers helped, but member involvement tipped the scales. Twice, when Michelle and Danny could not be present to assemble shares, an SOS to members brought the help they needed. The 2006 preseason meeting attracted sixty members willing to jump in for the season ahead.

Not all core groups perform regular tasks for the CSA; some are informal or advisory. Scott Chaskey, in his book *This Common Ground*, recalls that "it was a core group of ten families that actually planted the first seeds of the community farm. This group was the embodiment of the link between 'producer and consumer' (that word 'consumer' has an unfortunate echo; recently, at the suggestion of a progressive Danish farmer and thinker, Thomas Harttung, I was encouraged to substitute 'citizen'

for consumer). The group's role, similar to that of a board of directors, but predominantly advisory, changes with the seasonal fluctuations in the needs of the community."[6]

———————————•———————————

In building our own little institutions, like the CSAs, we are at the same time transforming ourselves into people who can listen to one another and take cooperative action. Changing isolated farm enterprises into community supported farms cannot happen overnight. We have all kinds of inertia, internal and economic, to overcome and a lot of new skills to learn. But our creativity is unlimited when we work together in an atmosphere of mutual respect. This next decade is a critical time for farming, both in this country and the rest of the world. A local food system based on direct links between farmers and eaters is the opposite of the global market touted by our government, which benefits only the Cargills, Nestles, and Altrias (Philip Morrises). Each new CSA is another little piece of liberated territory and a step toward the sustainable world that is our only possible future.

CHAPTER 6 NOTES

1. Frances Moore Lappé (keynote address at "World on Your Plate" Conference, Buffalo, NY, October 2006).
2. University of Massachusetts Web site, "CSA Farm Profiles," http://umassvegetable.org/food_farming_systems/csa/
3. Ibid.
4. Gloria Decater, "Ripening into Wholeness: The Story of Live Power Community Farm," printed in *Stella Natura: Working with Cosmic Rhythms* (Kimberton Hills Biodynamic Agricultural Planting Guide and Calendar, 2007), opposite July.
5. For the full program of the Angelic Organics CSA Learning Center go to csalearningcenter.org.
6. Scott Chaskey, *This Common Ground: Seasons on an Organic Farm* (Viking, 2005), 82.

7

LABOR

Principle of Fairness
Organic Agriculture should build on relationships that ensure fairness with regard to the common environment and life opportunities.

Fairness is characterized by equity, respect, justice and stewardship of the shared world, both among people and in their relations to other living beings. The principle emphasizes that those involved in organic agriculture should conduct human relationships in a manner that ensures fairness at all levels and to all parties—farmers, workers, processors, distributors, traders and consumers. Organic agriculture should provide everyone involved with a good quality of life, and contribute to food sovereignty and reduction of poverty.
— From "The Principles of Organic Agriculture," International Federation of
Organic Agriculture Movements (IFOAM) Basic Standards 2005

There are never too many hands to do all the work on a small farm and rarely enough dollars to pay for all the work that needs to be done. What makes a small farm a wonderful place to work is the great variety of jobs and the chance to be outdoors in all weather. Some of the jobs require technical training; other jobs are simple and repetitive, though skill and experience can greatly alter the pace. In our culture, when people eat, they rarely think about the many workers who have put their time and energies into planting, nurturing, harvesting, packing, shipping and distributing the food. Through membership in CSAs, more citizen consumers are learning about the effort that goes into what they eat, and the food actually tastes better to many of them when they feel secure in the knowledge that the workers up and down the food chain have enjoyed decent salaries and safe working conditions.

Member Work

According to U.S. Department of Agriculture (USDA) statistics, off-farm earnings help support 84 percent of the farms in this country. The unpaid labor of the farm family is essential. When the workload swells beyond what the family can manage, farms must shoulder the expense of hiring workers or take on trainees. CSA farms are no exception, but they have the exceptional opportunity to get help from a new source—the members.

Everything depends on how you ask the question. While CSA members and potential members, when polled, claim that they do not have time to participate, what people say does not always coincide with what they are willing to do. If a survey asks, "Do you want to work on the farm or take a job in the core group?" most people respond with a self-defensive

"No!" At least two other approaches work better. If, from the beginning, farmers and members discuss how to keep share prices as low as possible, participation is an obvious way to reduce cash payments for the food. No one refers to the work as "volunteering." If the CSA then organizes this work well, defining jobs carefully and getting members to sign up for specific tasks with limited hour requirements, members will find the time. If you as a farmer find yourself in the awkward situation of converting from no work requirement, appeal to the members for assistance, explaining that you cannot cope with all the tasks a CSA entails. Several CSAs have used the list of CSA jobs from this book (see p. 79) to successfully activate their members. The Genesee Valley Organic CSA (GVOCSA) sharers who come out to do farmwork usually discover that they enjoy it a lot. While they cheerfully pick and pack our vegetables, I feel like Tom Sawyer whitewashing his fence. (See Chapters 6 and 8 for more details on involving members.)

Family Work

Many CSAs are on very small farms, from one-half acre to ten acres in size, and with as few as half a dozen shares. The smallest CSAs are run by gardeners or homesteaders who are sharing their bounty with friends and neighbors. Calculations of hours and efficiency do not apply in the same way as in a commercial operation. When I first moved to a farm in Gill, Massachusetts, one of my goals was to keep out of the supermarket. For eight years I grew, canned, froze, and bartered for my food supply. One year I even grew wheat on a raised bed. We scythed it down by hand, fed it through an old combine to separate out the grains, dropped the grain off the barn roof on a windy day to get rid of the chaff, and collected enough grain for five loaves of the best-tasting bread I ever baked. Was that efficient? Are you kidding? Did that matter? Of course not. It was definitely worth the effort to me and my son and his whole fifth-grade class, who got to taste the bread with honey from our beehives.

cosmos and pigweed

sometimes the work
seems no longer to be blessed
by companion hum of honey
bee or caress of the
sun,
hawk whistling high circles
on the breeze
but becomes more
mosquito whine and
nettle sting, sweat
in the eyes.
and where the mind once envisioned
flowers like stars
amidst crystal abstractions of
fiber and light,
and the eyes rejoiced
at broccoli and cosmos,
kale, calendula, lettuce,
and fennel—
it all becomes pigweed,
potato beetles, and the prickly mystery
weed that tears open the palm
no matter how calculated the approach.

what to do then?
find a tool.
keep your sights close.
rest, quench thirst, satisfy hunger.
take a walk.
admire your progress.
sing. breathe. stretch.
make conversation with the pigweed,
        the potato beetle, the prickly mystery.
labor in love, not in vengeance, not in heat
        of contest. not to win. just to do.
remember the flowers.

—Sherrie Mickel, 1992

Farmer Katie Lavin and intern Patrick Keeler harvest Swiss chard at Peacework Farm. PHOTO BY ELIZABETH HENDERSON

Living on a farm blurs the line between life and work—it is a lifestyle choice that usually means less cash, fewer consumer amenities, and more physical labor. People should enter farming knowing this and should not be surprised and disappointed when they find they cannot earn as much as in most other professions or take paid vacations to exotic places. As in other small businesses, farmers have to pay for their own health insurance and plan for their own retirement. All the more reason, then, to learn how to work smart and to use your own labor (and that of anyone willing to get involved with you) as thoughtfully and efficiently as you can.

Farmers have been putting the family to work ever since Adam and Eve. Yet the very idea of unpaid family labor makes agricultural economists squirm. Rarely calculated or remunerated in dollars per hour, that family labor is basic to the whole economy of an integrated farm, where one of the main "products" is the people. Sure, some people who grew up on farms look back in horror at their childhood slavery and swear, "Never again." They barricade themselves in a life of office work or what-ever else is as far as possible from picking beans on the farm. But many more farm-raised people feel nostalgic about the hours spent close to parents, remembering the many sensual pleasures of sounds and smells that folded in with the work. Doing chores and being expected to take responsibility are good for children. When they are really needed, they rise to the task. Old-fashioned as it may sound, having to work builds character. How else are we going to raise fewer couch potatoes and more active citizens? Learning that physical labor is good for the body and for the soul and seeing parents involved in meaningful activity never hurt anyone.

There are limits, however, and the economic pressures on small farms have all too often forced farmers to overload themselves and their families. Many love relationships have turned sour under the strain of picking too many vegetables for market. Should we then conclude that the work itself is the problem? Or rather that we need to change the economics and learn how to organize the work better? In a conversation about the push to save people from the drudgery of small farms, Wendell Berry exclaimed:

Farming is a hard life. That's what these rural soci-ologists were talking about in the start. It's a hard life, therefore nobody ought to live it. What a remarkable conclusion. There are several steps that are left out. What causes the difficulty? Does freedom come out of it? Does family pride come with it, family coherence? Does some kind of idea of community come with it? Does some kind of idea of stewardship, of essential, irreplaceable, indispensable stewardship, does that come with it? Do ideas of affection or love or loyalty or fidelity come with it? The basic question is, how hard would you be willing to work in order to be free?[1]

There are volumes to be written about men and women working together on farms. Figuring out a fair division of work and responsibilities is extremely challenging. I realize this is a vast over-simplification, but years of observation have con-vinced me that most men and most women really do not think about the same things or in the same way. Ideally, this should be a source of strength for a working partnership where the masculine and femi-nine approaches complement each other. In reality, too often these differences are a source of friction. I have talked to farmers who run successful farms in partnership with close relatives. When I asked them what was most important for working together, their answers were similar: the ability to forgive and to place the family ahead of the farm.

On the mythical family farm, husbands and wives, parents and children, friends and neighbors all work together in cooperation and harmony. Mutual respect and appreciation govern relations between sexes and generations. When differences of opinion occur, the people involved take the time to sit down and negotiate a mutually acceptable resolu-tion. If only we could live up to this myth! In reality, all too often tempers flare; in the pressure of the moment, people who care deeply for one another say angry or thoughtless things and feelings are hurt. How can we move our reality closer to the myth?

At the Northeast CSA Conference in 1999, a diverse group gathered for a workshop on the topic "Women and Men Working Together." There were two of us who had left farms where relationships did not work out with partners, two young men who were about to start farming with their female part-ners, two young women who farm with men, a woman who has been farming for many years with her male partner, and a woman who is still farming with her ex-husband although they live separately. Everyone contributed to our lively discussion. We concluded by drawing up a series of recommenda-tions to help ourselves and others work more har-moniously and productively together:

- Don't let irritations and disagreements accumulate till an explosion occurs. When a difference occurs, note the context and skip the generalizations. Use conflict resolu-tion and hold regular meetings.
- Differences in sensibility can be a source of richness or of increasing irritation. Honor deeply your own perspective and the other person's.
- Balance expertise and nurture leadership skills. Take charge of separate areas and exchange these when possible to learn the perspective of the other role.
- Write a contract (the publisher Nolo has helpful models. See www.nolo.com).
- Resolve issues by the end of the day—don't go to bed angry or hurt.
- Learn good teaching skills—how to give enough information and then the space/time to absorb it.
- Do not overburden a partnership by always having the more skilled partner teach the less skilled. Turn to others for instruction. Our community needs to offer training in critical skills.
- We take this farming way too seriously. Build in play and social time. Have fun!

At Rose Valley, David and I divided work in a pretty traditional way. David took charge of machines, field preparation, and short- and long-term maintenance. He is a skilled carpenter and has done enough tinkering and repairs on machines to have become good at it. My father could not even change a fuse, so I grew up innocent of most mechanical skills. What I know, I have learned painfully as an adult. On the farm, I planned crop production, ordered seed, kept the books, ran the greenhouse, arranged for marketing (including participation in the GVOCSA core group), and oversaw picking, packing, and delivering. When needed, I drove the tractor for cultivation and rock picking. Both of us hoed and weeded and supervised apprentices. Twice a week, I worked with CSA members. David joined in on Sundays when we had a bigger crew than I could manage alone. Having clear lines of responsibility reduced conflict. In the spring, when David thought the ground was dry enough to disc, he followed his instincts and did not have to consult with me. I set the standards for quality in picking. David made no bones about my being fussier than he, but he accepted my standards and urged them upon our apprentices. In many ways our skills were complementary. What went wrong? I guess what so often goes wrong when the personal and the professional become inextricably entangled: a chance comment about the height of a shelf reverberates as a symbol of deeper levels of unhappiness and discontent.

At Harmony Valley Farm, near Viroqua, Wisconsin, Richard de Wilde and his partner, Linda Halley, did not have distinct lines of responsibility. In 1997, Richard told me that he and Linda were "interchangeable." He did more than half of the actual farming, the maintenance, repairs, fieldwork, and planting with their Stanhay planter. Linda helped him with crop planning, planted with their Planet Jr. seeder, did some of the tractor driving, and supervised the harvest crew. As the computer ace on the farm, Linda put together the weekly CSA newsletter. The two shared bookkeeping and took turns driving the delivery truck to Madison and serving as advisers to the University of Wisconsin in Madison. Linda left Harmony Valley in 2005, with Richard attributing the departure to conflicts over the farming. From my experience, I wonder how he can be so sure. What is certain, though, is that some of the members of their CSA will be very upset, as

ours were when I left Rose Valley. When you make your life as a farm couple public by recounting it in your CSA newsletter and inviting members to your farm, there is no way to prevent some of them from feeling cheated that the myth of the family farm has melted down into a soap opera. Richard reports that one member adviser consoled him, saying "half of us have been divorced, we understand!"

At Peacework Farm, there are four partners and our goal is for each of us to learn all of the skills so that no one will be indispensable. Inspired by *The E-Myth Revisited: Why Most Small Businesses Don't Work and What to Do about It*, by Michael E. Gerber, we drew up a chart of all the jobs necessary to run the farm. Each winter Greg and I took responsibility for a slightly different combination of tasks. As we have added two more partners, we have spread the responsibilities out further. The first two years, Greg took charge of sales to stores, farm maintenance, and accounts receivable. I was in charge of communications with the GVOCSA core, accounts payable, greenhouse work, production planning, and post-harvest handling. Together, we did planting, weeding, pest control, and work with CSA members.

When Greg and I started Peacework in 1998, neither of us had ever plowed or disked, spread manure, prepared ground for planting, or made permanent beds using a combination of chisel plow and spader. I was determined to learn how to do tillage and cultivation. Initially, we rented two almost new Ford tractors from our landlord, Doug Kraai, who was good at teaching us how to use them. (Little did he know that Greg and I had clocked in even fewer hours than those shiny tractors.) Both of us learned to spade with the five-foot-wide Celli that we purchased new for $5,000 with member donations to our capital fund. Greg took on the job of running the loader, while I did the spreading of compost with the manure spreader. Since I had done a little cultivating during my last year at Rose Valley, I did most of the cultivating for a few years. I used a basketweeder belly-mounted on our Allis-Chalmers G tractor, which I had bought at an auction, and a Perfecta field cultivator with S-tines that was easy to switch from one-row to two-row configuration. Greg did almost all of the mowing. The severe drought of 2000 drove him to learn how to design and manage trickle irrigation and then to teach the rest of us. When Ammie joined our team, she took on a lot of the mowing and pest control and replaced me in the packing shed and at quality control. In 2004 she learned to use the spader and in 2005 to drive the tractor with the water wheel transplanter. That year, Katie's first as a full-time farmer, she understudied with me in the greenhouse, with Ammie in the packing shed, and learned to cultivate with the G. She is taking the lead on our new enterprise—medicinal herbs. Now that Peacework has become a limited liability company (LLC), Ammie has taken over most of the bookkeeping. One of my goals for 2007 is to teach Ammie and Katie to do mechanical cultivation. Greg wants the three of us to take over some of the tractor maintenance. On CSA harvest days, all four of us, and our two interns, work side by side with the members.

Many of the many hands at Many Hands Organic Farm in Massachusetts belong to members of the Rawson-Kittredge family. Julie and Jack put all four children to work as they were growing up. Now that they are grown, Dan, the eldest, has settled on the farm to work with his parents. The three work together. Their system has evolved from a lot of yelling to "something we are happy with," according to Julie. Over the years they have figured out who does what well and what each wants to do. The result, Julie says, is that "we are stress-free and we all feel like being here every day." Dan works for a salary, while the parents make their living from NOFA and take only food from the farm. Jack takes a big role in budgeting, running the orchard, and some building and the homesteading aspects of their lives. Dan and Julie manage the vegetable production and do most of the animal care, raising chickens and hogs. While Ma oversees the daily crew, son tends equipment and makes hay. They trade meat with a woman who does payroll and taxes and pay a salary to a newsletter editor. There are ten working shareholders, mainly women Julie's age, who do eighty hours of work in exchange for a share,

Jean Mills transplants cucumbers at Tuscaloosa Farm in Alabama.
PHOTO BY CAROL EICKELBERGER.

earning from $6.50 to $9 an hour. Each year, Julie also trains four teenagers straight out of high school. Truly a diverse collection of workers.

It is interesting to compare the experience of Carol Eichelberger and Jean Mills, female partners. According to Carol, they share most work, but they divide responsibility according to what each cares more about or dislikes the least. Carol is responsible for crop planning, pest control, the greenhouse, and irrigation. Jean takes responsibility for the organizational work, weed control, and harvesting and writes the newsletter every other week. Together they do composting, tractor maintenance and repair, soil preparation, and green-manuring. Carol says they have average amounts of mechanical know-how, but she is pleased that their relatively new tractor runs well.

For a few years at Tierra Farm in New Hope, New York, Gunther Fishgold and Bruce Schader were two male partners. Their division of work responsibilities resembled what David and I did at Rose Valley. Bruce, who has been farming all his life, did the machine work and was responsible for farm operations. Gunther, with a background in environmental organizing and lobbying, took charge of marketing, recruiting, and communicating with CSA members, as well as the crop planning, and worked along with the apprentices on the list of jobs drawn up each week by Bruce.

In an e-mail message, Jean-Paul Courtens sketched out how many people it takes to run Roxbury Farm, which supplies over 1,000 shares and limited non-CSA sales:

> We have four apprentices and four hourly workers. The apprentices work about 40 to 45 hours a week and get paid $1,000 plus housing (a four-bedroom house with all the utilities paid for) and all the pork and vegetables they can eat. The apprentices work with us for about 5 to 8 months, averaging 7, while the hourly workers work with us between 5 and 7 months, averaging 6. Each worker works about 40 hours a week, so this makes for an estimated total of about 4,800 hours input from the apprentices and 4,000 hours from the hourly workers. Jody and I put in about 2,400 hours a year with a total of 4,800 of which 2/3 is spent raising crops, 1,000 hours of administrative work, and 1,600 hours during the off season creating a farm plan, meeting people and farm-related travel. Grand total is 13,600 hours to grow 30 acres of vegetables (which would make about 450 hours per acre of vegetables) and 40 acres of cover crops plus three sows and piglets of which 10 are raised as feeders.

Besides the farmer workers, we hire a bookkeeper for 2 days a week and a driver (in season).

Hiring Help

If you decide to hire workers, you enter a maze of state and federal regulations. Neil Hamilton's *The Legal Guide for Direct Farm Marketing* is a helpful resource for hiring and other issues; remember, though, that Neil is a lawyer and gives the most cautious interpretation of each situation. As Doug Bowne put it at a workshop on this topic at the 1997 Northeast CSA Conference, you are immediately answerable to six or seven agencies that do not have clear answers for CSA farms. "We are pioneers in a litigious system," Bowne warned. For those who have never had employees before, the first stop should be the Internal Revenue Service—(Circular A), the Agricultural Employer's Tax Guide. This document will lead you through the steps you must take. You can call (800) TAX-FORM, and on a good day they will send you what you need. The Economic Research Service offers the comprehensive "Summary of Federal Laws and Regulations Affecting Agricultural Employers, 2000," by Jack L. Runyan, which covers agricultural employers' federal safety requirements, migrant and seasonal farmworker provisions, and tax requirements. (*Agricultural Handbook No. 719*, July 2000, is available online at www.ers.usda.gov/publications/ah719/) In many states, the Cooperative Extension Service or the nearest office of the state labor department will be able to provide you with information. In New York, FarmNet, a nonprofit based at Cornell University but supported by tax dollars as well as contributions, provides counseling for farmers on financial as well as psychological matters. Other states may have similar services.

The simplest course for getting started might be to consult with a neighboring farmer who handles the paperwork for his or her own employees. Some states have a tax guide for new businesses. For all employees earning over $150 a year, the employer is responsible for Social Security and Medicare withholdings. The combined tax rate is 15.3 percent of gross cash wages (this does not include payments in kind, such as farm produce or lodging). The employer deducts half from employee gross wages and pays the other half. To determine how much money the employer must withhold for the federal income tax, new employees will need to fill out W-4 forms. The employer is then required to deposit all of the Social Security, Medicare, and withheld income tax at an authorized bank. A special deposit coupon (Form 8109) must accompany these deposits. By January 31 of each year, employers must file Form 943, Employer's Annual Tax Return for Agricultural Employees, and send a W-2, Wage and Tax Statement, to each employee. Each state also has its own peculiar tax requirements. Calling the state tax number should furnish the appropriate regulations and forms. Employers must fill out an I-9 form for each employee to document citizenship, legal alien, or visa status and keep it on file for at least three years. Federal law requires that all employers must also purchase workers' compensation insurance, which varies in price from state to state. Although the rates are set at the state level, the various kinds of farms have different risk pools. An uninsured employer is liable for the costs of medical treatment for a job-related injury or illness and risks fines for failure to carry insurance.

To fill out the many forms properly, you must have an Employer Identification Number (EIN), which functions like a Social Security number for tracking you through the system. To get one, fill out Internal Revenue Service (IRS) Form SS4. You can complete this process over the phone in a few minutes. Special laws regulate the hiring of youngsters under sixteen years of age who are not members of your immediate family. You must make sure they have working papers and that you do not ask them to do jobs that are forbidden by law, such as driving equipment. The fines for violating these regulations are very heavy. Putting your own children to work may not look like a bad alternative after all!

Finding really good, steady help is not easy. The pool of U.S. citizens willing to do farmwork is shock-ingly small. The ideal employee is a local person whom you can train and keep for a long time. At Rose Valley we were fortunate to find a friend, Greg Palmer, who lived nearby and who began as an apprentice, graduated to junior partner status, and then became my partner in Peacework Farm.

Few CSAs are large enough to support more than one or two full-time workers. Even the farmers are not getting paid for year-round work. At Harmony Valley Farm, where eight hundred CSA shares amount to half the farm sales, ten people work year-round in addition to the two farm partners. The employees are local people who have been with the farm for many years. The summer crew adds fifteen more workers, a diverse assortment of local teenagers, a few people in their thirties, some young Mexican guys who live in the area, and others who come from farther away. There have been Laotian Hmong, and one year a man came from Ghana to learn organic farming. As labor needs have grown, Harmony Valley has added ten H2A migrants. H2A is a federal program with stringent requirements for wages, housing, and transportation costs that allows farms to import legal workers. At the end of each season, the migrants must return home. This program is not popular with farmworkers because of its many weaknesses; for example, if a worker does not like conditions at the farm, he or she has no legal recourse and cannot simply leave and take a job at another farm. Despite these problems, Richard de Wilde says his workers are happy to go home and, without H2A, they would be afraid of never being allowed to return again. To provide them with health insurance, he has been able to find a company that allows them to drop and then pick up again when the crew comes back from Mexico.

A few of Harmony Valley's CSA members have worked for the summer, including one who is thinking of quitting his job as a college teacher and making a permanent move to the farm. The total payroll is $500,000 at a pay rate of $9 to $12 an hour. Smaller farms will be envious to learn that Harmony Valley also employs a full-time cook who, besides preparing meals, supplies recipes for the CSA newsletter.

At Good Earth Farm in Weare, New Hampshire,

with the help of one adult and three teenage workers, David Trumble supplied eighty families from three acres of vegetables and an acre of fruit. His experience in hiring workers has been very positive:

> We have had the privilege of employing a couple of dozen people over the years and maintain good friendships with all of them. They come back to visit us during college vacations, and have helped us babysit our child. We follow them to big events in their personal lives, or even just watch them in countless theatre productions at the local high school. We pay all of our workers a decent wage, deduct taxes, provide workman's compensation, take the time to teach them the skills they will need to be good farmers, and treat them like adults in the workforce. We are doing this with local labor, and we feel that this meets the definition of community better than importing interns from far away. We impact them, their friends, and their families. In addition, adults who stay with us for any length of time become like partners in that we seek out their advice and opinions in matters of agriculture.

The reality is that immigrant farmworkers do a large portion of the weeding, picking, and packing of vegetables in this country. U.S. labor laws specifically exclude agricultural workers from some of the protections granted to workers in other sectors, such as the right to organize and time-and-a-half for overtime. When small-scale organic farmers expand beyond the one- to five-acre size that a single farmer or family can manage, they discover that in most cases skilled farmworkers do this work faster and more reliably than reluctant teenagers or college students from the suburbs. In the role of boss, many CSA farmers retain their sense of social justice and seek ways to make the working conditions as decent as possible within the tight budget constraints that they face. The farmers at Roxbury place labor standards next to organic practices as their top priorities; Jean-Paul says, "We are interested in creating fair working conditions and providing fair wages to our workers as we feel this is as important as applying organic methods to our land."[2]

Several CSA farms in California have gone to year-round shares in order to provide steady work for their farmworkers. The T and D Willey Farms Web site features their farmworkers and urges customers to be grateful for their hard work that makes the food possible. The Web site explains the significance for the workers of year-round employment:

> Farm labor is hard work, and those who perform it do not receive adequate compensation or respect. The improvements we have made on our farm, though modest, are not generally available to farm laborers. Our farm is known in the area as a "good" place to work, and we are able to attract a high caliber work force as a result. While earning only minimum wage, our staff enjoy annual incomes roughly twice that of the average California farm worker. Year-round employment enables them to become stable members of our community. Educational opportunities of children are enhanced when their families have steady employment. After earning a permanent position on the farm, they become eligible for an annual bonus and paid vacation.

Julia Wiley of Mariquita Farm in Watsonville, California, explained to me, somewhat apologetically, that while they do not pay high wages to their workers, their farm provides year-round work and has been able to pay bonuses and give loans to several farmworker families, making it possible for them to build decent homes in the area.

Four partners share in owning and operating Full Belly Farm, a 250-acre operation in Guinda, California, where 800 shares make up only one-third of the income. Like Mariquita, Full Belly provides year-round work for many of the farmworkers they employ. Besides the partners, four or five other people live and work on the farm and fifteen additional local people work year-round. The harvest crew swells to forty. Full Belly also hires a full-time person to coordinate and handle the bookkeeping for the 450 shares delivered to Berkeley. The partners take turns overseeing the daily harvest. In 2006, Full Belly was honored as one of the recipients of the

91

Sustainable Agriculture Research and Education Program Patrick Madden Award for Sustainable Agriculture.

To help him farm the thirty acres he rents from the Stony Brook–Millstone Watershed Association in Pennington, New Jersey, Jim Kinsel hires a team of full-time seasonal workers. With over 2,100 shares, his Honey Brook Organic Farm is the largest CSA that I have been able to identify. Sherry Dudas, Jim's partner and farm planner, attributes the farm's impressive volume to their workers, who learned their skills on their own farms back in Mexico. Since 1993, Jim has developed a close working relationship with the Camacho family from Zacatecas, Mexico. The eldest brother, David, has become his indispensable field manager. David has kept his family farm and returns there in the winter to grow crops of chili peppers or pinto beans. A younger brother, Israel, is assistant field manager, and David's son and a brother-in-law also are regular workers at the farm. David and Israel are bilingual, so they are able to manage the work crew and communicate with Jim, who has mastered Spanglish.

Jim does the overall management of the farm, buying equipment, hiring, doing CSA outreach, ordering seed, helping the Watershed manage its 830-acre holding, and filling in on fieldwork when he has time. The Camacho brothers have primary responsibility for managing the harvest. At the height of the season, Honey Brook has ten field-workers and houses them all in a four-bedroom rented house. In sharing the house, the workers have devised a system of two teams that take turns cooking dinner and making lunch for the next day. The farm also provides them with a van. The starting salary is $8 an hour. Because the Camachos work only nine or ten months a year, Sherry says she has not been able to find an insurance company that would provide them with health insurance. In 2006 Sherry and Jim are starting a 401K retirement plan for David. When the other workers leave in November, Jim and Sherry do not know which of them will return the next spring. Jim reflects, "At one level, I'm tearing families apart—two of the brothers have wives—but I'm also bringing them together."

Interns Jeff Schreiber and Andy Simmons with a huge pile of the garlic they have harvested at Peacework Farm. PHOTO BY ELIZABETH HENDERSON.

When I asked Bill Brammer, whose CSA had grown to eight hundred shares in 1998 and was still expanding, whether his members did any work, he said, "Having all those people at the farm would drive me crazy." Then he told me that he employed a farm manager, 2 office workers, 4 foremen, and up to 110 field-workers, most of them Mexican. In other words, he and I supervised about the same numbers of people. If I had to be boss to that many, it would drive me crazy! So, you pays your money, and you takes your choice.

Interns

Taking on interns is an appealing way to get farm help for less cash (because of the legal meaning given to the term *apprentice*, the term *intern* is more appropriate in the context of CSA farming). To manage interns fairly and well, however, means assuming a major responsibility to teach and share what you know. If you do not want to spend time explaining why you are doing what you are doing on your farm, and just want to do a quick training and assign tasks, you should not entertain the idea of interns.

Over the past five years, the New England Small Farm Institute (NESFI) has taken the lead in developing high-quality resources for training new farmers.

The first piece it created was *DACUM Occupational Profile for On-Farm Mentor* (NESFI, 2001). Eleven farmers with experience training interns along with Russell Libby of MOFGA, who has run an intern program for years, spent two days listing all of the tasks and duties involved in mentoring. Next came *Cultivating a New Crop of Farmers: Is On-Farm Mentoring Right for You and Your Farm?* by Kathryn Hayes (NESFI, 2005), a "decision-making workbook" that farmers can use to figure out whether to take on interns and what resources their farms need to do it successfully. A series of work sheets leads you through the thinking and evaluation process. Finally, Miranda Smith built on five years of discussions and her own experience to write *The On-Farm Mentor's Guide—Practical Approaches to Teaching on the Farm* (NESFI, 2006), which completes the series. This book should be required reading for any farmer who trains interns.

In the draft version of "Internships in Sustainable Farming: A Handbook for Farmers," Doug Jones sums up the difference between hired workers and interns:

> Workers usually do specialized work in one area of the farm; they often have prior experience; they receive an hourly wage and usually do not live with you. State and federal governments have many regulations and officials assigned to protect workers from exploitation by employers. An essentially adversarial relationship is assumed to exist.
>
> With interns, on the other hand, you have a much greater obligation to instruct. They expect you to explain the "whys," not just the "hows." They deserve a diversified learning experience through a broad exposure to many different tasks, as well as through frequent discussion of the overall goals, methods, and systems of the farm. They are preparing themselves for a vocation, or at least learning how to grow their own food. Interns usually live on the farm, expect to interact socially with you, and may wish to learn a variety of rural living skills. . . . In most arrangements, interns receive a cash stipend that is not directly related to the number of hours worked. Hopefully, they will share some of your ideals and aspirations, and a mutual rather than adversarial relationship will prevail.[3]

[*Note*: This is actually a good description of how to achieve friendly and productive relations with employees as well.]

So, if you are willing and able to teach while you work, interns may be the answer to your labor shortage. The research that Eric Toensmeier did for the *On-Farm Mentor's Guide* proves conclusively that, however special the relationship may seem to you, the authorities will consider your interns or apprentices employees. *The On-Farm Mentor's Guide* provides the information you need to comply with the many laws and regulations that govern agricultural employment.

In my experience, when you find the right people, interns not only lighten your workload but also make the work itself more pleasant. Working with someone who is enthusiastic about learning what you are doing can transform an otherwise routine task. When you explain to your intern how a repetitive job fits in the context of the farm's systems, you keep alive for yourself the interconnections that are so essential to a sustainable farm. I still keep in touch with most of my former interns. Several have gone on to farm on their own, and six run CSAs, so our paths continue to cross in mutually beneficial ways.

Taking on interns has a broader importance as well. If we want CSA to continue to grow, experienced farmers must learn how to pass on their knowledge. "We need thousands, tens of thousands of new farmers!" a quote that could have come from Robyn Van En but was part of Paul Bernacky's presentation on apprenticeships at the 1997 Northeast CSA Conference. The practical skills of farming and,

"Interns expect you to explain the 'whys,' not just the 'hows.' They deserve a diversified learning experience through a broad exposure to many different tasks, as well as through frequent discussion of the overall goals, methods, and systems of the farm."

in particular, managing a CSA farm are not part of the curriculum of many schools. A few institutions are starting to catch up with the encouraging demand for this kind of learning. In the meantime, those of us who are doing it need to train an entire generation from scratch in local, sustainable food production and all the skills involved.

Most of the people who are interested in learning to farm come from nonfarm backgrounds. They have a double set of lessons ahead: learning how to do physical labor and learning how to farm. Shane LaBrake, who runs a two-year internship program at the Ecosystem Farm sponsored by the Accokeek Foundation in Maryland, divides these differently into three sets of challenges: physical, mental, and emotional. As Cass Peterson put it in her letter to potential interns, "It takes time to learn some of the basic farming skills that most people nowadays think of as 'unskilled labor.' It isn't unskilled labor. Hoeing requires agility and practice. Harvesting requires judgment and speed. Marketing requires communications skills and experience."

Getting into physical shape is only the first necessary step. To use your body without hurting yourself and perform tasks with the least amount of energy take practice and experience. Cooperating with other people on physical tasks adds another layer of complexity, an instinctive awareness that most people are not born with. When I bend to pick up a board, I can sense instantly if my companion understands what is required to move that board in the easiest way. If you grew up in the suburbs as I did, you might need a few pointers to master what farm kids take for granted as common sense.

Shane's goal is not just to train farmworkers but to empower people to farm on their own. To do that, he believes, requires the acquisition of skills and self-confidence. The first year of his training focuses on skills, physical and mental. He has designed the second year to provide the interns with management experience and more confidence in their ability to recognize and solve problems. Long hours, low pay, fatigue, droughts, floods, and pestilence—the emotional challenges of farming come along with the territory, built in to every real-life farm. Probably the

only way to know if you will be up to meeting them is to try it out. At least through an internship, you get to do it on someone else's dime.

In recruiting interns, it is essential to give a realistic picture of what working on your farm will be like, and to dispel the many romantic notions abroad about the idyllic farm life. The Angelic Organics brochure inviting interns stresses the team discipline involved in farmwork:

A farm is a weaving. Everything that happens on it affects everything else on it. If time is lost because of a late start, or using an ineffective tool, or because a communication is misunderstood, the work still has to be done some time; it doesn't just go away. It will have to be done in the afternoon or in the evenings, or we will have to hire more help. Otherwise, the weeds will get away from us, or the harvest won't be completed before the rain, or transplanting will be delayed and the crop will be impaired, causing our CSA members to receive shabby or inadequate produce.

Cass Peterson does her best to scare away the unrealistic:

Farm work is hard. It involves long hours at times, as well as sore muscles, insect bites, sweat and dirt, and the duress of cold, heat, and rain. Not many jobs require real physical strength, but you will need a good measure of endurance. If you are expecting summer camp, complete with weenie roasts, hay rides, and romps in the ol' swimming hole, please consider doing something else with your summer. We strive to have a good time, and we do like to kick back at the end of the day with a cold beer. But our success depends on a certain obsessiveness about what we're doing and a serious adherence to demanding schedules.

To make an internship a success for both farmer and intern, both sides need to be clear on expectations and responsibilities. *The On-Farm Mentor's Guide* provides suggestions on recruiting, examples of interview questions, application forms, and contracts. Many farms ask that candidates fill out application forms, including essays on why they think they want to farm, work references, and a date for a personal interview. Sam Smith, who worked with four interns a year for over twenty years, stresses fostering a sense of teamwork, giving the interns specific areas of responsibility so they have the experience of making some decisions, taking the time to walk the farm together every week and share observations, drawing up a weekly work plan, and holding weekly study groups on important topics. He and Elizabeth had a sort of final test for their interns (and themselves!)—they went away for a week, leaving the interns to run the farm on their own. In my experience, interns who have not farmed before need six to eight weeks to develop a clear idea of where they want to focus. At this point, well into the season, it is a good time to ask them to draw up a learning contract, outlining what skills they especially want to acquire. Mid-season and again at the end of the season, you can do a mutual evaluation of how well both sides have fulfilled the contract, the intern through learning and the farmer through teaching.

Many of the farmers who hire interns worry about what to pay. The law requires minimum wage and all the other legal requirements for any workers, including Social Security, Medicare, and workers' compensation insurance. That does not compensate the farmer in any way for the value of the education provided. (Ironically, you are less likely to get in trouble if you charge the interns than if you pay them!) To keep the pay at an amount a farm can afford, the farm can charge the interns for room and board at rates set by state laws. There can also be a charge for instruction, but this payment must be counted as part of the farmer's income.

There are many internship listings and programs around the country through which a farm can advertise for candidates and potential interns can seek placements. Appropriate Technology Transfer for Rural Areas (ATTRA) has assembled the most comprehensive list of both individual farms and organizational programs. If you call their toll-free number, (800) 346-9190, they will send you a copy in short order. Their listing includes CSA farms. OrganicVolunteer.org also has a national list, though most of the people who use it are seeking short-term placements.[4]

Many CSA farms accept interns, but there are very few programs that are dedicated to training CSA farmers in particular. At the University of California Santa Cruz, thirty-eight apprentices live in tents while completing an intense six-month program with the Center for Agroecology and Sustainable Food Systems in ecological horticulture, which includes running a one hundred–member CSA. At the instigation of John Biernbaum, Michigan State University, in 2006, initiated the College of Agriculture and Natural Resources Institute of Agricultural Technology Organic Farming Certificate Program—forty course credits in organic farming and specialty crops, including credit for a year of experience with the Student Organic Farm CSA. In the first year, fifteen students learned to manage year-round CSA production through their forty-eight–week share program, divided neatly into semester-sized increments of sixteen weeks to meet the needs of a university community.

Shane LaBrake runs the two-year internship at the Accokeek Eco-Farm; interns at Sauvie Island Organics in Oregon spend two growing seasons doing hands-on work and the winter in between taking classes and planning a project for the second year at the farm; Sebastian Kretschmer at the Camphill Village Kimberton Sankanac CSA provides biodynamic training and is a central figure in the development of the On-Farm Mentoring Network; Happy Heart Farm in Colorado offers a three-year program; the Michael Fields Agricultural Institute Internship Program gives a seven-and-a-half-month training, combining classroom study

> Finding really good, steady help is not easy. The pool of U.S. citizens willing to do farmwork is shockingly small. The ideal employee is a local person whom you can train and keep for a long time.

Farmer Katie Lavin and intern Andy Simmons set out plants with waterwheel transplanter. PHOTO BY CRAIG DILGER.

with hands-on projects. A group of the students at Michael Fields started Stella Gardens CSA, which is still run as part of the training program in biodynamic agriculture. (See the "CSA Resources" section, p. 284, for more details.)

At the Food Bank Farm in western Massachusetts, when Linda Hildebrand left, Michael Docter decentralized his management structure, dividing the farm into several distinct areas of responsibility. He encourages interns to stay for three or four years, increasing their pay from $700 to $1,200 a month plus room and all the food they want from the farm and the farm's store. Each intern takes charge of a different area. Michael says that this arrangement is more stressful for him, since every few years he loses good people; but this works out better for interns who want to move on to run their own farms. Two recent interns are taking over Ol' Turtle Farm in Massachusetts, and another two are starting up their own CSA farm in Minnesota. For Spanish-speaking would-be farmers, many of them currently farmworkers, the Agriculture and Land-Based Training Association (ALBA) in Salinas, California, provides six-month courses for twenty to thirty farmer

apprentices that cover organic growing practices, business and management skills, and English as a second language. ALBA's mission "is to advance economic viability, social equity, and ecological land management among limited-resource and aspiring farmers." After the first six months, farmers can continue using Rural Development Center land with as much as five acres for up to three years. They can market their produce through ALBA organics and through its CSA. Maria Inés Catalán, a graduate of this program, has been running her own CSA, Laughing Onion Farm, for nine years. (See more about Catalan in chapter 20, p. 235.)

The Collaborative Regional Alliance for Farmer Training (CRAFT) is a model program, the creation of a group of farmers in eastern New York and western Massachusetts, many of whom run CSA farms. The CSA Learning Center at Angelic Organics coordinates the Upper Midwest replication, there is an eastern Massachusetts group, and Sam Cantrell at Maysies' Farm in Pennsylvania coordinates the Sustainable Agriculture Internship Training Alliance, another similar program uniting the efforts of a dozen farms. According to Katy

Smith, the program arose from the two hours a week (from six to eight on Saturday mornings) that Jean-Paul Courtens committed to sharing his knowledge with her while she was an intern at Roxbury. In Katy's words, "I became increasingly aware what a big burden this was for Jean-Paul, every week to prepare two hours of information just for me. . . . I feel very fortunate! After two years, I was able to go out and start my own business, which has been successful. But it was draining for him." To lighten the load, they began visiting other farms in the area. In the winter of 1995, these farms met and agreed to exchange farm visits to provide all of their interns with a broader exposure to the variety of farming techniques and designs even among organic and biodynamic farms. Sam Smith, one of the founding farmers, writes that a second, but equally compelling purpose has been "to enable the apprentices to experience the importance and power of forming close associations and a supportive community among peers."[5]

Each of the thirteen farms participating in CRAFT hosts a tour, and the farmer speaks on a favorite topic. Sam reports: "For the 1997 season, the topics are: tractor maintenance and farm safety, integrating livestock and crop production, hands-on experience at a dairy farm, cover crops, tillage, and water management, getting started, soil structure, financial management, seed production and saving, orchard management, greenhouse operation, and soil fertility and testing." The sessions last a half-day every week or two: they tried full-day and two-day sessions, but those turned out to be too exhausting. The program opens and closes with general gatherings for introductions and evaluation. From all reports, CRAFT has been a great success in enriching the learning experience of the interns while strengthening the sense of community for both them and the farmers. One of Sam's interns observed, "It has been reassuring to discover that my farmers, Sam and Elizabeth, are part of a much larger circle of farmers sowing, cultivating and sharing their beliefs."

Over the five years of the On-Farm Mentoring project at NESFI, we asked ourselves repeatedly what would be the most effective way to improve the

Interns Who Graduate as CSA Farmers

The fine CSA farms created by some of our former interns are a source of great satisfaction to the older generation of CSA farmers. There are some outstanding examples: Dan Kaplan took Brookfield Farm out of the red and turned it into a model CSA after interning with Sam and Elizabeth Smith. Katie Smith, who works with her husband, Chris Cashen, at The Farm at Miller's Crossing, studied with Jean-Paul Courtens. With her husband, Erik, Birgit Landowne took over Kimberton CSA Garden from her mentors, Barbara and Kerry Sullivan. Sam and Elizabeth Smith could pass on their farm with confidence to Don Zasada because he had studied with their student Dan Kaplan. Here are brief profiles of farms run by former interns at Peacework:

During the 1996 season, Deb and Stew Ritchie interned with me and my former partner at Rose Valley Farm. Today, they own Native Offerings Farm in Little Valley, New York, a 180-acre farm. With three full-time employees and a lot of volunteers, they run a CSA with 250 summer shares, 100 winter shares, 180 fruit shares, and they also raise pigs and cattle. The steady income the CSA provides helped them persuade a bank to give them a mortgage on the farm. Their finest crop, however, is their three children. Sundays are family day, and a major goal of theirs is two days off a week. Deb and Stew see their farm as an opportunity for their children. Their management style is remarkably integrated and thoughtful, aimed at reducing risk. And to reduce stress, Stew meditates and Deb does yoga.

In 2002, Rebecca Graff and Tom Ruggieri started Fair Share Farm in Kearney, Missouri, on land belonging to her father. Rebecca left a job managing volunteers for a San Francisco affordable housing nonprofit to learn organic farming at Peacework, with the intention of settling in Missouri. The winter before she joined us, GVOCSA member Tom Ruggieri had given up his career as an environmental engineer to seek more meaningful work. At the first distribution that spring, he cast his eyes on Rebecca, and his life was transformed. By 2006 they had 75 members who pick up their shares in Kansas City or Liberty, Missouri, or at the farm. As at Peacework, everyone helps with the farmwork or takes part in the core committee. Sometimes Rebecca and Tom call us with questions, but, just as often, we call them.

skills of mentors and ensure that would-be farmers get the training and support they need to enter farming. The guides to assessing one's readiness and to mentoring emerged from these discussions. We also concluded that we needed an official mentoring association that could provide the framework for farmers to develop curricula, and establish professional standards for farmer mentors, in the model of the biodynamic training programs in Germany and England. In January 2006 a group of experienced mentors and others convened to launch the association, ratify bylaws, and select a steering committee. Over the years to come, we will see how this association evolves and whether it meets the founders' high hopes and expectations.

We can experience farm labor as an ecstasy of oneness with the Earth. Josh Tenenbaum, a member of GVOCSA, wrote, "When we dig our hands into the brown, damp soil, does not the entire Earth tremble? When we plant seeds and eat green, living food, does not all the Sun's light intermingle with our own? We see that planting and weeding, building a community around sustainable agriculture, is the most fundamental peace work, no matter what our political beliefs or ideas."[6] Or we can experience this labor as spirit-crushing, unrelenting drudgery. The challenge is ours to make of it what we will.

CHAPTER 7 NOTES

1. Wendell Berry (interview conducted by Elizabeth Barham in 1996 as part of her work on her dissertation in rural sociology. Elizabeth generously shared her notes with me and allowed me to use this quote)
2. Jean-Paul Courtens, "Why Do I Farm Organically?" Jean-Paul generously provided me with a copy of this article in which he recounts the life experiences that led him to become an organic farmer and reflects on his values. You can find it at www.satyacenter.com/why-farm-organically.
3. Doug Jones, *Internships in Sustainable Farming: A Handbook for Farmers* (NOFA-NY, 1999), 2.
4. Several regional organizations have listings of farms that take interns: NEWOOF (Northeast Workers on Organic Farms), www.smallfarm.org, NOFA-VT, NOFA-NY, Pennsylvania Association for Sustainable Agriculture (PASA), MOFGA, and others.
5. Sam Smith, "Collaborative Apprentice Training," *CSA Farm Network*, 2 (1997): 45.
6. Josh Tenenbaum, "Global Effects of CSA Participation" (unpublished article, 1991).

8

SHARERS ON THE FARM

*The only possible alternative to being either the oppressed
or the oppressor is voluntary cooperation.*
—Enrico Malatesta

Time-study experts will tell you that the most efficient harvesting system gets the most vegetables cut, washed, and packed in the shortest amount of time by the smallest number of people. If you aspire to this definition of efficiency, you will have a tight, experienced crew do all of your harvesting, as they do at the Food Bank Farm in Massachusetts and at Wise Acre Farm in California. However, you might want to consider other values, such as community, education, and participation. If you rank these higher than efficiency and are sociable by nature, you may decide to involve your sharers in the harvest. My experience is that members who work on the farm feel a deeper connection with the project and learn much more about the realities of growing food.

As Robyn Van En put it,

The weekly harvest days are a great time to engage CSA members in hands-on activity. Except in emergency situations, the regular farm crew should always accompany the members. It is a wonderful opportunity for them to bond with the farm or garden, as there is little to compare to mornings in a field, the crops heavy with dew, and the birds singing. Members who feel welcome and take the time from their regular work to help with the harvest often become regular volunteers. Deep, philosophical conversations can arise while everyone is vigorously picking string beans. The community spirit is palpable. Many hands make light work and create an unavoidable feeling of accomplishment."

If your project does plan to have volunteer or mandatory help with the farmwork, make member participation as easy as possible. At an orientation meeting or in the newsletter, list all the jobs that need to be done. Indicate the days of the week and hours of harvesting, critical times in the season for weeding, setting out transplants, and the like. Let members know which jobs are "down and dirty" and which are stand-up with an apron on.

At Genesee Valley Organic CSA, we sign members up for "Special Vegetable Action Teams" (SVAT) for jobs such as planting onions, putting up the pea fence, and harvesting winter squash or garlic.

Every Community Supported Agriculture project has the opportunity to involve members. Some CSAs do not ask members to work at all. More frequently, CSAs offer their members the option of a reduced fee for working a certain number of hours per season. Only 22 percent of the CSAs Tim Laird surveyed require that their members pitch in; many more solicit volunteer help. According to Laird's study, work-hour requirements range from three to eighty hours, with an average of twenty. The completion rate for required work was 70 percent, with four of the fourteen farms claiming 100 percent and

GVOCSA members put up pea fencing at Peacework Farm.
PHOTO BY TOM RUGGIERI.

the others ranging from 5 to 90 percent. Some of the farms do not require work but instead offer some crops on a pick-your-own basis. Where members come to the farm for pickup, inviting them to pick for themselves is a popular approach for high-labor crops, such as peas, beans, berries, and cherry tomatoes.[1] The Genesee Valley Organic CSA requires work as part of the contribution for every share, as do Blackberry Hills Farm in Wisconsin (one workday per member), Casey Farm in Rhode Island (eight hours per share), Holcomb Farm in Connecticut (four hours per individual share and fifty hours per organizational share), Many Hands Organic Farm in Massachusetts (four hours per share), and Valley Creek Community Farm in Minnesota (one harvest or delivery per share). The Genesee Community Farm in Wisconsin and Seeking Common Ground in New York function more like worker cooperatives, with each member working three hours every week. Many farms offer working shares through which members pay for their shares by doing farmwork.

At Quail Hill Farm in Southampton, New York, members pick most of their own vegetables. A video of Quail Hill on a picking day shows a remarkable tableau of dozens of people harvesting asparagus and spinach. When interviewed in front of the camera, the members express their delight at the opportunity to pick for themselves. To prepare members for this harvest, the farm crew holds three field orientation days in the spring. They teach members where to find the crops, how to pick or cut, and how to walk around a bed instead of through it. Veteran members are usually willing to help newcomers, but the crew is always available during harvest hours to answer questions. Maps, clear markers, and signs help prevent chaos.

One CSA farm has a labor credit system offering members $6 per hour toward the price of their share. West Haven Farm at the EcoVillage in Ithaca, New York, gives free shares to members who help harvest for two months of the season. In exchange for a rebate and "as much produce as you can pick yourself and carry home," Harvest Farm in Pennsylvania solicits help with mailing, the newsletter, distribution coordination, share packing, driving, and six hours of labor a week or biweekly on the farm. Live Power Community Farm in Covelo, California, delivers its shares to San Francisco, over three hours away. Members rarely come to the farm to help work, but they have full responsibility for distribution in the city. Ruckytucks Farm in Stillwater, New York, gave members the choice of working but sent those who didn't show up a bill for the hours missed. New Town Farm in Waxhaw, North Carolina, asks working members to pay the full $375 for a share and then gives them a $50 rebate when their work is completed.

Most of the workforce for growing and harvesting the 950 shares at Vermont Valley in Wisconsin consists of fifty-six working members. Farmer Barbara Perkins oversees this remarkable system. Having majored in sociology, she enjoys sizing up the strengths and weaknesses of each recruit and assigns them the jobs they can execute most effectively. Each member-worker signs up for a regular four-hour work shift for the entire twenty weeks of the harvest season. Five mornings and three afternoons a week, groups of members work at the farm, doing harvesting, washing, and bagging and also other tasks, such as transplanting, weeding, trellising, and mulching. They do no mechanical work. Barb reports that the quality of their work is amazingly high. People with no background in farming quickly learn under Barb's skilled training. The description of this system on the Vermont Valley Web site emphasizes that the work is "hard," repeating this phrase at least three times. Yet 50 to 60 percent of the workers return year after year, and one man is in his ninth year. The eighty hours of work are in exchange for a standard share priced at $470, or $5.88 an hour. Barbara's husband, Dave, tells me that the Internal Revenue Service (IRS) allows for payment in commodities without requiring Federal Insurance Contributions Act (FICA) fees.

Glen Eco Farm CSA in Virginia lures members to volunteer for work with these terms: "Work share participants last year enjoyed breaking from their routines to work in a pleasant and peaceful outdoor setting, to socialize with each other and ourselves, and to connect with people and the source of their food. With work valued at $5.50 per hour, a share can be paid off in full with 52 hours of work (2 hours per week), in a 26-week season." The sign-up form lists work choices, including bug squashing.

During our first year of operation, our little core committee of David and me (the farmers) and two consumers decided to require farmwork as an experiment. We were not sure people would like it, but we wanted to try. We reasoned that people who could afford to buy their way out of work were already being served by the local food co-op and health food stores. To keep CSA prices lower, members would help administer the project and provide labor, especially for picking and packing. For people who could neither afford to shop in stores nor do the work because of illness or physical disability, we organized a small buying club. The coordinator received produce in exchange for taking orders, collecting payments, and using her home as a pickup point. After eighteen CSA seasons, with a membership that has grown to over 300 households, we can say confidently that requiring farmwork has been a great success, enabling us to keep the share price low—$14 to $20 a week on a sliding scale. Members consider the farmwork a benefit. Their end-of-the-season evaluations are unanimously pos-

> Most members consider the farmwork a benefit. Their end-of-the-season evaluations are unanimously positive about only two things: the quality of the food and the farmwork.

itive about only two things: the quality of the food and the farmwork.

Our CSA distributes twice a week. At the beginning of the season members sign up for a particular day for farmwork, distribution work, and weekly pickup. Two mornings a week, groups of sharers drive to the farm. We figure one helper for every ten shares is adequate for the amount of work required and for vehicle space to transport the food from the farm to the pickup center at the Abundance Cooperative Market in Rochester. Each full share entails three 4-hour sessions at the farm and two 2 1/2-hour sessions at the distribution center.

We prepare sharers for the farmwork with detailed suggestions on what to wear and what to bring, depending on the weather. Both our Members Guide and the GVOCSA.org Web site include this information. Sharers arrive dressed in boots and hats, carrying water bottles, sunscreen, and bug repellent. Some show impressive foresight, coming with work gloves, gloves for warmth, and a third pair of rubber gloves for washing vegetables; rain jackets, winter jackets, and a change of clothes if they get wet. Some of the women tell delighted stories about their reception at a coffee shop where they stop on the way home covered with mud. My heart sometimes sinks when I wake up to a CSA workday of particularly trying weather. One snow-covered October morning, I was starting to mumble an apology for the snow to the first comers when a second carload arrived. Kim Christopoulos leaped out, exclaiming, "Yeeha! It's too slippery to ride a motorbike. Let's get picking!" No apologies were needed after that.

After signing up, each sharer receives a copy of the work schedule for the entire season. Anyone who wants to change work dates is responsible for trading with someone else and reporting to the farmwork schedule coordinator. The CSA e-mail Listserv helps with these trades. A few people are usually willing to be on call at short notice for emergencies. Frequently,

While farmer Elizabeth Henderson digs, members pull and bunch carrots at Peacework Farm. PHOTO BY HIROKO KUBOTA.

extra people come or members bring friends and relatives. The attendance record has been phenomenally good. Only once in eighteen years has no one shown up on a workday. On that occasion, one CSAer suffered a detached retina the night before; the car the other two were driving lost its brakes on the way to the farm.

We plan carefully for each farm workday, aiming to keep everybody busy at a relaxed pace. Members drive a full hour from Rochester to arrive at 8:00 A.M. at the farm. We schedule from nine to fifteen workers a morning, depending on what we are harvesting. As people arrive, we divide them into two crews—field-workers and packing shed workers. Usually, we spend two to three hours picking, washing, and packing the food for that day's shares. With few exceptions, totally unskilled people can learn how to make bunches of greens, dig carrots, and pick beans, peas, corn, peppers, tomatoes, or berries. The hardest part is getting the count right. We count out rubber bands in advance so we don't have to try to count as we pick. We (the farmers) usually harvest vegetables that require experienced judgment—asparagus, lettuce, broccoli—the day before or in the morning before the CSAers arrive. After everything is picked and stored in the cooler, we spend the last hour or so weeding or hoeing. In the fall, sharers clean and sort onions, garlic, potatoes, and squash. We've agreed with our insurance company that we will not use chemical pesticides on the farm and that only the farmers will use machines or ladders.

New CSAers who have never been on a farm before, or even gardened, are often nervous that they will make a mistake or step in the wrong place. We reassure them by explaining as we go along where things are and why we have planted the way we have. Most of the tasks are so simple and clear that people quickly realize they will do all right. After a year or two, some of our sharers have become old hands who need only the slightest direction and can be trusted to do tasks carefully. Obviously, they don't get the work done as quickly as a skilled crew, but our sharers have been remarkably conscientious and willing. I can think of only one man who was an exception. I noticed him doing a very sloppy job, mixing as many weeds as Swiss chard leaves in his bunches. I pointed out to him that he was going to eat that food. A light seemed to go on in his brain. He replied, "Oh, I see what you mean. You are professionals, so we need to take pride in this work."

I have learned not to make any assumptions and never to laugh at members' questions. After I had explained to a group that we were going to weed a bed of carrots, one woman asked me, "What is a weed?" She had never gardened, and caught up in the complexities of her life as a young doctor and mother of two children, she had never even thought about growing plants. Suppressing my amazement, I showed her again which were the weeds and which were the carrots, and observed from a distance as she slowly, but ever so carefully, pulled out the right ones.

Valerie Vervoort, a member of Zephyr Farm CSA in Stoughton, Wisconsin, appreciates the socializing that can occur while doing jobs that can seem like boring stoop labor when done alone: "The bean field is where you really get to know people. There

> We reasoned that people who could afford to buy their way out of the work requirement were already being served by the local food co-op and health food stores. To keep CSA prices lower, members would help administer the project and provide labor, especially for picking and packing.

are so many beans on one bush, you're sitting on a bucket picking them forever. You get into politics and issues, everything from whether you like the governor to the role of men and women in society. And when the group is small, it can get pretty confessional about relationships."[2]

At Happy Heart Farm in Colorado, about 20 percent of the members work three hours a week in exchange for a 50 percent reduction in the share price. Dennis Stenson describes sharer help: "When members work with you, there is just a rhythm to the dance, rather than a frenzy, and the working members share that dance with us and our interns. Supervision and planning play key roles in maximizing the blessing of those helping hands; they are working with us rather than for us. Time must be spent with each new task and with each new worker to show how, where, and when, rather than to tell. Constant follow-up is also essential, because anything taken for granted will be a mistake."[3]

The 1992 season at Genesee Valley Organic CSA put to the test our members' agreement to share with us the risks of farming. Along with the rest of the Northeast, we suffered from a cold, wet growing season. To make things worse, our county was hit with disastrous rains, which began with five and a half inches in one afternoon, followed by twenty-five inches in thirty-five days. We averaged three-quarters of an inch of rain a day, the kind of aberrant weather that is predicted to accompany global warming. Half of our crops drowned outright, while the rest eked out a third to a half the usual rate of production. In mud up to our boot tops, we asked the core committee at the end of July if they would prefer us to give back the money and close down for the rest of the season. They answered unanimously that they didn't care if we gave them only one item a week; they were determined to stick with us. The sharers were more worried about us than about how much food they got. They called,

Leslie Shaw, member of Peacework Farm, models perfect farmwork attire: many layers, gloves, and waterproof boots. PHOTO BY ELIZABETH HENDERSON.

wrote encouraging letters, and even offered to donate money. They showed up on time for their work and slogged with us through the mud. After a morning with us, Sean Cosgrove wrote:

> Driving back from the farm today, we all commented on how much we enjoyed our visit—even in spite of the rain. I wanted to thank you again for your hospitality, for sharing your knowledge about growing things and for the work you are doing.
>
> I imagine that things may be looking a little bleak right now, but the CSA project is so important. I've been despairing a lot lately about our food system and the choices we are given—so it is reassuring and heartening to me to see that there are alternatives . . . albeit struggling ones. Besides the delicious lunch, the fine company, and the

pleasures of working a little bit on the farm, you gave me some hope today. So—I'm sending some hope back your way along with many thanks. I look forward to returning to the farm again—hopefully under brighter circumstances.

Our sharers kept us going. The 150 people who attended our end-of-the-season dinner gave us something not many farmers get to experience from the people they feed—a standing ovation.

Some of the farmers surveyed by Tim Laird expressed a fear that people would not join a CSA if required to work. Other farmers told of their disappointment at the low level of participation in voluntary farmwork or of the nagging necessary to get members to fulfill their commitments. Switching from no work to a work requirement could be quite difficult; however, when members participate, their sense of commitment to the farm and appreciation for the farmers' work increase dramatically. A study of member satisfaction in five CSAs in the mid-Atlantic area by K. Brandon Lang shows that working members are more likely to be satisfied by the CSA experience.[4] As Dennis Stenson puts it, "We've found over the years that it is the working members who really get all the deeper economic and social concepts that CSA is about. Their families become the long-term, committed, enthusiastic, word-of-mouth sales reps we need to keep up with the annual fluctuation of those who come and go."

After her first work experience, Pat Mannix, one of the founding members of the GVOCSA, wrote this deeply felt appreciation:

> For four hours we worked. We picked broccoli, corn, beans, greens, and herbs. We washed and packed them for transport. We hand-weeded and hoed, making a way through the clumps of earth for tiny seedlings. The sun beat down, the mosquitoes bit, the sweat ran. My skin itched, my shoulders ached, my cheeks burned, my dry throat scratched. It was wonderful!
>
> During the week as I prepared meals, I noticed a different feeling, a change in perspective. I found myself preparing the vegetables in a

"Bean concentrate": Members and farmers pick beans and share stories at Peacework Farm. PHOTO BY TOM RUGGIERI.

loving, respectful manner. I planned with a passion so nothing would go to waste. I began to compost. When, for the first time, I ate what I had harvested, it was both an awakening and yet a deepening of the mystery. I clearly understood how this food was becoming a very part of me; its life had been sacrificed that mine could continue and this was all right. It was as it was meant to be and this made it sacred. The old liturgical declaration—fruit of the vine, work of human hands—took on new meaning. I understood that the Earth was alive and that it gave and sustained other life. I knew that the vegetables and myself were both children of it, joined in a wonderful kinship. Food would never be the same for me again.

For most farmers, the experience of community and support is worth the effort to involve members in the farm. After Julie Rawson's first season with seven working shareholders doing eighty hours each, replacing her four dispersing children (Dan, the eldest, has since returned to farm with his par-

ents), she wrote: "I get a sense of confidence from so many people having ownership in the day-to-day operation. Interestingly enough, their presence made the experiences for the paying shareholders, whose work commitment was only four hours, more meaningful. It was with great peace of mind that we left the farm for a week in August and left the working shareholders in charge of the entire management."[5]

Andrea and David Craxton at Roots and Fruits in Dalton, New Hampshire, wrote this appreciation of their members' work:

"We had a lot of return customers, which translated to lots of experience and expertise in field work. Increasingly, our operation is becoming a well-oiled machine. The old-timers quickly instructed and supported the newcomers as they adopted the procedures that we follow. We soon became a big, happy, growing tribe of seventeen families. Our Friday morning harvests were, for many, the highlight of the week. We can't say enough about the enthusiasm and good will as everyone pitched in and made our CSA a delicious success."[6]

For me at Peacework Farm, the intimacy that comes from sharing our farming transcends our business connection. Working with our customers is also enriching socially. I enjoy the opportunity the CSA provides to get to know the people who eat our food. While our hands perform routine tasks like picking and weeding, we are free to have long and intense conversations. Sharers get to know one another as well: surprising exchanges occur between an engineer and a welfare mom or a nurse-nutritionist and a newspaper reporter working together. Seeing the children at regular intervals year after year, we have the pleasure of watching them grow and develop. We don't just sell our sharers food. We were adopted by this growing community of friends who consider Peacework "their farm" and us "their farmers."

Children on the Farm

Children are welcome at most CSA farms. Willow Pond Farm in Maine has a special garden by and for the children. Marty Grady, an appreciative parent wrote:

> When we go to the farm, we take a picnic so we don't have to rush out and rush back. The first place we go when we get there is the children's garden. Each week my children monitor the progress of the garden and their favorite pumpkins. They're really involved in the whole growing cycle. Jill is very open—she really trusts the kids and their exploration. They're free to roam the fields . . . and to use the bathrooms! The kids just soak it up out there. It's great!

At Peacework, children are free either to participate in the farmwork or to play in designated areas: the sandbox, the swings, or the packing shed on rainy days. We have clear rules against touching equipment or wandering into sheds and buildings. Although we have had as many as nineteen children at a time (for a birthday party), no one has ever vio-

Member Susanna Hamer harvests potatoes at the age of 2 and then at 6. PHOTO BY ELIZABETH HENDERSON.

lated these rules. We encourage families with small children to sign up for designated Children's Days, for which we schedule an extra adult or two to do child care. On those mornings, our friend Roland Micklem, a retired science teacher, a naturalist, and a writer, leads the children on nature walks. Some families prefer to come an extra time so that the parents can take turns between child care and farmwork.

To start off a work morning with children, as part of our introduction circle, one of us gets the children's attention and explains our rules (see chapter 10, pp. 131). Shy on their first visit to the farm, by their second trip, children have made it their own turf. Most children divide their time between work and play. Children as young as four or five have helped for the entire four hours with a break or two for snacks. Few children like weeding, but many take part in washing and sorting jobs, and pick beans, peas, and corn and pull carrots and beets with enthusiasm. One morning when at the last minute a father could not make his work assignment, his wife persuaded their two sons, four and six, that they could replace him. The two boys pitched in willingly—they counted rubber bands, washed lettuces, and began picking beans. About halfway down the row, they turned to their mother and said, "You *know* we've done more work than Daddy would have done!"

Another morning, the presence of a large group of children inspired us to dig potatoes. If you have ever run a crew of teenage potato gatherers, you know that getting them to work can be harder than picking the potatoes yourself. The CSA children, aged two to six, had never seen a potato harvest. As the digger brought the red potatoes up from the ground and dropped them in a row, the children shrieked with delight. From a distance one might have thought we were running one of the wilder rides at a county fair. The children ran after the digger, snatching those potatoes up as fast as candy out of a piñata. Since then, we try to schedule Children's Days to coincide with potato harvest.

Caretaker Farm includes clear guidelines for parents in its CSA Handbook:

> The farm is a wonderful place for children to explore and learn about plants, animals, and where and how their food is grown. It is also a place to learn respect for the land that feeds and delights us. While we want the children to feel at home on the farm, we also want them to be safe.
>
> Parents are expected to know where their children are at all times.
>
> We have created some guidelines to be followed:
> 1. The Ponds. Children must be accompanied by an adult swimmer when they visit the ponds.
> 2. The Electric Fence. The fence is on at all times to contain livestock and to protect them and the gardens from coyote and deer.

Things to Consider in Deciding Whether to Ask Members to Work:

- Social inclination of farmer
- Distance from distribution
- Flexibility for cost sharing
- Harvest less efficient, but gain in long-term members (efficiency in member retention)
- Opportunity for exchange of information and forming friendships

If Members Work, You Need:

- To make sure farm is safe—no rusty nails hidden in the weeds, and so forth
- Clear directions to farm
- Clear signs on farm
- Specific work descriptions in newsletter and Web site to prepare members for work shift—what to wear, what to bring
- Weather report
- To provide good tools
- Flexible schedule—way for members to substitute dates on CSA Web site or e-mail Listserv
- Choice of special projects

Children on the Farm:

- Provide clear, simple rules
- Be clear on who is responsible for the children at all times—parents or child care worker
- Offer jobs children enjoy—pulling carrots and beets, picking cherry tomatoes, digging potatoes
- Breaks and rest periods
- Snacks
- A children's garden
- Guided nature walks

> All wires are "HOT," including the temporary plastic fence. It is not dangerous to touch, but it isn't pleasant either. We are most concerned about babies and toddlers who are too young to know. Please educate your children to respect it.
> 3. The Animals. For the protection of the children and of the animals, please do not enter the animal pens, the chicken houses, or the

pastures unless accompanied by a farm staff person.

4. The Farm Buildings. Barns, workshops, tractor sheds, and greenhouses are off-limits to children unless accompanied by a farm staff person or an adult.

5. The Children's Garden. This is a special place for children when they are visiting the farm. Families who wish to help may adopt a bed for the summer.

Weather conditions that discourage grownups may have the opposite effect on children. The truth is, if we will admit it, we all like to play in the mud. Children were especially appreciative of the opportunities afforded by 1992—the Year of the Big Waters. Nathan, son of Pam Hunter, the minister of the Lutheran Church that hosted our pickup site for several years, was two that summer, toddling along on his hind legs but just barely. Pam was apprehensive as he walked from the swing to join us at the beds where we were picking. No more than three feet up the row, Nathan did a perfect pratfall face-first into the mud. He struggled back on his feet, pushed back his rain hood, and laughed. The worst had happened. Pam relaxed. At the end of another rainy morning, three older children lingered near the packing shed, reluctant to get in the family car. They crowded around me, faces streaked with mud, asking excitedly, "Do you get this dirty every day?"

The families in our CSA who are homeschooling consider participation in the farmwork a significant addition to their children's education. Ann Bayer wants her four children to understand where their food comes from. Having grown up in the city, she does not want her children to believe, as she did, that food comes from the back room at the grocery store.

Members of other CSAs around the country have expressed how important the farm connection, the pesticide-free food, and the education about how food is grown are to their children. Many report that their children eat more of the vegetables after they've been to the farm and helped pick them. In their sociological study, "Factors Influencing the Decision to Join a CSA Farm," Jane Kolodinsky and Leslie Pelch did telephone surveys of members and nonmembers of three Vermont CSAs and came to the puzzling conclusion that having children makes people less likely to join CSAs. These "scientific data" contradict the real-life CSAs I have known, in which members with children make up the majority and many parents say they have joined *because of* their children. A member of Harmony Valley wrote this comment, which I think typifies parental sentiment about CSAs: "The biggest benefit is that our kids could see where our veggies come from (at Strawberry Days), and I believe this encouraged them to believe that veggies are a wonderful gift and therefore would eat them."

———————————•———————————

On too many farms in this country and abroad, the exploitation of children's labor deprives them of their childhood and of the chance to go to get an education. This sad reality, however, should not keep us from involving children in farmwork in healthy, nurturing ways. By working along with adults, farm children learn how to be farmers. If they never get to dig potatoes or pull carrots from the ground, children from nonfarm families will never know that their calling is to join the next generation of farmers, of which our culture is in such desperate need.

CHAPTER 8 NOTES

1. Timothy J. Laird, "Community Supported Agriculture: A Study of an Emerging Agricultural Alternative" (master's thesis, University of Vermont, 1995), 54–55.

2. Valerie Vervoort, *Country Journal* (May/June 1998), 26.

3. Dennis Stenson, *Seasonal News* 1, no. 1 (1994), 3.

4. K. Brandon Lang, "Expanding Our Understanding of CSA: An Examination of Member Satisfaction," *Journal of Sustainable Agriculture*, vol. 26, no. 2 (2005), 61–80.

5. Julie Rawson, "Working Shareholders," *CSA Farm Network*, 2, 74.

6. Andrea and David Craxton, *CSA Farm Network*, 2, 83.

MONEY MATTERS FOR CSAs

There can be nothing sacred in something that has a price.
—E. F. SCHUMACHER, *Small Is Beautiful*

Besides all the wonderful earthy and human qualities of CSAs, they must also function as viable small businesses. Throwing all the receipts in a shoe box might work for the first year, but long-term survival requires economic management systems for budgeting, pricing, book-keeping, and contractual agreements with members. An essential aspect of CSA sustainability is providing a decent living for the people doing the farmwork, and adequate cash flow to cover both the annual costs of production and the continuing reproduction of the farm—that is, investments for maintenance, improvements, transfer to the next generation, and retirement.

Keeping in mind Trauger Groh's image of the CSA farmer freed from economic considerations, it is still important to me to run a farm that functions successfully as a business. I want my farm to serve as a demonstration to my farming neighbors, many of them very conservative people, that ecological farming is a practical possibility. I realize that the conventional farmers I know consider my organic CSA to be a sort of special case, but at the same time, they recognize it as a creative approach to marketing and admire my ability to get the cooperation of my customers. Although what I am doing is different, it is somehow still within the range of acceptable local farming practices, and that is a great advance over how it was viewed ten years ago. While the ideal may be to break the usual market connection between a particular

weight or count of vegetables and the dollars members pay, and to inspire support for the entire farm, we cannot jump over our own heads. Moving toward a mutual aid economy, what the World Social Forum calls a "solidarity economy," and transforming all of our members into active "co-producers," to use the term of Slow Foods founder Carlo Petrini, will take time and a lot of education to reach a solid footing where farmers and farm members understand and support each other's needs.

The U.S. government's cheap-food policy is an obstacle to charging the true cost of producing food. While in other countries citizens pay up to 35 percent of their incomes for food, middle-class U.S. consumers are accustomed to spending only 12 to 13 percent or less. According to United States Department of Agriculture (USDA) statistics, off-farm income supports 84 percent of the farms. Selling directly to the consumer saves some farms, and finding a group of loyal customers provides a refuge from the ruthless battering of the global marketplace. People determined to stay in farming have been finding many creative ways to sell directly to the consumer: farmers' markets, farm stands, or mail-order sales through catalogs or the Internet. In one sense, CSA is simply the latest wrinkle in this long-term trend.

According to the 2001 CSA Survey, a smaller percentage of CSA farmers than non-CSA farmers are dependent on off-farm income. Converting all or part

Share Prices

Regarding sign-ups for next year: We raised our price $15 per share. We are offering a $15 discount to anyone who signs up by December 10; as you can probably tell from how far ahead of time we are dealing with 1996, we want to sell our shares early. (You might also note: We are already preparing fields for 1996, so this forward planning does come out of the farming process.) Several of you suggested that we raise our price to cover the cost of production. I know that price would have prevented many shareholders from signing up again; it would have simply been too high. That's known as "pricing yourself out of business." The price we are asking might cause us to eventually farm our way out of business, but at least it will keep us going while we try to figure out a way to make Angelic Organics truly sustainable. We'll include a 1996 sign-up form with a future newsletter.

For 1996, we are creating different categories that shareholders can sign up for beyond the basic box price. We think this is an effective way to give you a choice about the long-term future of Angelic Organics. This method does not force a big price increase onto shareholders, but it does create a clear opportunity for people to pitch in to make this farm continue. The different levels of support are: seed $15; cotyledon $25; seedling $50; vine $100; pumpkin blossom $500; giant pumpkin $1,000. You'll get this information when you get your form later in the season, but I'm including it in now, so you can mull it over.

We understand that, for many, renewing a share at the basic share price is a considerable financial stretch. If you want to help out Angelic Organics, but money is not your way to do it, another way you can pitch in is to encourage your friends to join Angelic Organics for next year—sometime this fall would be best. We hope it seems more like fun than work for you to get your friends to join Angelic Organics; then you get to compare recipes, carpool to our field days, trade vegetables, and keep your friends healthy! For us, it is very costly and time-consuming to persuade newcomers to join our CSA, when we have our hands (and hearts) full of farming. Most people seem stuck in the paradigm that anyone they buy anything from is just another business. If we were just another business, first of all, we wouldn't be in business, and secondly, we'd have a boring newsletter. If you think your friends would like what we do, please encourage them to join—this makes a huge difference to Angelic Organics. Note: we only want people to join who are aligned with the CSA model; it's impossible to please people who think of us as a store. We're a farm, through and through.

John Peterson, Angelic Organics, Caledonia, Illinois, used with permission.

---

of a farm's cropland to growing for a CSA is, for many farmers, an attractive marketing alternative. As more and more people catch on to the fact that you get what you pay for, those who really care about food have become more willing to pay higher prices for food that is genuinely high quality, fresh, and locally grown. By setting a realistic price for CSA shares and cutting out the middlemen, farmers can eliminate debt and stabilize their incomes. Those who adopt CSA successfully find that it offers more than just a way to survive.

Share Pricing and CSA Budgets

CSAs fall roughly into three groupings according to how closely price is linked to the amount of produce in a share. Farms that produce mainly for sharers stress support for the entire farm rather than the price per item of food.

A very few community farms, such as Kimberton CSA in Pennsylvania and Temple-Wilton Community Farm in New Hampshire, break the price connection completely. These farmers submit a budget to the members, who then pledge what they can afford to contribute to the expenses of maintaining the entire farm. If the first go-around does not cover the costs, members are urged to pledge more until the budget is either met or altered to adjust to their ability to pay. Regardless of how much a member pays, each takes as much food as he or she needs.

A second group of CSAs goes partway toward

2001 CSA Survey

Farmer assessments of the CSA operation's effects on the following aspects of their farm as a whole.

	n	Greatly Undermines	Undermines	Change	Improves	Greatly Improves
				Percent of Respondents		
Financial ability to meet annual operating costs	293	2.4	7.5	16.7	48.8	24.6
Farmer compensation	289	3.1	10.0	32.5	39.5	14.9
Financial security for farmer including health insurance, retirement, etc.	285	8.1	16.1	53.0	17.5	5.3
Financial ability to build and maintain physical farm infrastructure	281	3.9	10.7	43.4	35.9	6.1
Farmer stress level/quality of life	285	2.1	14.4	26.7	41.4	15.4
Maintenance or improvement of soil quality	290	1.0	3.1	43.4	33.1	19.3
Workload for the farmer	284	3.9	22.2	40.5	27.5	6.0
Compensation for other workers	257	2.3	9.7	51.4	31.9	4.7
Workload for other workers	251	2.4	9.6	59.8	24.3	4.0
Community involvement	284	2.1	3.9	29.6	40.5	23.9

breaking the price/produce connection. These farmers present a budget to the core group for discussion. To meet the budget, these CSAs divide the total figure by the number of shares. CSAs concerned about economic diversity permit sharers to pay on a sliding scale. Instead of charging by share, a few farms charge so much per adult member and feed the children for free.

Even subscription-style CSAs weaken the price connection, since they neither price items individually nor necessarily maintain share size from week to week. The farmers charge a given weekly price for a share without divulging how they arrive at that price. Heavy weeks may compensate for light ones to reach a yearly average.

Trauger Groh, of Temple-Wilton Community Farm and now retired, is the eloquent spokesman for CSA as part of a new kind of "associative economy." By Trauger's logic, farmers are too busy farming to be able to bother with earning money, so other people must support them. He defines an associative economy as "an economy where the motive

In 2001, the CSA Survey found that the median price for a full share was $400, with a price range from just a few dollars to $4,000. The median price for half shares was $200, with a range up to $1,200.

of our actions is the need of others, and that is what a farm is about—the motivation of the farmer is the need of the members for produce, and the motivation of the group is the need of the farmer." In another example of alternative economic thinking, Jean-Paul Courtens explains why he farms organically and strikes at the heart of the usual business ethic. "It is my hope that the old paradigm of profit can be replaced with service and sustainability as part of an organic approach." Or as Karen Kerney, the artist for this book, who runs a mini-CSA in Jamesville, New York, puts it, "I believe when you grow food—you grow food. You do not make money. The money is made in the harvest and handling." As I see it, harvest and handling is part of growing the food; perhaps it is helpful to view the entire process from seeding to selling as a highly skilled craft, which should be rewarded fairly like any other comparable skill, such as carpentry, masonry, or midwifery.

CSAs also vary in how public they make the budget. The earliest North American CSAs, following

Massachusetts CSA Farm 2005 Income and Expenses	
Item	**Dollar Amt**
(Annual Operating Expenses)	
CSA Revenue ($500/share x 106 shares)	
(Assumes discounts paid for)	$53,000.00
Wholesale Revenue	$ 9,047.80
Winter Share Revenue	
($100/share x 85 shares)	$ 8,500.00
Gross Revenue	
(from 3.5 to 4 acres cash crops)	$70,547.80
Rent	$4,600.00
Seasonal Labor	$5,545.00
Fertility (Manure, Bagged Fert, Potting Soil,	
Straw Mulch, Cover crop seed)	$4,178.50
Seeds (Incl. seed potatoes and	
sweet potato slips)	$2,720.96
Equipment (Truck & Tractor Op.	
Expenses: Oil, Filters, Ins., Repair)	$2,170.99
Fuel	$2,212.42
Biodiesel Processor (not done)	$ 489.55
DLCT land improvement	475.00
Combine	750.00
Pumps & tanks	954.19
New greenhouse & cellar fan	2,271.57
Subtotal, Capital Expenses	$4,940.31
Total Expenses for 2005	$30,582.29
Farmer's total cash income for 2005 (Gross Revenue less total expenses) (about $20/hour for a nominal 2000 hour work year or $22.50/hr. if durable purchases are counted as "asset income"—although asset depreciation ought to then also be deducted from this, for which I have no good estimate.)	$39,965.51
Debt Summary (no outstanding debt)	0.00

even with our core group because we did not feel it was necessary given the small percentage of our farm sales represented by the 31 shares. Of the CSA farms Tim Laird surveyed in 1993, 26 percent marketed exclusively to their members, and, eight years later, the 2001 CSA Survey found that only 16 percent of the farms were 100 percent CSA. The Kimberton and Temple-Wilton CSAs are among that 16 percent. They analyze their expenses at length with their farm members. From the beginning, they have conceived of their projects as community farms in which all members share tasks, responsibilities, and financial contributions according to ability. Temple-Wilton farmer Anthony Graham reminisces:

> Back in 1985, out of our discussions with Trauger, we decided on our approach. We asked members of the farm community for a pledge rather than asking them to pay a fixed price for a share in the harvest. We realized that the members of our community had a wide range of needs and incomes and that one set price was not necessarily fair for every family. . . . Our approach works. It requires honesty and good will, but it works. The last four or five years, our annual budget meeting with the farm members has only taken about 45 minutes. It's fast, up front, and everyone understands it by now.[1]

Not many CSAs have taken the path of asking members to share the entire financial support of the farm. Like Rose Valley, most farmers initially view the CSA as an alternative marketing approach for the farm. Michael Docter of the Food Bank Farm takes the pragmatic attitude that members are more likely to be concerned about the value of the food they receive than about the finances of the farm. He contends that to become widespread, CSAs must produce food efficiently and provide good value to their members. Many other CSA farmers would appreciate more membership involvement and support, but feel they must move in that direction cautiously so as not to scare members away.

Only gradually, as the Genesee Valley Organic CSA became a genuinely supportive community and

the model of Indian Line Farm, recruited members by circulating a brochure that included the budget. In the Northeast some CSAs continue this practice, but it is rare in other parts of the country. When Rose Valley Farm launched the GVOCSA in 1989, we did not share budget information with the general public or a larger part of our marketing, did we begin sharing more information with members about our farm as a business. In 1996, when we cut back on production and dropped most other markets, we printed our farm budget for the first time in the Genesee Valley Organic newsletter. Since 1998, when Greg Palmer and I founded Peacework Farm, the GVOCSA has consumed over 90 percent of our veggies. Every January we discuss the annual budget proposal with the core group and use that as our basis for setting share prices and numbers. In the March newsletter we print the budget proposal for the coming year alongside our actual expenses for the previous year.

Whether you make it public or not, the farm budget needs to include certain standard expenses. The county extension office or the Farm Service Agency can supply you with budget examples. The Internal Revenue Service (IRS), in Schedule F (Form 1040), "Profit and Loss from Farming," provides a handy guide. Setting up your bookkeeping with this form in mind will aid you in doing your taxes. Here is the IRS's list of legitimate farm expenses:

- car and truck expenses (Form 4562)
- chemicals
- conservation expenses (Form 8645)
- custom hire (machine work)
- depreciation and Section 179 expense deduction not claimed elsewhere
- employee benefit programs other than on line 25
- feed purchased
- fertilizer and lime
- freight and trucking
- gasoline, fuel, and oil
- insurance (other than health)
- interest
    a. mortgage (paid to banks, etc.)
    b. other
- labor hired (less employment credit)
- pension and profit-sharing plans
- rent or lease
    a. vehicles, machinery, and equipment
    b. other (land, animals, etc.)

- repairs and maintenance
- seeds and plants purchased
- storage and warehousing
- supplies purchased
- taxes
- utilities
- veterinary, breeding, and medicine
- other expenses (specify)—Includes investments in land improvement, such as tiling of fields, gifts and contributions, legal and professional fees, advertising, dues and seminars (including certification), pest control, and others.

Budgets in CSA brochures usually divide expenses into labor, capital, and operating costs. Labor includes salaries for the farmer or farmers, assistants or interns, additional hourly help, and insurance and tax costs. Farms that exchange shares for work would include this expense under labor. (See chapter 8 for more on working shares.) I have seen only a few budgets that build in a pension fund for the farmer. Operating costs may be divided into separate sections for farm, share distribution, and CSA administration. The published budget for Happy Heart Farm in Colorado for 1996 included a shortfall to be made up by fundraisers. Under capital expenses, CSAs include the purchase of tools, equipment, fencing, construction costs for coolers, and other durable materials. Large expenses, such as tractor purchases and greenhouse construction, are usually amortized or factored in over several years. A few CSAs that provide meat, milk, or egg shares as well as vegetables break out their expenses into separate enterprises. *The CSA Handbook* provides useful work sheets for calculating both farm expenses and share prices (see the "CSA Resources" section in this book for ordering information).

Note that the IRS does not include a salary or health insurance for the farmer. In official business tax-think, the farmer receives the difference between gross revenues and expenses—the "profits."

Farmer Earnings

As Jill Agnew of Willow Pond Farm puts it, "Most of the food in this country is provided by people working for substandard wages. Paying people a living wage for work in agriculture is also critical to the CSA philosophy." Published budgets usually include the farmer's salary along with other labor costs. In a few cases, CSA members have been shocked at how little the farmer earned and have insisted on raising share fees to provide a living wage. Both Tim Laird and Daniel Lass discovered that many of the CSAs they surveyed did not include farmer salaries or capital expenses in their budgets. Among Laird's respondents, the average salary was $11,225, with a range from $0 to $30,000.

CSAs with two hundred shares or more that have matured into well-established community farms are paying their farmers $30,000 to $45,000 a year with such benefits as health insurance and even retirement funds. While these earnings are not spectacular, CSA farms are doing well compared with other kinds of farms, and a higher percentage rank as serious enterprises rather than hobby farms. According to the 2001 CSA Survey, 63 percent of CSA farms have gross farm incomes over $20,000, compared with 30 percent of U.S. farms as a whole with gross incomes over $24,999.[2]

In addition, the 1999 CSA Survey found that the 28 percent of the CSA farms that have core groups had an annual median income from CSA shares that was $10,000 higher than the other farms. The core group farms sold more shares and charged a higher price. The 1998 study, "The Economic Viability of CSA in the Northeast," by Daniel Lass, Sumeet Rattan, and Njundu Sanneh, which analyzed data collected over three season from a varying group of CSA farms, showed similar results and concluded that the core group provides a structure that allows the farmer to benefit financially from the goodwill of the shareholders.

With the long-term economic viability of the farms in their network as their primary goal, Équiterre has done two studies, in 2000 and 2004, closely examining the financial data from carefully selected coop-erating farms. In the publication of the first study of eleven farms, "A Guide for the Management of CSA Farms" (December 2002), the authors present five sample budgets based on the actual farms in the study. Since the sample group of farms was small, 2000 weather conditions in Quebec were poor, and the study lasted only one year, no great conclusions result. This study is useful, however, as an economic snapshot of five very different CSA farms:

1. a generalized average of the farms in the study with 144 shares
2. an experienced, established farm with little debt with 300 shares;
3. a market garden on rented land with 45 shares
4. a farm typical of the members of the Fédération d'agriculture biologique with 80 shares, representing only 16 percent of the farm's production
5. a more mechanized farm with a high level of debt

The second study, "Four Viable and Enviable Economic Models of CSA" (December 2005), conducted during a season when weather conditions were more or less average, provides convincing data showing that the ten farms are covering their costs of production and even generating a small surplus for future investment or the repayment of debt. While the eleven-farm average in 2000 came to a loss of $46 Canadian per share in 2000, in 2004 the average was a net profit of $17 per share. Farmers who read French would learn a lot by examining the detailed model farm budgets and accompanying summaries of the farms' resources in land and equipment.[3]

From 2002 to 2004, John Hendrickson, a researcher and a Madison Area CSA Coalition (MACSAC) board member, worked with a group of nineteen vegetable farmers in Wisconsin to collect data on sales, labor, and business costs. The goal of "Grower to Grower: Creating a Livelihood on a Fresh Market Vegetable Farm" was to help the farmers compare their farms and figure out how to improve their economic viability.[4] The conclusions highlight the economic benefits of CSA: "CSA appeared to help stabilize income. . . . Other marketing strategies are subject to the vagaries of the marketplace and weather." The CSA farms also had higher ratios of net cash earnings to gross expenses. Of nine farms with a ratio over 50 percent, eight of them had CSA programs. Hendrickson surmises that "crafting a budget and having cash in hand at the beginning of the season may result in more careful spending on CSA farms." However, the hourly earnings for these farmers are not good news. The largest and most efficient farm paid the farmer a grand $11.36 an hour. The smallest farms, with less than three acres in crops and most of the work done by hand, rewarded the farmers at the rate of $4.96 an hour.

Farmer Tracie Smith in New Hampshire describes her struggle to make a living at the work she loves: "The year I graduated, I started my first year of CSA with 15 members; now, 5 years later, I am up to 90 members. I am still struggling to make a real living wage, earning less than $10,000 a year by the time all the expenses are taken out, but those around me see the farm as successful in terms of quality and being known and respected. I do feel we should be able to make a fair living wage, enough to buy the land we grow on, which I am not." In her September 5, 2005, newsletter, she expands on her efforts: "To keep the farm going, I spend 60 to 80 hours per week for 6 months, and 20 hours per week for the months of March and April, along with many hours of planning, ordering, and advertising throughout the fall and winter. I have become more efficient each year with the farm tasks, have invested in many tools and supplies to make it more efficient, and the soil is getting better every year. . . . There is hope. I am not going to give up, but need to reconsider everything I am doing because I work too much and am not getting paid enough myself, never mind making enough to pay other people to help."

Capital Investments

To avoid taking it out of their own hides or borrowing from banks for major improvements or expansion, several CSAs have appealed to their members, cashing in on the substantial social capital that a farm can accumulate through good service. As Signe Waller and Jim Rose of Earthcraft Farm wrote in a fund-raising letter to the farm's most loyal members: "To be a sustainable operation, Earthcraft Farm needs a subsidy from a source other than ourselves." Along with the annual budget, they included "Items to be Purchased from Sustainability Fund," which was a list of soil amendments, livestock, and equipment that will contribute to building the long-term fertility of the farm. Happy Heart cheerfully invites people to sign up as "Friends of the Farm," holding out the "opportunity to provide the needed capitalization funding." Its brochure lists four levels for donations: Root—$25, Stem—$50, Leaf—$100, and Flower—over $500. Another Colorado farm, Blacksmith Ridge, had a similar scheme. In 2004, Brookfield Farm in Massachusetts was able to raise $150,114 by appealing to members to pay for a new barn and farm center. Three hundred forty people responded with pledges! Members of Common Harvest Farm in Minnesota paid for seven years of shares in advance to provide the farm with funds to build a barn and purchase fuel-efficient vehicles. In 1997, Angelic Organics invited sharers to invest in a limited liability company to purchase thirty-eight acres of additional land for the farm. (See chapter 5 for more examples of member contributions to land purchases and conservation easements.)

The most radical (or perhaps the most conservative) aspect of Peacework Farm is our total avoidance of borrowing. If we have the money to buy something, we pay in cash; if not, we wait. The need to make mortgage or other debt payments can add an unbearable level of pressure to running a farm and have a profound effect on all management decisions. Avoid it if you can.

Calculating the Share Price

How to set a fair price for shares is one of the most puzzling questions for CSA farmers. The cheap-food policy in this country exerts a sharp downward pressure on any attempt to sell food at a price that will sustain farms. Rather than apologizing for their share prices, Tom and Denesse Willey declare right up front: "Your money goes directly to a family farm. Every dollar of your share goes directly to the farm. A similar purchase at a supermarket may return a farmer as little as 20 cents on the dollar. By participating in T and D Willey's CSA you are supporting the farm's community of year–round employees who are able to be paid steady and reasonable wages. Your financial support is a vote for organic, family-based farms as the backbone of our agriculture production system." After struggling along for their first three years without enough money to cover all their living expenses, Barbara and Kerry Sullivan at Kimberton CSA asked for a substantial increase in their share price. Barbara describes the change:

> After we raised the cost of the share about four years ago, the membership had a whole different feeling. We ended up with the serious, committed people who really appreciated the garden for what it is. We lost the people who were just looking for cheap vegetables. . . . It was the best move we ever made. We could not have survived until we made that decision. You can start out the way we did, but sooner or later you've gotta have enough people who are really committed to making the CSA work. It's a real community garden now. We could go, and the Kimberton CSA Garden would still go on. This is a real people garden.[5]

A few years later, the Sullivans did leave and Kimberton continues to thrive.

I keep discovering more ways to calculate the share price. CSAs that make their budgets public usually divide total expenses by the total number of shares to get the price. Some CSAs divide shares into various sizes or qualities, such as full and half-shares, single-person shares, barter shares, gourmet shares, macrobiotic shares, or winter and summer shares. Jean Mills and Carol Eichelberger divide their season into spring/fall and summer shares and charge more for the summer weeks

when the burning heat of Alabama makes production more difficult. Dutchess Farm in Castleton, Vermont, offers more northerly Vermonters a five-week share of vegetables most backyard gardens won't produce until much later in the season. CSAs often price half-shares at more than half of a full share because of set costs per share and the additional work involved in making up a greater number of smaller packages.

Farms with other markets tend to calculate the share price based on the market value of the produce. Subscription schemes charge a fee based on what the market will bear. The Willey Farm in California values their vegetables at the wholesale price but adds to that the value of the nonfood items included with a share—the special service, the newsletters, on-farm events, and so forth. At Rose Valley in 1989, I began by adding the farmers' market price of the shares on one date each for early spring, summer, late summer, and fall. I then divided by four to get an average weekly price. Through the first season I used this average price to guide the quantities of vegetables I allocated for the weekly shares. Since the early weeks were lighter, I compensated by adding more food later in the season. Marianne Simmons, a core member who has overseen distribution for sixteen years, reported that there was occasional grumbling about the price, but only among the wealthiest members. When she pointed out that her teenage son earned about as much for an hour of babysitting as the farmworkers, the grumbles subsided.

Some farms set their price based on the average weight of the produce they will give each week. They determine how much they value the produce per pound and then multiply by the number of pounds they intend to provide. At the end of the year, they can then report that members paid only $0.79 or $0.99 per pound for their organic produce all season.

Yet another way of calculating the price is used by a few multifarm CSAs: unit pricing. The purpose of this approach is to ensure that the various farms are compensated fairly for their contribution to the shares, since some vegetables require more work

Genesee Valley Organic CSA members harvest greens. PHOTO BY MARILYN ANDERSON.

than others. The farmers agree on a way to convert all produce into a standard unit, for example, a bunch of beets equals a pound of tomatoes equals a pint of strawberries, and assign a value to the standard unit. They then multiply by the number of units to get the share price.

For a few years, Caretaker Farm tried calculating how many mouths they could feed and divided their estimated annual costs by this number to arrive at the charge per adult. Children were thrown in for free. In 1996–97, each adult paid $310 toward the farm's budget of $72,657.[6] Sam Smith says that this approach led to misunderstandings, and in 1998 they went back to a regular share price.

Jim Rose and Signe Waller of Earthcraft Farm CSA in Indiana had a novel system. Instead of a share, the farm sold members a cluster of 180 units, which they could use in any combination they chose over the course of the season. Members allocated their units as they wished—a dozen one week, none the next—or gave a few to a friend. The farm's weekly menu assigned unit values to the produce. A

unit might represent a pound of beans, two pounds of collards, a bunch of green onions, a pint of raspberries, or one bok choy, depending on the supply available. As Signe explained to me, this system worked only for the section of their CSA that picked up their shares at the Earthcraft farmers' market stall. For the Indianapolis members, the farm packed the usual weekly bags. Signe says they felt real differences in member responses; the more-local members were more satisfied, loyal, and likely to stay with the project.

The 2001 CSA Survey found the median price of a full share to be $400, with 50 percent of the farms charging from $340 to $500, and the top price $4,000. The few farms with the top share prices are coming closer to supplying all the food needs of their members, including meat, eggs, and other products year-round. Nineteen of the sixty-five CSAs in Laird's study offered half-shares, with an average price of $244. Most of these farms charged more than half the price of the full share to cover set costs per share, such as bookkeeping, postage, and other overhead. The median price for half-shares was slightly lower in 2001, at $200, with 50 percent charging from $200 to $320, and $1,200 the top price. Intervale Community Farm and Quail Hill Farm charged an extra-high fee for half-shares in part to discourage people from buying them. In the 1995 season Quail Hill eliminated half-shares altogether and settled on a single-share size.

Member participation can influence the price of shares. Some CSAs offer members the opportunity to reduce their share price by working on the farm. The going value of members' time runs from $5 to $6 an hour (though legally, to avoid the Wal-Mart syndrome, it should equal minimum wage). Meeting Place Farm in Ontario, Canada, cuts $50 off the full-share price for ten hours of work and $25 off the half-share price for five hours. New Town Farms also cuts its price by $5 per hour of work, but asks for twenty hours. Salt Creek Farm in Washington gets away with only $3.50 an hour. The Genesee Valley Organic Community Supported Agriculture (GVOCSA) keeps share prices lower by requiring that all members work either on the farm or in administration of the project. I calculated one year

that if we had to pay for all the work our members do, we would have to raise the share price by $200. (See chapter 8 for more on working members.)

To attract lower-income members, a few CSAs ask for payments on a sliding scale. Rising River Farm CSA in Washington has a sliding scale of $425 to $500 for a full share and $212.50 to $250 for a split share. The GVOCSA began with a scale of $5 to $7 a week and gradually went up to $14 to $25 in 2006. Members decide for themselves how much they can afford to pay, with no attempt made by the CSA to verify their income. On the CSA brochure and at the orientation sessions, core members explain that the actual value per week is $17 and ask members to pay at least that, if they can. People who pay more understand that they are subsidizing those who pay less. Almost every year, by some miracle, the higher payments exactly balance out the lower payments.

Introducing a sliding scale in 2006, Local Harvest CSA gave this explanation:

> Susan and Simone, Local Harvest's founding mothers, have always been committed to the three

---

2006 Payment Plan
Call Alyssa Lopez at (212) 894-8094 with any questions. Or stop at
Food Change: 252 West 116th Street (Between 7th & 8th Avenues)

< $15,000*	$250	$50.00	$25.00	$25.00	$25.00	$25.00	$25.00	$25.00	$25.00	$25.00
$15,001 to $34,999	$300	$76.00	$28.00	$28.00	$28.00	$28.00	$28.00	$28.00	$28.00	$28.00
> $35,000	$400	$104.00	$37.00	$37.00	$37.00	$37.00	$37.00	$37.00	$37.00	$37.00

*Food Stamp Payment Plan　　$250
　Cash Deposit of $30.00 due by May 1st
　First Food Stamp payment due June 8th

Pay $19.25 with Food Stamps every 2 weeks during the course of the season from June to December.
Members that make all 13 Food Stamp payments will get the $30.00 deposit back at the end of the year.

*Please note: Proof of income is required for subsidized share.

primary components of farming sustainably—building and preserving our land and environment for future generations, building and caring for our community, and pricing our products to provide us with a living wage. With steadily increasing costs this last item is becoming more challenging. While seeing the need to raise our vegetable share prices to reflect our real costs, we remain committed to the belief that everyone should have access to fresh, wholesome vegetables.

In response to this dilemma, we decided to offer you a price range and you choose where you fit in according to your own budget limitations. The lower dollar amount reflects our real costs to meet the basic needs of our family and the people who live and work with us—shelter, food, clothing and transportation. The higher dollar amount represents our costs to reach a more sustainable goal that includes medical insurance and a retirement plan.

Susan reported to me that, to her relief and surprise, most members paid at the highest level!

Scholarship funds are another way to provide for low-income members. Many CSAs solicit contributions from members on their annual commitment forms. Vanguarden CSA in Massachusetts explains their request: "We at Vanguarden are serious about ensuring that no one is denied fresh, healthy produce due to financial constraints. Your contributions make this possible. Thanks." Twenty of Vanguarden's 106 shares go to "Co-operative Economics for Women" at half price for distribution to low-income families. Chris Yoder raises the $5,000 to support these shares from the other members of his CSA.

Rochester churches, to which some GVOCSA members belong, have made several generous contributions to the scholarship fund. A portion of the price of every *FoodBook* (the Rose Valley Farm crop description and recipe book) has also gone for scholarships, so that in 2006 ten members paid as little as $8 a week. Some of the payments were in food stamps. The difference between the payments and the average cost of $17 a week came out of the

scholarship fund. (See chapter 10 for more on food stamps and Chapter 20 for more on involving low-income people.)

To entice early sign-up or enlist member help in recruiting others, some CSAs give a discount on the share fee. If you are considering this, you might ask some of your members whether they need these "bonuses." You may find they are willing to sign up earlier or recruit friends without a financial incentive. The core group of the GVOCSA decided against a discount for members who paid in full before the season because they felt it was unfair, awarding a financial break to the people who needed it the least. Angelic Organics gives a $100 discount, and Full Belly Farm in California gives a free share to members who are willing to host a distribution pickup site in their garage or on their back porch.

Evidence suggests that price is not a serious hurdle for the majority of CSA members. In 1995, Gerry Cohn did a study of consumer motivations in joining CSAs in California. With the help of four CSA farmers, he created a four-page survey with questions about why people join, preferences on share size and pricing, alternative sources of produce, level of interaction with the farm, core group participation, and household demographics. He received 235 responses from members of ten CSAs. He found that most of the members had fairly high incomes and education levels. Price was not an important question in deciding whether to join a CSA. Of those members surveyed, 64 percent were willing to make donations toward scholarships for low-income shares.[7]

Researchers Jack Cooley and Daniel Lass set out to determine whether membership in a CSA actually saved consumers money. In their survey of 250 members of four CSAs in the Amherst area of western Massachusetts, they found that nearly half believed that they were paying about the same amount of money or more than they would for the same produce in a local store. To test if this were true, Jack found out what was in the shares each week and then priced those items in three different local groceries: a national food chain, a regional food chain, and a local store. Only the regional food chain carried organic

Sales and expenses over one year on a 1.5-acre market garden	
Gross sales	
CSA (45 members)	$14,415
Direct wholesale	$9,869
Total sales	$24,284
Annual cash expenses	
Bank service charges	$48
Hired labor	$4,400
CSA crops purchased	$300
Fuel—for equipment (tiller, mower, vehicle)	$200
LP for greenhouse	$320
Greenhouse supplies	$590
Insurance	$310
Memberships/dues	$130
Miscellaneous	$301
Organic certification	$520
Postage, printing, and reproduction	$160
Repairs	$580
Seed	$820
Soil amendments	$520
Supplies	$1,490
Taxes	$400
Telephone and utilities	$940
Total expenses	$12,029
Net cash income	**$12,255**

FROM JOHN HENDRICKSON, "GROWER TO GROWER"

Sales and expenses over one year on a 4.5-acre market garden	
Gross sales	**$46,460**
Cash expenses	
Hired labor	$5068
Seeds	$3,361
Property taxes	$2,558
Fertilizer	$2,206
Services	$2,010
Fuel	$1,983
Farm and vehicle insurance	$1,716
Greenhouse supplies	$1,300
Maintenance	$1,285
Phone	$650
Electricity	$630
Small tools and misc. supplies	$590
Communications (printing, copying)	$531
CSA supplies	$435
Employment taxes	$250
Bags	$222
Office supplies	$122
Total cash expenses	**$24,917**
Net cash income	**$21,543**

FROM JOHN HENDRICKSON, "GROWER TO GROWER"

produce. Out of the eighty items, fourteen could not be found in the stores at all, so the researchers estimated a price by multiplying the conventional price by 1.5122, which they found to be the ratio of the average price for all organic items to the average price for all the conventionally grown items. They did not include herbs and flowers in their cost comparisons. Two of the farms charged $450 for a full share; one provided 24.7 pounds of produce per week and the other, 27 pounds. Both of these farms used machines for planting and cultivation and irrigated a large portion of their fields. A third farm, a smaller operation with no mechanization or irrigation, provided 8.5 pounds a week for $250. When compared with the retail price for organic produce, the members of the first farm saved $548, of the second saved $682, and of the smaller CSA saved $149. Even compared with the price for conventional produce, the CSA prices were cheaper.[8]

A cost-conscious member of the GVOCSA did a similar comparison with similar results. In 2000, while on the staff of Just Food, Corinna Hawkes compared CSA prices with those of several New York groceries and delis, a supermarket, and an organic food store. Even during the disastrous food conditions of that fall, members of Threshold CSA saved money when compared with the organic food store prices.[9]

A few farms are using this kind of comparative

Sales and expenses over one year on a 16-acre market garden	
Total gross sales	**$250,000**
Cash expenses	
Automobile expenses	$3,800
Chemicals	$800
Custom hire	$150
Equipment purchases	$43,000
Employee benefits	$7,000
Fertilizer	$2,600
Trucking	$500
Fuel	$2,400
Insurance	$2,900
Mortgage	$7,300
Other interest	$500
Hired labor	$40,000
Equipment rental	$1,900
Land rent	$1,900
Repairs	$2,400
Seeds	$5,000
Misc supplies	$16,000
Taxes	$1,900
Utilities	$2,400
Office supplies	$2,000
Marketing	$3,300
Subscriptions	$300
Training	$900
Professional services	$1,200
Total cash expenses	**$150,150**
Net cash income	**$99,850**

FROM JOHN HENDRICKSON, "GROWER TO GROWER"

market information in advertising the benefits of joining their CSA. The Cate Farm brochure boasts that members will receive a $700 value for only $450. Dutchess Farm touts a money-saving value of $430 worth of produce for only $320.

While providing good value is laudable, some researchers suggest many CSA farms may not be meeting the goal of providing adequate compensation for growers. In "The Economic Viability of Community Supported Agriculture in the Northeast," a three-year study of CSA economics,

Daniel Lass concludes that farms could be charging more for what they provide.[10]

Only members who actually make use of all or most of their share truly appreciate the competitive value. For those who don't like to cook or who eat out often, the vegetables may amount to expensive compost. Before the Lass-Cooley survey, almost half of the respondents did not realize they were saving money. This perception may be a result of paying a large sum for the share all at once instead of paying in dribs and drabs, as most people are used to when they shop. One year, at the GVOCSA sign-up meeting, a lawyer member made a loud demonstration of paying for his entire share up front. He insisted that I be summoned to witness his largesse. To my astonishment, I discovered that he was paying at the bottom of the sliding scale. When I laughingly pointed out to him that the share's actual value was $3.50 higher a week than what he was paying, he promised to make up the difference. People are funny about money.

Bookkeeping

Using a computer program like Quicken or QuickBooks may make bookkeeping easier, though the electronically impaired might not agree. For the Genesee Valley Organic CSA, two members of the core group share the bookkeeping job: For many years, Dennis Lehmann kept meticulous records by hand, while Judy Emerson used a Lotus spreadsheet computer program; at the end of each season they compared the results. When Dennis moved with his wife and children back to South Dakota to be closer to his mother and Judy confessed to being tired of the job, we had some nervous moments wondering whether we would ever find another CSA member willing to do the treasurers' job. Amazingly, a librarian and a trained accountant, Nora Dimmock and Susan Stoll, have replaced Dennis and Judy. They keep track of member payments, dues, and administrative expenses. We keep the books for Peacework Farm as a separate business. (See chapter 11 for other kinds of records farms should keep.)

Dan Kaplan at Brookfield Farm in Massachusetts offers a set of software programs for all CSA bookkeeping, from share prices to row feet for each crop (www.brookfieldfarm.com).

Jill Agnew has a simple way of accounting for member payments. At the farm, she keeps an index box with a card for each member. The members are on the honor system to keep track of their own payments and pay at whatever interval is convenient. When she receives checks, Jill records them on a receipt pad and thus has a double-entry system. She says she has never had anyone who failed to pay.

Harmony Valley Farm in Wisconsin has been able to arrange with its bank to have automatic monthly transfers from members' bank accounts. The annual CSA contract includes an authorization form for the transfer.

Contracts

The CSA brochure or the annual contract or commitment form is where the members first encounter the price of their share. The typical contract has spaces for name, address, and phone number; weekly or seasonal fee for the share and for any additional shares, such as eggs or flowers; preferred pickup day and site; and signature and date. Some contracts also reinforce the message that joining a CSA means agreeing to share the risk with the farmers. The contract of Common Ground CSA in Washington spells out this mutual obligation in detail:

> I understand that I am making a commitment to Common Ground Community Supported Agriculture and recognize that there is no guarantee on the exact amount of produce I will receive for my share. The CSA harvest is protected by sustainable and organic practices such as organic soil improvements, crop rotation, and irrigation. I will share both the rewards and the risks of the growing season along with the other members and the growers. It is my responsibility to pick up my share

within scheduled pickup hours. If it is not picked up, I understand that it may be donated elsewhere.

Crop Insurance

Diversifying your crops and recruiting members to share the risk with you is the most cost-effective form of crop insurance. When you buy crop insurance, you contribute to the enormous profits made by insurance companies. Even the new "agri-lite" policies do not match up well with a small-scale, multi-crop farm. In 2005 the companies that sell crop insurance to farmers made $927 million in profits and also received $829 million from the government to cover their administrative fees. Meanwhile, the insurers paid farmers only $752 million.[11] If CSA is nothing else, it is an alternative to making the insurance companies richer.

Start-up Expenses

"If you want to make a small fortune in farming, start with a large one," suggests traditional farmer wisdom. Precisely how much you need to start up a CSA depends on a multitude of factors: whether you have to buy the land; the buildings already in place on or near the ground you plan to farm; the level of intensity of the methods you plan to use; whether you will need irrigation; and the amount of volunteer labor available. Estimates range from $11,000 to $250,000. Choosing the level of technology that is appropriate for you, your skills, and your preferred lifestyle is one of the most important decisions you will have to make.

The $11,000 figure comes from *The Rebirth of the Small Family Farm: A Handbook for Starting a Successful Organic Farm Based on the Community Supported Agriculture Concept*, by Bob and Bonnie Gregson. Anyone thinking about entering farming should read this charming little book of practical pointers based on the Gregsons' own lives. They list the equipment and materials you would need for a two-acre market garden where most of the work is done by hand. The income they estimate is from sales of gourmet produce to middle- or upper-class customers.

In 1998, CSA Works—the consulting firm of Michael Docter, Linda Hildebrand, and Dan Kaplan—suggested an investment of $15,000 to $20,000 for the farm implements you would need to cultivate five acres or more using mechanical weed control systems. They recommended the purchase of an Allis Chalmers G tractor with cultivating implements and a Planet Jr seeder. Farmer Anne Morgan (Lakes and Valley CSA, Minnesota) cautions that only farmers with mechanical skills should start with old equipment; she suggests that you would do better buying a tractor that has a dealer close by. The $20,000 figure does not include buildings or the larger tractor a farm would need if you were planning to do your own plowing and make compost on a large scale. While exemplary in many ways, the Food Bank Farm is not self-sufficient in terms of on-farm fertility; if that is one of your goals, your equipment costs will be higher.

Jean-Paul Courtens sketched out an equipment list and rough costs for the level of capitalization in place at Roxbury Farm. Not including the land or livestock, he estimated an investment of $250,000 for barns; greenhouses with a vacuum seeder or soil block machine and flats; three tractors; a manure spreader; tractor implements for tillage, cultivation, transplanting, seeding, and haying; a harvest truck or flatbed wagon; a delivery truck; a root crop washer; irrigation equipment; a walk-in cooler; a bucket loader; and a forklift. I think Jean-Paul was pricing new equipment. At Rose Valley Farm we had

Sample Equipment List for a 1.5-acres market garden*		
Item	Purchase price	Current value
Hoophouse and related supplies	$1,700	$1,000
Used walk-behind tractor w/ rotavator	$2,700	$1,600
Used mower	$250	$100
Used Walk-in cooler (6x6x4)	$900	$700
Garden cart	$350	$150
Miscellaneous garden tools, harvest crates, and irrigation lines	$750	$300
Totals	$6,650	$3,850

*This data comes from a participating market garden.

FROM JOHN HENDRICKSON, "GROWER TO GROWER"

Sample Equipment list for a 5-acre market farm*		
Item	Purchase price	Current Value
Tractors and Vehicles		
35 hp tractor with loader	$8,500	$6,500
All Terrain Vehicle (ATV)	$2,500	$2,000
Pickup truck	$5,500	$5,000
Implements		
3pt rotary mower	$1,000	$800
3pt tiller	$1,400	$500
3pt digger (field cultivator)	$250	$250
3pt chisel plow	$250	$250
3pt tool bar and clamps	$150	$100
3pt sprayer	$750	$500
6ft grain drill	$50	$0
Manure spreader	$50	$25
4 x 6 trailer	$500	$100
Lely spring tine cultivator	$1,400	$500
Bedding (mulch) chopper	$350	$350
Irrigation		
5.5 horsepower pump	$550	$300
Lay flat hose (~3000 ft.)	$1,000	$100
Sprinkler heads and couplers	$500	$100
Greenhouse		
1000 sq. ft. greenhouse	$4,500	$1,000
Benches	$200	$0
Heater	$650	$100
16' x 96' hoophouse	$1,400	$500
Hand tools, harvesting		
Planet Jr. seeder	$250	$100
Misc. tools	$500	$250
Misc. shop tools	$1,500	$500
Computer and printer	$200	$0
50 harvest tubs	$600	$200
Saw horses	$100	$0
Wash tubs	$300	$30
Pressure washer	$500	$300
Totals	$35,400	$20,355

FROM JOHN HENDRICKSON, "GROWER TO GROWER"

Brookfield Farm Equipment List

Vehicles	Repl. value
1985 Ford F350 Box Truck	$7,500
1988 Chevy S10 4wd truck	$3,000
1965 Chevy 3/4 ton truck	$750
1978 Chevy 3/4 ton truck	$500
1978 Chevy dump truck	$500

Tractors

Massey Ferguson 35	$3,000
Case 1410	$6,000

Machines/Tools

Pittsburgh 6' disc harrow	$500
JD KB 8' Harrow	$1,000
JD 3 Bottom Plow	$1,000
Bezzecchi Fertilizer spinner	$200
JD Corn Planter	$150
Mechanical Transplanter 2-row transplanter	$400
Iron Age Potato Planter	$500
Pittsburgh 3pt hitch cultivator	$200
6' Buddingh basket weeder	$300
Lely Tine Weeder	$1,000
Lilliston Cultivator	$500
5-row dibbler	$100
Hardi 6' boom sprayer	$300
Potato digger	$200
Potato digger	$200
potato picker	$200
Bush-hog	$400
IH 160 manure spreader	$100
New Idea 467 manure spreader (jointly owned)	$1,700
2 wagons	$500
6' utility trailer	$250
Kohler 14 hp irrigation pump	$1,500
B & S 5 hp irrigation pump	$200
800' 2" aluminum irrigation pipe – hook & latch	$400
600' 2" aluminum irrigation pipe – friction	$300
800' 3" aluminum irrigation pipe – hook & latch	$600

Vehicles	Repl. value
2000' 3" layflat irrigation pipe	$500
30 Nelson sprinklers	$160
Root Washer	$750
Propane greenhouse heater	$800
30' Germination pad	$100
8 flexinet electric fences	$200
8 farm gates	$80
Lawn mower	$150
Weed wacker	$150
Husqvarna chain saw	$200
Hand tools	$400
Workshop tools	$500
	$37,940

Supplies

50 – 20 bu. wooden bins	$500
220 Tomato trays	$1,100
400 3/4 bu. waxed boxes	$150
500 1/2 bu. waxed boxes	$100
50 1 3/4 bu plastic produce bins	$250
80 1 1/9 bu plastic produce bins	$400
500 Pro Trays 72s – greenhouse trays	$100
50 10-row germinating trays	$10
1100 carrier trays	$200
electric fencing supplies	$1,500
4000 tomato stakes	$400
	$4,710
	$42,650

Leased equipment (from Dan Kaplan)

farmall cub	$2,500
side dresser for farmall cub	$500
allis chalmers "g"	$3,000
allis chalmers "g"	$3,000
5 row planet-jr seeders	$500
tine weeder – allis g	$300
5' Buddingh basket weeder	$500
	$10,300

a comparable level of capitalization, but by doing the construction ourselves and purchasing mainly used implements, we kept the costs at less than one-quarter of Jean-Paul's estimate. Like many other small farms, we also made our purchases gradually, adding one or two key pieces of equipment each year as we could afford them out of current earnings.

Bill Brammer of Be Wise Ranch expanded his operation every year for twenty-one years, starting from three-quarters of an acre and a rototiller to over three hundred acres with several tractors. A 1998 acquisition is a machine that enables him to roll up and reuse drip irrigation tape five or six times. By combining drip irrigation with transplants, Bill has reduced water use from three acre-feet to one acre-foot for some crops, a major saving

Malek and Lisa, members of Cadet-Roussel Farm in Quebec, Canada, pick up their shares at the Bakery Le Fromentier in Montreal. PHOTO COURTESY OF EQUITERRE.

when water costs $640 per acre-foot. One of the advantages of getting larger, Bill says, is that the increased volume of sales has provided the capital to purchase adequate equipment. Although he depends on compost to improve the rocky, clay soils, mainly used by other farms for cattle ranching, he waited until 1997 to purchase a manure spreader to mechanize the job.

To make good use of equipment demands skill, experience, and careful planning. The more experienced you become as a farmer, the more ways you will discover to make your farm more efficient, more conserving of energy (both purchased and human), and more in harmony with the natural forces on and around your farm. A workshop at the Upper Midwest Organic Conference in 1998 compared the economic "facts" and "ratios" of two CSA farms. Earthcraft Farm added a column with its own figures and sent me a copy. They compared numbers of shares, acres, gross and net farm income, farmer person days, total person days, and payroll expenses. John Hendrickson reports that in the "Grower to Grower" project this

ratio exercise enabled the farmers to learn a great deal, and many of them put the numbers to practical use on their farms right away. Training in holistic management offers a powerful set of tools for improving decision making and earnings on any farm. Workshops and conferences where CSA farmers and active members share openhandedly with one another are of tremendous value. We can all run our farms so much better if we pool our best discoveries.

Despite the lofty goals of CSAs, there is no evidence yet that they guarantee a comfortable income for farmers. Two small studies found that about half the farmers reported increased incomes when switching from previous marketing arrangements to CSA. For nearly as many, income remained the same. But almost all the farmers felt less insecure and expressed optimism that their incomes, though still low, would grow as the CSAs matured.[12] This hope is borne out by the farms profiled in Groh and McFadden's *Farms of Tomorrow Revisited*: While not making out like bandits or bankers, after seven more years of development, the farmers were doing better financially. The 2001 CSA Survey similarly found that CSA gave farmers an optimistic view of their financial future. See the table on p. 111.

In 1993, Jack Kittredge interviewed a farmer with a twenty-five-member CSA who shared these reflections on his farm's economics:

> Small scale makes sense ecologically. I wish there were more CSAs. A local one went out of business this year because he owed too much money for land, mortgage, and equipment. A successful CSA needs to grow cautiously. It needs to be more of a lifestyle choice than a way to make money. I couldn't possibly do this if I owed money. We own our land and house with no mortgage. Taxes are less than $500 a year. I haven't figured out how most farmers survive, especially if they want to live like what I perceive to be the "average American."[13]

Would You Please Add a Little Glitter to That Squeaky Ripper?
(John Ponders Metal and Muscle)

We have thirty-three machines on this farm. They have names like discer, hiller, harrow, digger, mulch layer, field cultivator, cultipacker, barrel washer. The machines aerate, pulverize, plant, scrub, dry, hill, till, weed, rip, mix, smooth, cool, shuttle. They have shanks, knives, bearings, teeth, scrapers, brushes, lathes, belts, wheels, springs that bend, break, screech, corrode, wear in, wear down, wear out. They need greasing, sharpening, inflating, welding, patching, riveting, tightening, drilling, shimming, propping, oiling, jerry-rigging.

This is a little farm, and that's a lot of machinery propping it up, and occasionally dragging it down. Most of it is used. Some is ancient, like our Iron Age (actual brand name) potato planter—originally designed for hitching to a horse. Much is cast off from farms that specialized and expanded, and some is residue from farms that folded.

The vast majority of work here as measured in hours is through hands-on physical labor. I keep looking for "labor-saving" devices to create more balance between technology and the human back. I read the classifieds and watch the sale bills. I drive to Michigan and up into Wisconsin in the hopes of locating obscure specialized equipment that will ease the sheer physical demands of bringing a weekly box of produce to your neighborhood from some packets of seeds the previous winter. Imagine all the different actions involved in that metamorphosis—procuring the ingredients for the greenhouse soil, mixing, sieving, planting, watering; preparing the fields: composting, tilling, transplanting; tending the crops: weeding, irrigating, thinning, foliar feeding; harvesting: discerning, cutting, picking, digging, peeling, separating, hanging, drying, trimming, snapping, bunching, lugging, conveying, washing, sorting, divvying.

I have an affinity for machinery. One machine can sometimes do the work of twenty laborers. It will never sneak out of the field in the middle of the morning for a nap. It's usually where I left it. If it breaks I can almost always fix it. And I've never seen a drunk chisel plow.

On the other hand, the 656 Farmall tractor has never made me lunch, I've never flirted with the wheel hoe, and I've never enjoyed an Amazake White Russian with the tiller.

(I was going to delete these last two paragraphs, but Jill and Cindy insisted that I leave them. Editing talent is another quality that my machinery lacks.)

Fieldhands have other redeeming qualities that give them a special status here. They plant and water the seedlings, clean the packing room, yank weeds, discern ripeness, maintain machinery, stock our stand, and fill and deliver your boxes week after week with vegetables they harvested, cleaned, and graded almost totally by hand.

And as I learned Saturday night, the workers dress splendidly for a farm party and laugh 'till their sequins drop. My tillage tools reposed obediently, drably, in the full moonlight of the farmyard, as the farm help spun, glittered, and danced.

—*John Peterson, Angelic Organics, Caledonia, Illinois. Used with permission.*

We should not expect to divorce CSA farms from the world context in which they exist. As Jim Rose and Signe Waller said in their "Five Year Retrospect and Prospect": "The entire food system, from the most personal level of the shopping and eating habits of advertising-hounded consumers to the macroeconomic level where world food prices are controlled by a few corporations, constitutes an environment hostile to small and middle-sized farmers. It is almost a mission of insanity to undertake to farm as more than a hobby, to actually try to make a living at farming." Miraculously, the CSA connection is helping farms to do just that! And in times of crisis, the social capital CSA farmers accumulate can turn into hard cash. When raging flood waters destroyed more than half of Harmony Valley's crops in August 2007, members called farmer Richard deWilde to express their concern and sent check—totaling roughly $45,000—to keep the farm going.

CHAPTER 9 NOTES

1. Steven McFadden, "Community Farms in the 21st Century," *New Farm* (January 2004).
2. 2002 Census of Agriculture, www.nass.usda.gov/census_of_Agriculture/index.asp.
3. Équiterre, "Quatre modèlés economiques viables et enviables d'ASC" (December 2005).
4. John Hendrickson, "Grower to Grower: Creating a Livelihood on a Fresh Market Vegetable Farm," www.cias.wisc.edu.
5. Trauger Groh and Steven McFadden, *Farms of Tomorrow Revisited*, Biodynamic Association, 1997, 155.
6. Ibid., 178.
7. Gerry Cohn, "Community Supported Agriculture: Survey and Analysis of Consumer Motivations" (research paper, University of California at Davis, 1996). Available from Gerry Cohn, 127-B Hillside St., Asheville, NC 28801.
8. Daniel Lass and Jack Cooley, "What's Your Share Worth?" in *CSA Farm Network*, vol. II (Stillwater, NY: CSA Farm Network, 1997), 16–19.
9. Corinna Hawks (study presented at Just Food 5th Annual CSA Mini-conference, February 2001, and published in the Just Food *CSA Toolkit*).
10. Daniel Lass, Sumeet Rattan, and Njundu Sanneh, "The Economic Viability of CSA in the Northeast" (study, University of Massachusetts 1998).
11. Gilbert Gaul, Dan Morgan, and Sarah Cohen, "Farmer Insurance Firms Reap Billions in Profits," *Washington Post*, October 16, 2006.
12. Timothy J. Laird, "Community Supported Agriculture: A Study of an Emerging Agricultural Alternative" (master's thesis, University of Vermont, 1995), 78–80; and Rochelle Kelvin, *Community Supported Agriculture on the Urban Fringe: Case Study and Survey* (Kutztown, PA: Rodale Institute Research Center, 1994.
13. Jack Kittredge, "CSAs in the Northeast: The Farmers Speak," *The Natural Farmer* (Summer 1993)," 12.

LEGALITIES

Character is the internalization of responsibility. What we are talking about when
we talk about a local food system or CSA is a food system that relies more on
character than it does on legal, bureaucratic, or commercial procedures.
—WENDELL BERRY, quoted in *Safe Food News*

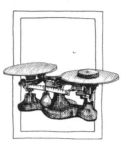

Like many other farmers, Community Supported Agriculture (CSA) farmers tend to have an aversion to elaborate legal structures and government regulations. KIS is what we prefer—keep it simple. In this book we can offer only very general information on the legal aspects and implications of CSA. Laws vary from state to state and county to county. You may need to consult a local lawyer. Better yet, recruit one as a sharer and barter for the advice. First be sure to find out, from other farmers or area cooperatives, which lawyers have experience with relevant issues. This chapter will at least give you enough information to formulate the right questions.

CSA Legal Structures

A CSA can adopt a variety of legal structures. Each group should determine which form is most appropriate. Some CSAs are "sole proprietorships" or partnerships; in other words, both farm and CSA business are the property of the farmers. Other CSAs separate the CSA from the ownership of the land. The land may be held as a sole proprietorship, a partnership, or a corporation, while the CSA is an unincorporated association or is incorporated as a nonprofit corporation. Groups of farmers can organize as farmer-owned cooperatives, most of which are corporations. There is no set structure in the law for food co-ops or buying clubs, so groups of consumers can choose the corporate structure that suits them best in forming a CSA, including a member-owned cooperative. Institutional CSAs usually hold both the land and the CSA as part of a nonprofit corporation. Each form has advantages and disadvantages. The details of these legalities will vary from state to state.

Robyn Van En sketched out the following information based on Massachusetts state law:

Sole Proprietorship
- Not regulated by the State Statute; consequently no legal help is necessary to establish it.
- Proprietor is responsible or liable for debts and obligations of the business or property.
- Proprietorship income can be shown on a schedule of the proprietor's individual tax returns.
- Proprietor pays half of Social Security and Medicare, employees pay other half.
- Basic bookkeeping is necessary.

Partnership
- Similar to sole proprietorship, but requires registration of partnership with county clerk. In forming a partnership, it is wise to write a partnership agreement, which includes how the

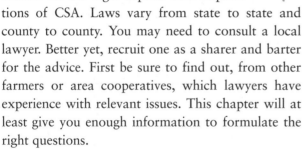

partnership will be dissolved. *The Partnership Book* (Berkeley, CA: Nolo Press, 1995) contains all the clauses and instructions on how to form your own partnership.

Limited Liability Partnership

- General partner runs the business, and has rights and liabilities of any partnership.
- Limited partners have no management powers, and are liable only to the limit of how much they have invested. Limited partners are basically investors.

Unincorporated Association

- Not regulated by the state; no filing fees necessary.
- Administering individuals are responsible and personally liable for obligations and debts.
- All labor may be contract labor, and contractors are responsible for own taxes and insurance.

Incorporation

- Filing fees are necessary (usually about $500, plus possible legal fees to process the paperwork). All activity is governed by the state.
- Creates a legal entity, separate from stockholders, that can hold a mortgage or long-term lease.
- Legal suits would be directed against the assets of the corporation rather than individuals. The owners (stockholders) are not personally liable.
- Separate books must be kept and separate tax returns must be filed.
- A Board of Directors must be elected or selected to manage the administration of the corporation, which might include hiring and firing of employees.
- Stockholders and shareholders invest with the expectation of a financial return on their investment.
- Takes tax deduction for compensation paid to employees, including shareholders who are employees.

Farmer-Owned Cooperative Corporation

The Capper-Volstead Act of 1922, a Federal law, authorized the formation of farmer cooperatives. The USDA Rural Development office (Stop 3257, 1400 Independence Ave., SW, Washington, DC 20250; 202-720-8381) will send a packet of information on how to form a farmer cooperative, with sample legal documents, policies, and a history of co-ops in the United States.

- The structure and liability limits are similar to a corporation as described above. The minimum number of farms required for incorporation varies under the statutes of different states. Most states require that a cooperative have officers, a name, an annual membership meeting, and a mailing address.
- Five underlying principles distinguish cooperatives from other types of private businesses:
  1. Ownership is by member-users.
  2. Control is on the basis of one vote per member, or on volume provided.
  3. Operations have an at-cost (nonprofit) objective.
  4. Dividends on member capital are limited.
  5. Education is necessary for understanding and support.

Nonprofit Corporation

- The procedure for organizing a nonprofit is very similar to forming a corporation, including the filing of fees. An important difference is the ability of nonprofits to receive grant funding.
- Investors do not expect financial return on their investment. There is no capital stock or stockholders.
- Nonprofit tax status is required to be eligible for charitable donations, which are tax-deductible to donors.
- Assets can be transferred only to another nonprofit should the organization cease operation.
- If the nonprofit qualifies for 501(c)(3) status, all activity is exempt from federal and state taxes.

The unincorporated association or sole proprietorship forms are the most appropriate for the first year or so while a CSA gets itself together. Many farms, while remaining under the operation of a family, incorporate to limit liability.

The members of the Hudson-Mohawk CSA invited Janet Britt to be their farmer and then chose to classify their CSA as her sole proprietorship. In an

On her distribution shift, member Sharon Hamer checks off names of members as they come to pick up their shares. PHOTO BY MARILYN ANDERSON.

article in *Harvest Times* (Winter 1992), Janet explains the decision: "The people who had come together to get our farm started were willing to work hard to organize it, and to have the trust and faith to buy a share, but were not ready to assume legal liability for the project. I was willing to assume responsibility because I wanted our CSA to start and had few personal assets to lose in the unlikely event of legal problems." While legally a sole proprietorship, the group agreed to a philosophy of cooperative ownership, which meant that Janet owned all capital investments but agreed to leave those assets with the CSA if she were to leave.

Four farmers are partners in Peacework Farm, which produces most of the food for the Genesee Valley Organic CSA. In preparation for signing a long-term lease on our land with the Genesee Land Trust, we formed a limited liability company (LLC) under New York State law. An LLC is similar to an official partnership. The LLC has to pay an annual fee to the state and we must have officers, keep minutes of our annual meeting, and have a written oper-

ating agreement. To confound the gods of hierarchy, Katie Lavin is the Chief Executive Officer (CEO), Ammie is the Chief Financial Officer, Greg is Chief of Strategy, and I am Chief of Operations. In our agreement, we defined LLC members as the farm managers, with no provision for outside investors, and agreed to make all major decisions by consensus. We had to declare each partner's capital investment and decide how to divide profits. Our personal assets are protected from any liability action against the farm. The LLC itself does not pay taxes but passes the farm's profits or losses on to the LLC members and we pay taxes as individuals.

The GVOCSA is an unincorporated association that does not make any profit. It is a buying club that passes all the money it collects in food payments to the farm. An annual membership fee covers administrative costs. From the point of view of the various authorities, the GVOCSA contracts with the farm to purchase the crop and is responsible for the picking and distribution of the food that it owns. When members work at the farm, they are not working as farm employees. We do not have to fill out nine hundred W-2 forms. The farm business and all farm equipment belong to the Peacework LLC, and only the cooler used for distribution belongs to the GVOCSA.

Qualifying for nonprofit status can be tricky. State and federal regulations govern the kinds of organizations that can qualify and the activities that are acceptable. Failure to make a profit does not in itself qualify a farm to be a nonprofit corporation. An established nonprofit can more easily include a CSA among its activities than a new CSA can set up its own nonprofit. Donations to a nonprofit must be spent on acceptable activities such as educational programs and religious or scientific activities open to the general public. A CSA whose primary goal is to feed and educate its own members would have a hard time qualifying as a 501(c)(3): The benefits have to go to society at large, not to a limited group. A nonprofit that adopts a CSA as fitting with its mission might serve as the funds administrator for the CSA and require a small fee for the service. It might also request that the CSA handle some of the required paperwork. The GVOCSA fits very well with the community food security work and urban-rural linking done by Rochester Roots, but the organizations, by mutual consent, remain separate.

Insurance

Most CSAs carry standard liability insurance. As separate coverage, liability can be very expensive; as part of a farm insurance package, the price is more reasonable. You should try to get a liability policy that includes a stated level of medical expenses paid out without a lawsuit. Some CSAs have additional liability as a special form of "pick-your-own." The rates for "pick-your-own" will be lower if you specify that you do not use synthetic pesticides and members do not use equipment, horses, or ladders. Take the time to explain to your insurance agent how your CSA differs from a regular pick-your-own operation, and provide documentation, such as a copy of your brochure or newsletter. If an accident should occur, your agent

will then not be able to plead ignorance about the nature of your operation.

Pick-your-own coverage will allow members to help harvest and to use hand tools. Keep a first-aid kit handy. Encourage members who work regularly to get a tetanus shot to reduce the danger from puncture wounds.

If children are welcome at your farm, you must have clear farm rules. Unless you are providing child care, parents or guardians must understand that they are responsible for their children. At Peacework, when children arrive, we round them up, get their attention, and explain our four rules:

1. Children are not allowed in the sheds or barns or on any farm equipment unless one of the farmers is with them.
2. Children must not go near the Gonargua Creek without an adult.
3. They must learn to recognize poison ivy and stay out of it but if they think they have touched it, tell one of the farmers, who will help them wash off the toxin (they all look tense and serious by this time).
4. Have fun! (laugh of relief).

Silver Creek Farm has a pond where farmers and members can swim. Molly Bartlett makes sure all visitors know that no one is allowed to swim alone, that children must be supervised, and that there are set times when everyone can swim. (See "Children on the Farm," p. 106, for more on this topic.)

In this machine age, children love to ride on farm equipment. Satisfying this desire is tremendously seductive. It makes for happy children who express themselves in shrieks of delight (if you give them a good rough ride) and warm gratitude. Enter the insurance grinch—no affordable policy exists that would protect your farm should an accident occur.

If you are going to welcome outsiders to your farm, you must make every effort to reduce hazards. Clean up junk piles; pick up rusty nails, old tines, pieces of equipment, and anything lying around that can poke, pinch, or cut your visitors. Install a few warning signs, such as Employees Only on the

131

equipment or tool shed. Distribute a map of the farm to CSA members with restricted areas clearly indicated. An insurance agent Robyn consulted said, "Forewarning members is your best protection."

Health Insurance

When you succeed in creating agriculture that is truly community supported, you won't need health insurance, because your community will bring forth the contacts or funds you need. Unfortunately, most of us have a way to go before we reach this level. In the meantime, the CSA must decide whether to factor health insurance for the farmers into the annual budget or leave it up to them to pay for themselves.

The legal structure of the CSA will determine how health insurance is handled. If your CSA is a sole proprietorship or a partnership, the farmer or farmers will have to purchase insurance for the self-employed. You may be able to find a group in your area that you can join, such as a company that allows small businesses to join together for group rates. If the CSA incorporates—as either a for-profit or nonprofit corporation—and hires the farmer as an employee, it should include health insurance as part of the benefits package. If the farmer is an independent contractor, responsibility falls back on the individual.

Food Stamps

Receiving authorization from the United States Department of Agriculture (USDA) to accept food stamps in payment for produce may help attract more low-income members to your CSA. Before they became CSAs, many farms that sold at farmers' markets or through farm stands had received this authorization, and when they became CSAs, they continued accepting food stamps without problems. However, the USDA has turned down a few CSAs that had included a full explanation of CSA (weekly shares, payment in advance, and so forth) with their application for food stamp authorization. In a 1994 policy memo, the USDA stated that CSAs are not eligible on the basis of collecting payment up front at the beginning of the season. CSAs that are nonprofits and deliver the food within two weeks of receiving payment remain eligible for food stamps, as are CSAs that operate like farm stands. If a CSA is willing to accept payments every two weeks, the USDA is more likely to grant authorization for food stamps.

Meanwhile, electronic benefit transfer (EBT) complicates small farm acceptance of food stamps. Instead of issuing food stamps, the USDA gives recipients a plastic card, much like a credit card, that they swipe through the appropriate electronic equipment at the cash register. For major food chains, which already had such receptors, EBT entailed only a minor retooling. For small farms, purchase of this equipment is expensive, though grant money has helped some farms. An alternative is for the CSA to associate with a farmers' market or a local nonprofit that is staffed and equipped for food stamps. If your farm receives food stamps only from time to time, you can register for a special account with JPMorgan Bank. With the arrangement, you do not need an EBT machine; instead, you have to call the bank for authorization each time you accept food stamps.

To apply for food stamp authorization, call or write your nearest regional USDA office for an application form. The agency usually takes a few weeks to process your application. The reauthorization process that USDA requires every few years is more elaborate than the original application. Reauthorization forms provide no category that fits a fresh market farm. The local representative responded to my queries by advising me to fill out the forms as best I could. Many moons later, JPMorgan sent me a mailing explaining the bank's voucher process. Once you receive authorization, include the food stamp payment option in advertising your CSA to encourage participation by lower-income people. To accommodate lower-income members, you will also have to allow more frequent payments. Food stamps are issued monthly, and the regulations stipulate that recipients can make advance purchase transactions at two-week intervals only.

———————————•———————————

Robyn concluded her notes on CSA legalities with some reflections on future growth and possible threats to the spirit of CSA:

> As I write, CSA is still a little-known alternative, but the concept has tremendous potential to engage masses of people. Genuine CSA, where members

have opportunities to meet the farmers and production crew, will be difficult for agri-business to co-opt. But like the word 'natural,' which we see plastered all over many things, natural and otherwise, the concept is vulnerable to the more money-minded to put their spin on it. Increasing the scale of CSAs will be necessary to feed the millions of people in urban centers. The challenge is to retain the integrity, quality, and underlying philosophy, as well as the community-building functions. I believe the answers will come; it is all unfolding as we go. In its grossest terms, CSA is still just a marketing technique. Each community is responsible to develop and perpetuate the more subtle values and benefits that are near and dear to many of us.

TO CERTIFY OR NOT TO CERTIFY?

Principle of Health
Organic Agriculture should sustain and enhance the health of soil,
plant, animal, human, and planet as one and indivisible.

Principle of Ecology
Organic Agriculture should be based on living ecological systems and
cycles, work with them, emulate them, and help sustain them.

Principle of Fairness
Organic Agriculture should build on relationships that ensure fairness
with regard to the common environment and life opportunities.

Principle of Care
Organic Agriculture should be managed in a precautionary and responsible manner to
protect the health and well-being of current and future generations and the environment.
—From "The Principles of Organic Agriculture," International Federation of
Organic Agriculture Movements (IFOAM) Basic Standards 2005

What the USDA National Organic Program (NOP) guarantees as organic is only a small portion of the full scope of organic agriculture, as articulated in the IFOAM Principles above. For thousands of people in this country and millions around the world family- scale organic farming is a way of life based on cooperation and harmony with nature, building more socially just and sustainable communities; producing healthy, safe, nutritious, and minimally processed food; reducing food miles, energy waste, and pollution; and trading on fair terms both with neighbors and with people in distant regions. Organic farming is part of the striving for peace among humans and between humans and all the other creatures of the earth above and below ground. The USDA "organic" seal, on the other hand, is a marketing tool, and thus the decision whether or not to certify one's farm as organic is primarily a business decision.

While organic certification is not a legal necessity for a CSA, if your farm is in fact organic or biodynamic, you should consider it carefully for several reasons. Certification will help recruit new members who are seeking organically grown food and do not yet know the CSA farmer personally. At its best, the certification process can be the occasion for an annual reflection on your farming practices and progress in building soil health and biodiversity, as well as for thorough record keeping. Farmer- or farmer/consumer–run associations sponsor many certification programs, so membership brings access to lively and informative newsletters and farmer networks. The negative aspects of certification can be the cost and the paperwork. Some farmers do not want to be regulated, even if they agree with the regulations.

A long twelve years after the passage of the Organic Foods Production Act (OFPA), the USDA, in October 2002, rolled out the NOP, which has transformed the landscape of the organic market. Whereas previously farmers had the choice of whether or not to call their farms organic, the NOP took control over the "organic" label. Today, if you want to call your produce organic, you must be certified by a USDA-accredited organic certification program. By all sources of statistics, that deprived a good half of all CSA farms of the organic label. Farmers reacted in different ways: Some certified their farms for the first time; some dropped their certification; still others joined Certified Naturally Grown, an alternative program. (See page 139 for details.) As far as I have been able to determine, no one has done a study to find out whether members really care if their CSA farm can use the government's "organic" seal.

Certified Organic

My partners and I revisit our decision to certify every year. Loyalty to the Northeast Organic Farming Association (NOFA) certification program after many years in the struggle to create the organic label is certainly a factor. My friendship with my partner Ammie dates back to meetings of the NOFA/Massachusetts (NOFA/MA) certification committee in 1984 or 1985. Together, we helped initiate the NOFA/MA organic certification program. My farm at that time, Unadilla, was one of the first to be certified. The process of creating NOFA's certification standards was highly participatory and exciting. Once our committee had a completed draft of the standards, we circulated them among members of the organic community—both farmers and consumers—and held a series of open meetings around the state. Anyone could come and argue for or against any of the standards, or propose additional ones. Participants expressed passionate views on

issues: Should farmers be allowed to use any botanical pesticides? Should the standards allow any chemical residues in the soil or on crops? Who should pay for testing? Should we include organic materials that had to be shipped thousands of miles, such as bat guano? Should we allow any sodium nitrate? A particularly fierce debate occurred over whether to allow farms to have both organic and conventional fields. (We finally agreed to allow "mixed" farms if their transition was a one-way path toward complete organic management.) The annual revision of the standards gave farmers and consumers a way to learn together and to improve our systems constantly. In the pre-NOP years of certification, I looked forward to the annual inspection as a great opportunity to walk my farm with someone from whom I could learn more about organic growing and marketing.

The NOP has changed a lot of that. The NOP accreditation has been very expensive in time and money for small nonprofit certification programs such as the NOFAs'. The NOP's interpretation of *conflict of interest* forced certified farmers off the boards that run these programs, despite carefully structured procedures that had kept farmers from making decisions on farms or handlers in which they had financial interests. The NOP put an end to the educational work of certification inspectors, turning them into regulators, pure and simple. And the codification of national organic standards eliminated the annual participatory revisions that gave a lot of life to the certification process.

Nevertheless, the standards for organic farm management in the OFPA are basically solid. These standards are based on those used by the so-called "patchwork" of certification programs that the NOP replaces. When organic farmers from around the country compared all of our programs back in 1989, we were surprised and pleased to discover that we were 90 percent in agreement and that the differences were largely procedural. Many hundreds of volunteer hours, the conscientious efforts of farmers and consumers thinking together, went into the formulation of these organic standards, going back thirty years to the original guidelines provided by IFOAM.

The wording of the OFPA section on materials, which came from the members of the Organic Farmers Associations Council (OFAC), and the work of Lynn Coody, in particular, articulates the experience of certification programs in their struggle to develop criteria for judging what kinds of materials have a rightful place in organic agriculture. If respected in the implementation of the NOP, these criteria will ensure that accredited organic certification programs will approve only ecologically sound materials. The Organic Materials Review Institute (OMRI), established in 1997 with the support of most of the existing private certification programs, uses these criteria as the basis for its materials list.

The NOFA-NY certification program has kept farmer involvement alive through a farmer advisory committee, and the staff is truly responsive to farmer concerns. Even though the NOP considers NOFA's limited liability corporation its agent, the program still *feels* like it belongs to us. Despite what Certified Naturally Grown proclaims on its Web site, the paperwork is less burdensome now than it was before the NOP existed, and the program has worked on reporting systems that are time efficient with farms such as ours that produce hundreds of varieties of crops.

Debate continues to rage nationally on whether the NOP has destroyed the integrity of the organic label. There is no doubt in my mind that USDA's habitual role as the Department of Agribusiness and the pressures of greedy Big Food corporations could force this to happen. Had certification been left up to the USDA, the degeneration might have been instantaneous. The OFPA, however, set up a "public/private partnership," leaving certification in the hands of the existing certifiers. Most of these programs survive and function in a professional and conscientious fashion. Admittedly, there are a few upstart certifiers with dubious qualifications. In its role as accreditor, the USDA should ferret these out, but its failure to perform evenhanded and high-quality accreditation is one of the most serious problems. Fortunately, the organic movement and its many allies among nonprofit consumer, environmental, and faith groups have thus far prevented serious lowering of the pro-

duction standards. Watch-dogging the USDA has required constant vigilance, but as Wendell Phillips said back in 1852, "Eternal vigilance is the price of liberty."[1] The outcome of this struggle is highly political and will depend on the amount of public involvement and the ability of the organic movement and the organic industry to maintain a unified front. In the 2004 campaigns to keep the provision for 100 percent organic feed and to cancel out NOP's "directives," this unity has held. An irrevocable crack opened in 2005 with the change to the OFPA engineered by the "pay-to-play" members of the Organic Trade Association who were determined to keep synthetics in processed foods. For the time being, we continue to certify Peacework Farm.

Under the OFPA, the basic certification document is an organic farm plan, recorded in the initial application form. Created by the farmer, the plan describes soil and livestock management, pest and disease control, and use of water resources and sets out goals for continual improvement in these areas. If a farmer has never gone through certification before, filling out the application is a rigorous and time-consuming process. It probably takes a good two weeks of steady work to amass the necessary documentation of rotations and cropping field by field, listing all materials used for fertility and pest control, verifying the farm's borders for buffer zones with neighbors who use chemicals, and so forth. The farm must maintain records of all harvested crops, an audit trail from seed to consumer, for the inspector to review. In subsequent years, the update is a lot simpler.

Depending on where you farm, certification can be an expensive proposition. Private programs charge a basic fee, inspection expenses, and a percentage of sales. With federal funding, many states provide cost share of 70 percent of the annual certification fee up to $500. Tax dollars support a few state-run programs. The first round of USDA accreditation of certification programs was free; the second round will be very expensive. In the winter of 2007, the NOP sent the certifiers bills for $15,000 to $20,000—and that will surely affect fees. A CSA may well decide that the expense is too high for the farm budget, especially in the first years.

Tim Laird writes that organic certification was an issue for both of the CSAs for which he farmed. The Intervale Community Farm considered certification crucial for recruiting the fifty or so new members it needed every year. Quail Hill Community Farm decided to certify only after a few years of operation. These are the reasons Laird lists:

- We wanted to support NOFA-NY financially beyond membership dues. As a CSA we added the certification to our budget, so the community as a whole, and with some consciousness, knows that by joining the CSA they are supporting not only our farm, but the movement of organic agriculture as well.
- As farmers, we wanted to feel more a part of the organic farming community. Becoming a certified farm is a natural means to achieve this.
- We wanted to draw on technical support that NOFA-NY offers.[2]

Long-term planning and careful record keeping are part of the orderly management of a CSA farm, whether certified or not.

The following records are useful to keep:

- seed and plant purchases
- purchases of soil amendments, pesticides (including botanicals, purchased beneficial insects, pheromones, etc.), farming supplies
- date, amount, and purpose of pesticide applications
- dates of equipment purchases, maintenance, and repairs
- livestock and feed purchases and sales
- date, amount, and purpose of livestock health treatments
- field maps showing rotations, soil amendments, date of plantings, date and amount of harvest
- regular soil tests
- use of water, water tests
- compost-making ingredients, dates of pile creation and use

Because the certification process demands this kind of detailed record keeping, belonging to a certification program can be a useful prod to complete paperwork that not all farmers love to do. Joining a certification program by becoming a member of a farmer advisory committee or a review board is an excellent way to learn more about organic practices. Most certification programs go through an annual review: Participation ensures that the particular needs of CSA farms will be taken into account.

Certified No More

We don't have statistics on the number of organic farms that dropped their certification when the NOP was implemented in 2002. A few CSA farmers have told me that they left the certification program because they did not think they could afford to switch to certified organic seed or they could not make compost according to the NOP regulations and did not want to lie about their practices. Michael Ableman, who had never certified Fairview Gardens, prominently declared his intention to go "beyond organic." After 20 years in organic agriculture, John Gorzynski of Gorzynski Organic Farm in New York made what for him was a very painful decision to drop certification once the federal program went into effect. John said:

The NOP degraded what the word *organic* had always meant. By allowing synthetics like streptomycin and tetracyclin for fire blight on apple and pear trees, the program could continue to be degraded with the inclusion of more synthetics. Allowing broad-spectrum antibiotics to be sprayed on trees, exposing the whole environment, air and water, is totally ludicrous under the word *organic*. To me, it is as bad as GMOs and sewage sludge. I had been enthusiastic about having a national program before I saw the regulations. I did not see how we could save the meaning of the label. I wanted to dissociate myself from it. To throw twenty years of my life away to make organic mean something—it was not an idle decision. It really ripped me apart.

137

This next question was how close I would remain with the organic community, NOFA and all my friends. I decided that the real organic still exists. There is a discrimination thing going on in my mind all the time—are you the corporate organic that just wants to make money off it, or are you one of the people who put all the effort into it all these years?

John subsequently renamed his farm Gorzynski Ornery Farm.

Participatory Alternatives to Certification

The elaborate apparatus of certification and accreditation that has emerged over the past thirty years is adapted to the globalizing marketplace. Farms that are selling directly to customers, whether individuals, restaurants, or local stores, do not need a third party to certify their practices. Besides that, most of the organic farms in the world are too small to afford certification. These farms provide for the subsistence of the farm family and may have some excess to sell for cash. Around the globe, organic agriculture groups of varying kinds have been developing programs that are better suited to these small, direct-sales farms. While each alternative program is unique and specific to its social context, they all have core principles based on sustainable, ecological practices, social justice, equity, and gender balance. Most projects have some documentation, but the touchstone for the organic guarantee is farmer integrity and trust between farmers and their customers. Shared goals are to empower even the poorest and smallest-scale farmers to become active contributors to and beneficiaries of local sustainable development and to offer continuous education to farmers and other stakeholders in the system. Paperwork and expenses are kept to a minimum so that even illiterate farmers can participate. The focus is on direct sales through farmers' markets, co-ops, CSAs, *teikei*, farm stands, or farmer-controlled stores.

There are some inspiring examples. Over the past five years, in the south of Brazil, the Rede Ecovida de Agroecologica has spread its decentralized network to producer associations, including over ten thousand producers, consumer cooperatives, and agricultural technicians. Instead of inspections and reports, Rede Ecovida asserts that its organic guarantee is the result of trust "generated by a huge process that begins as increased awareness of the producer (farmer, processor) related to produce without destroying, as in nature." The Ecovida Network has formed twenty-one nuclei, each consisting of six to eight local groups, which include farmers' associations, nonprofits, and consumer co-ops. Each group works in its own way to improve farmer and consumer awareness and technical facility in organic production and to build local markets. Instead of inspections, groups, which include other farmers, a technician, and some co-op members, walk around the farm and discuss progress, problems, and new discoveries with the farmer.

Stakeholders in the organic sector in New Zealand formed OrganicFarmNZ to provide a credible guarantee for the smallholders for whom existing certification was too expensive. After only two years, two hundred of the two thousand smallest farms had organized into twelve regional groups, or "pods," of seven to forty farmers who inspect one another's farms. They use a recognized national organic standard. Elected regional groups administer the program, which has the support of the Soil and Health Association and the New Zealand government. By working in groups, the farmers learn from one another and form relationships of trust that lead to joint marketing and purchasing as well as providing an organic guarantee.

Since 1992 the Institute for Integrated Rural Development (IIRD) in Aurangabad, Maharashtra State, India, has set up organic bazaars where 1,250 organic farmers sell their produce. Forty-five eco-volunteers, women who have completed a three-year training program in organic farming, visit the farms once a month to disseminate technical information and assure adherence to organic methods. Strengthening the social position of female farmers is a major focus of the IIRD program.

In Uganda, where traditional farmers still use basically organic practices, membership in the National Organic Agricultural Movement of Uganda (NOGAMU) has spread rapidly to eighty-two associations involving twenty-five thousand farmers. Farms that join NOGAMU can market their products as organic as soon as a local marketing officer inspects them for compliance with the national organic standards. The system is informal, with no documentation, but it has earned a high level of trust from Kampala shoppers at the NOGAMU retail outlets.

Disgruntled with the complexities of organic certification, a new CSA farmer in the state of New York, Ron Khosla had made an effort in the late nineties to revive the Natural Foods Association (NFA) certification program. With the implementation of the NOP, he realized that the NFA did not have the resources to qualify for accreditation. As he tells it, he was in a panic at the thought that his farm and those of friends and neighbors would no longer be allowed to use the organic label. In the summer of 2002, a small group of these farmers decided to create a new label—Certified Naturally Grown (CNG).

Using his computer skills, Ron fashioned a Web site–based program with simple procedures. To qualify for the free trademarked label, farmers fill out an application and a declaration of understanding. Then another farmer inspects their farm and fills out a short inspection form and a check sheet. Each participating farm commits to doing one inspection. The application, the declaration, and the inspection check sheet for each farm are posted on the Web site. As standards, CNG took the basic clauses from the National Organic Program and added a requirement for spot checks for chemical residues paid for by the program. There is no mandatory fee; the program invites "free-will donations" from farmer members. Alice Varron, the administrator, tells me that farmer payments average $54 each.

CNG struck a responsive chord; within a year, over two hundred farms all over the country had signed up. The number has continued to grow; by the fall of 2006, CNG had accepted over five hundred farms and posted their applications on the CNG Web site. Ron Khosla explains the value of the CNG program:

> The CNG program encourages and supports LOCAL un-certified organic farmers who would not otherwise participate in any sort of certification program, giving them a low or no-cost (and legal) eco-label for their produce now that they are no longer allowed to call themselves "Organic." Having a publicly available and acceptable set of standards associated with the label also keeps farmers "honest" (their words!) in terms of growing practices. When you publicly agree to do something . . . and are "checked on" by your neighbors, you are less likely to want to slip. . . . The CNG program also turns off a lot of the negative energy that was getting focused on the USDA NOP among small farmers. Local farmers that joined the CNG program tend NOT to trash the USDA program so much anymore . . . they don't have to feel the same level of frustration and anger towards the "government and industry" that took away a term that they had been using for 25-plus years in some cases.[3]

Owing to its extremely rapid growth and lack of staff and funding, CNG has not had the resources to ensure farmer-to-farmer inspections or to find consumer groups to inspect isolated farms. In March 2007 less than half of the farm listings on the Web site included inspection forms.

The Fredericksburg Area CSA Project, a cooperative venture uniting five farms, is a good example of the farms that have signed on with CNG. It provides shares to three hundred families in their part of Virginia. In 2006 it celebrated its tenth year. Among the five are two larger family farms, Canning Farm and Mount Vernon Farm; an oversized family garden, Rock Run Creek Farm; a one-acre plot where Heidi Lewis has made a home for Summerbeam Garden, growing vegetables, herbs, and flowers on land belonging to the Eitt family; and a community green space, Downtown Gardens, a nonprofit devoted to education for local youth. A grain, cattle and timber operation, Canning Farm dedicates six acres to the CSA. Mount Vernon Farm supplies the group with grassfed beef, lamb, and pork. Rock Run Creek Farm's main purpose is to feed the Ngohs' four sons; the overflow, and a specialty crop of shiitake mushrooms, goes to the CSA. Its brochure states that all its growers operate under the guidelines of CNG, but only Summerbeam appears on the CNG Web site.

The same realization—that the NOP might damage the livelihoods of the half of its organic farmer members who never certified—led NOFA-NY to provide an alternative to its own USDA-accredited certification program: the Farmer's Pledge. I was one of the small group of organic farmers who composed this pledge. My main motivation was my desire to call attention to the full range of values that most organic farmers and their customers share that go way beyond the NOP. Unlike CNG, the Farmer's Pledge has no inspection at all. Pledge signers agree to allow customers to visit their farms once a year, but the basis for the pledge is the personal integrity of the farmer. Peacework and other CSA farms, such as Roxbury, Whistle Stop Gardens, and Sisters Hill Farm in New York have signed the pledge. Other NOFA chapters are providing this service to their members.

Each pledge signatory receives a laminated version to display at the farm or farmers' markets. The list of farms adopting the pledge is posted on the NOFA-NY Web site and is published in an annual listing of organic farms in the state. The pledge has been published in the NOFA-NY newsletter and in *The Natural Farmer*, the NOFA regional newspaper covering seven states. Each farm pays $50 for the laminated copy and for the listings on the Web site and in the booklet. Since the pledge brings no monetary premium, it seems less likely that anyone will cheat.

The great strength of the pledge system lies in the breadth of values it encompasses. NOFA hopes it will help people recognize the numbers of small farmers who have a strong ecological approach to farming, who are treating and paying their workers in a socially responsible way, and who are committed to once again making farming an integral part of their communities. With additional publicity,

the pledge will help small-scale organic farmers distinguish their products in the marketplace from industrial-scale organic food, so that informed consumers can vote with their dollars for the kind of farms they prefer.

Neither CNG nor the Farmer's Pledge has gone beyond the first rungs of the ladder to participatory organic guarantee systems. But their rapid adoption by farmers, and the willingness of food co-ops and farmers' markets to display these labels, indicates that there is room in the marketplace for additional values.

Since 1999, I have been working with Michael Sligh of the Rural Advancement Foundation International (RAFI-USA) and Richard Mandelbaum of CATA (Farmworkers Support Committee) on creating social stewardship standards for organic agriculture. These standards are based on the complementary principles of fair pricing for the farmer and just working conditions for farm and co-op workers resulting in a win/win/win scenario in which workers, farmers, buyers, and ultimately consumers all benefit. The process aims for transparency and full participation by farmworkers, farmers, store workers, and buyers. The long-term goal of the Agricultural Justice Project (AJP) is to transform the existing unjust food system. We envision a food system that is based on thriving, ecological family-scale farms that provide well-being for farmers and dignified work for wage laborers and that distributes its benefits fairly throughout the food chain, from seed to table. As a first small step toward this ambitious goal, in the 2007 season, the AJP implemented a pilot project for a social justice label, Local Fair Trade, in the Upper Midwest to allow family-scale farms in their trade with food cooperatives to differentiate their products from industrialized organic. It will be interesting to see how consumers respond.

In 2006 the AJP, Organic Valley, Equal Exchange, and other organizations launched a national Domestic Fair Trade Association. The rising membership in CSA farms shows that there are people in the United States who are willing to trade fairly for their food.

Certified or not, CSA farms benefit from

The 2008 FARMER'S PLEDGE © ™

Knowing your farmer is the best assurance that the food you buy is responsibly grown; grown with methods that recognize the inherent implications of the web of life in all our individual actions. **Northeast Organic Farming Association of New York (NOFA-NY)** believes that farmers should work in harmony with natural forces and leave the little piece of the world over which they have stewardship in better condition than when they found it.

To further enable consumers to identify the farms they want to support with their food dollars, NOFA-NY has established a Farmer's Pledge, separate and distinct from USDA Certified Organic. Farmers and market gardeners who adhere to the following pledge have signed an affidavit which they display for customers and neighbors to view.

This pledge is based on the integrity of the farmer/gardener. Those who sign this pledge agree that consumers may inspect, by appointment, their farm/garden to judge the truthfulness of this statement. NOFA-NY does not investigate or make any guarantee that the individual farmer is complying with the Farmer's Pledge. This pledge arises from the expressed need of growers who have a fundamental disagreement with the usurpation and control of the word "organic" by the USDA, and those farmers who want to pledge to an additional philosophical statement about their growing practices.

WE PLEDGE THAT IN OUR FARMING, PROCESSING, AND MARKETING WE WILL:

Build and maintain healthy soils by applying farming practices that include rotating crops annually,
    using compost, cover crops, green manures, and reducing tillage;
serve the health of soil, people and nature by rejecting the use of synthetic insecticides, herbicides, fungicides,
    and fertilizers;
reject the use of GMOs, chemically treated seeds, synthetic toxic materials, irradiation, and sewage sludge
    in our farming, and all synthetic substances in post harvest handling;
treat livestock humanely by providing pasture for ruminants, access to outdoors and fresh air for all livestock,
    banning cruel alterations, and using no hormones, GMOs or antibiotics in feed;
handle raw manure and soil amendments with care;
support agricultural markets and infrastructures that enable small farms to thrive;
conserve natural resources including the atmosphere and climate, by reducing erosion and pollution of air, soil
    and water through responsible farming practices;
maximize the nutritional value of food and feed by practicing careful post harvest handling;
practice minimal processing for all food products to preserve the natural nutritional value of food: NO use of
    irradiation, ultra-pasteurization, excessive heat, synthetic preservatives, or GMO processing agents or
    additives and include all ingredients on labels;
ensure food safety by using potable water for washing crops;
reduce the ecological footprint of farms and homes by limiting energy use and converting to renewable
    sources of energy;
reduce food miles by selling produce locally and regionally;
create beneficial habitat for wildlife and encourage biodiversity;
help preserve farmland;
share and develop farming skills and know-how;
use ethical business practices;
pay a living wage to all farm workers and acknowledge their freedom of association and their right to
    collective bargaining;
treat family members and farm workers with respect, and ensure their safety on the farm;
work in cooperation with other farmers and with the neighboring community to create a more sustainable way of life;
encourage the distribution of unsold but edible food to people who need it;
sustain the land in healthy condition for future generations.

NORTHEAST ORGANIC FARMING ASSOCIATION OF NEW YORK, INC.
PO BOX 880, COBLESKILL, NY 12043 • 607-652-NOFA • WWW.NOFANY.ORG

belonging to networks with other organic farms. In many parts of the country, organic farming associations, such as the NOFAs, Maine Organic Farming and Gardening Association (MOFGA), Washington and Oregon Tilth, Carolina Farm Stewardship Association, and the Biodynamic Farming and Gardening Association, which initiated certification, also serve as educational networks for farmers, providing workshops, conferences, newsletters, and contacts with other growers. In the absence of a supportive land grant university, extension office, or other government programs, these farming associations have been the primary sources of information on organic farming.

Each CSA will have to make its own decision about organic certification, weighing the advantages and disadvantages. When we reach that ideal state in the marketplace when the label on every food item details where and how it was produced, all the materials used by the producers, and how much energy was expended in production, and when every buyer is sophisticated enough to read and understand those labels, we won't need certification anymore. We have a way to go yet . . .

CHAPTER 11 NOTES

1. Wendell Phillips (speech, January 28, 1852), *Speeches before the Massachusetts Anti-Slavery Society* (Boston: R. F. Wollcott) 13. These words have been attributed to Thomas Jefferson but are not found in his writings. Some attribute them to Patrick Henry.
2. Timothy J. Laird, "Community Supported Agriculture: A Study of an Emerging Agricultural Alternative" (master's thesis, University of Vermont, 1995), 27–28.
3. Ron Khosla (presentation, Workshop Reader for Alternative Certification Workshop, sponsored by IFOAM and Movimiento Agro-ecologico de Latina America et el Caribe [MAELA]), April 2004, Torres, Brazil.

COMMUNITY AND COMMUNICATIONS

Community is formed by people who are acting in cooperation with one another.
—JEANNETTE ARMSTRONG, "Sharing One Skin"

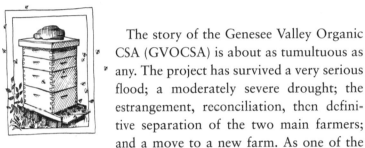

The brief history of Community Supported Agriculture (CSA) in this country dates back only to Indian Line Farm's Apple Shares in 1985. We are still in the pioneering phase: Anyone who ventures onto this frontier should expect to share the risk. CSAs that are nineteen, thirteen, or even nine years old are the venerated elders of this movement. What is the glue that holds them together? Despite traumatic upheavals—moves to new pieces of ground, irreconcilable splits within core groups, divorces, and financial crises—many of the early CSAs are holding together and moving forward. One common trait is their unabashed readiness and capacity to borrow and apply one another's best inventions. Their example and the power of the CSA concept are attracting new people with exciting ideas. What will give these new CSAs the persistence of witchgrass rhizomes or make them spread as irrepressibly as galinsoga weed?

Surprisingly, as I look over the history of these twenty years of CSA, no particular scale of operations or distance from members emerges as the ideal for forging community ties. Members have demonstrated passionate loyalty in both action and words to farms that follow the community farm model and to subscription farms; to farms that ship shares over three hours by truck, well beyond the "localvores," one-hundred-mile limit; and to CSAs with over two thousand shares. What seems to count most is not the size or miles but the vividness of the shared experience.

The story of the Genesee Valley Organic CSA (GVOCSA) is about as tumultuous as any. The project has survived a very serious flood; a moderately severe drought; the estrangement, reconciliation, then definitive separation of the two main farmers; and a move to a new farm. As one of the farmers involved, I appreciate most the way CSA softens the business side of farming. Selling vegetables directly to people I know shelters the farm and me from the impersonality and indifference of the global marketplace. Supermarket produce managers I have encountered care more about saving a penny on a head of lettuce than about what happens to my farm. CSA members, on the other hand, write us "vegetable love letters," such as this one from Colleen Fogarty: "How often I think of you farming the land at Rose Valley with awe, admiration, and profound gratitude. At the first distribution this year, I was elated to have fresh greens—lettuce, spinach, arugula—with two colors of radishes. It was such a cold spring, it seemed nothing would ever grow, and the miracle of those first vegetables stayed with me many weeks. More words couldn't really convey my thanks to you for coaxing such wonderful food from the ground with such love and care." These words counterbalance a lot of aching muscles and mosquito bites. The quality of my relationship with the members moved me to want to stay with the CSA even when I had to leave Rose Valley Farm.

As I peruse end-of-the-year surveys, I find two main attractions have helped keep GVOCSA members: the quality of the vegetables and the farmwork. These are some typical quotes from the surveys:

"The quality was excellent! Once I got the hang of it, we ate almost all of it every week and the selection was great. It taught me to cook with things I wouldn't have tried."

"I enjoyed working at the farm—especially the variety of jobs you have to do. I enjoy the warm atmosphere and especially that the children are welcomed."

"Being in a CSA reminds us of how much work and energy go into producing the food we eat. Working at the farm, even if only once a year, helps us not to forget the source."

To the question, "Is the CSA concept worth the trouble?":

"Yes, it is significant to be involved—even a little—with growing food in a healthy way and seeing it through from the ground to the kitchen. I am grateful to have the opportunity. And especially grateful to the core group for their efforts."

To the question, "What did you dislike the most?" one member answered: "Getting up early in the morning and working four hours outside." To the question, "What did you like the most?" the same member responded: "Getting up early in the morning and working four hours outside."

When I asked a dozen of the longest-term members why they have made CSA part of their lives, they always start with the vegetables, but then go on to deeper aspects of the work, the community, agriculture, and food economics.

Marian Vaeth, who joined in 1990, said, "I want to support organic agriculture. I love the produce and the people who are in the program. One year I wasn't a member, when my husband was sick—I missed the produce and I missed the people."

Suzanne Wheatcraft, an active core member since 1990, said,

I believe in some of the things that GVOCSA does that maybe the other CSAs don't. For example, you should not be able to buy your way out of farmwork. It's an important connection to what we are all about (this is the big we—humans). We eat food. Food grows in the ground. Farming makes that happen. Also, I like that pickup is always already on my calendar every week. I like that I don't spend a lot of time making food choices. I like that I am not trying to get the lowest price possible from some poor farmer sitting at the farmers' market on a cold, rainy Saturday morning trying to just get rid of all the produce they already picked, packed and transported, because there are no customers that week. I hate to spend money where I see no value—I never buy new clothes, I don't go to the mall, I re-use my plastic bags. But, I do believe that there is a big difference between price and value. The CSA offers great value!

Fred Miller, also a 1990 recruit, had cool things to say about some of the other members, but he stayed for the food: "I like the idea of going out on the farm and helping grow the food. I didn't have a place of my own to farm. Farmwork is good clean work on an organic farm. No problem with pesticides, herbicides, and I like the farmers too."

Brenda Mueller, a founding member, explained that she has multiple chemical sensitivities and needs to eat natural foods, free of pesticides: "The CSA and you have been flexible, finding work for me that I could do. Also, it is convenient to get all my vegetables in one place, even with the limitations of time."

Eloise Schrag, class of 1990, said, "A big part is my commitment to farmers. Both Dennis and I grew up on farms. Our parents were farmers. Having children, I like the idea of organic vegetables. I grew up gardening and canning and preserving food. I like the quality and the taste. And I like paying the farmers directly."

GVOCSA newsletter editor Rick Griggs wrote: "I continue with the GVOCSA and in the Core because it serves as both touchstone and creative release. Being part of an active expression of compassion for our mother Earth and her capability to sustain us helps keep me grounded in a callused world that takes so much for granted. Associating with others possessing a diversity of thoughts, experiences, and attitudes counterbalances my native linear, logical, business-oriented nature, rounding off the hard edges that life can hone, coaxing expression from the safe, deep silence in which it must reside."

Retaining Members

The GVOCSA is not strong on convenience or service to members. The work requirement has certainly discouraged some people from joining. When new recruits offer to pay more instead of working, we send them to the Porters' CSA across town. Tight parking and a two-hour pickup interval create the most discontent. Especially for people reluctant to ask for help, the necessity of being at

the same place at the same hour every week to get the vegetables is a constraint. The core does an annual phone survey of members who quit and tries to make adjustments. These minor inconveniences do not rank high in the reasons for leaving, though one former member gave the pickup time as her reason for switching to another CSA. The main causes of member attrition are:

- Moving out of town
- Having another baby
- Starting own garden
- Summer plans—long period of travel or frequent trips out of town
- Prefer to choose own menu
- Too much food

David Inglis reports that he counts on losing 15 percent of his members every year because of the local population's transience. He added divorce to my list of top reasons. Jennifer Bokaer-Smith's former members gave similar answers. She also had a few members join

---

The Tornado at Wolseley Community Garden

On Friday, I experienced my first tornado—not an experience I care to repeat! It touched down on the east side of the garden and took out a swath of eight or so trees. It touched down again in a large field just beyond my yard and flattened a perfectly round area. If you didn't know otherwise, you would think a million-pound space ship had landed there. It touched down again in my neighbor's yard and pulled up a few of her oak trees. Monsoon-type rains accompanied it. It was a complete whiteout, but I couldn't hear the rain because the wind was making jet-plane-landing noises. Of course, you don't get a storm like that without a little bit of HAIL thrown in. There wasn't a great deal of it, but the stones were large.

All this happened in about thirty minutes Friday night. I was sure I was going to wake up Saturday to total devastation. I went to sleep composing a letter to all of you. It went something like this: "Dear Members: I know $250 for one bag of produce is a bit steep but . . ."

On Saturday I had to get up at 5 A.M., load up and head for the St. Norbert Farmers' Market. I could not bring myself to even look at the garden. When I got home I had a power nap and headed over to a friend's house for dinner. When I returned, I couldn't stop myself from going out to the garden. It was a nightmare! Everything was smashed into the ground and covered with mud. I started working on the letter to you again.

Sunday was spent in a depressed haze. I tried not to look at the garden. On Monday, my forty-eighth birthday, I woke up to a gentle rain. It continued for several hours and then the sun came out. I took a walk in the garden and, although the plants had plenty of holes blasted through them, they were standing up again with clean and shining faces.

You will notice a lot of holes in your greens. Just eat around them.

—*Sandra Conway, from Wolsley Community Garden Newspage, Gardenton, Manitoba, July 4, 1996. Used by permission.*

Living Close to the Ground

I am on my knees again. It will be like this until the frost. After we finish the planting, we fall to our knees, pulling weeds and grass, thinning carrots, hilling potatoes, picking peas and beans. Some mornings, I walk down to the garden, and Rose Ann has disappeared among the rows. I know she is there, because her car is parked by the garage. But she is submerged in the green, until I see that familiar hat brim rise above the leaves.

The world is a different place, on my knees. It is smaller, and slower. My vista is all of twelve inches as I look hard to distinguish a mature snap pea from the curled new leaves on the vine. I do not, I cannot, look down the hundred-foot row, or I will lose sight of these peas ready to be picked.

On my knees, I meet the tiny green grasshopper who by fall will be brown and angular, as long as my index finger. The small Trichogramma wasp tickles the hair on my wrist as she flies to the nearest flower. I am happy to greet her because she will devour the aphids and deposit her eggs in leaf-eating caterpillars. I disturb the worm who works hard to break through our clay soil. I see the grass blades shiver, and glimpse the brown and black bull snake slithering away, as silently as he can.

As I inch my way down the row, my knees are saved by the bright blue foam pad that I kneel on, but my left hand gets raw as I lean on it each time to shift my weight and push along. I can feel how the earth changes every ten feet or so. It is soft and loamy here, and my hand is grateful, but the dirt is like rocks there, so I must put on my left-hand glove to save my skin. In this section are the thick, dried stems of the winter squash, in that section smooth river rocks and small chunks of quartz left behind centuries ago when the last glacier left.

On my knees, I begin to see with my hands. I can feel the plump stems of the foxtail grass and, without looking, distinguish it from the rounder, smoother stem of the dill. Lamb's-quarters are hard and square, pigweed fat. I've pulled so many of these stems, that all I need do is reach into a clump and feel their texture and shape.

As I pick the shelling peas, I lift the vines that have fallen over, and the coolness under the vines gives me relief from the burning sun. The soil is still wet under the cabbage leaves, while the rows open to the sun are dry and dusty. I feel the wind shift, and the air becomes cooler. Rain coming on. When I'm on my knees, the grass bends over my shoulders and protects me from the rain.

In 1764, Hargreaves invented the Spinning Jenny, making it possible to spin thread twenty-four times faster than with the conventional spinning wheel. In 1890, Frederick W. Taylor systematized work and production at Bethlehem Steel. Taylor was working at "The Steel," as it is called in the Lehigh Valley, as a consultant. He studied the work of the men who shoveled coal into the numerous blast furnaces. He redesigned the yard to save steps, made use of shovels of different sizes and shapes for different materials, and planned the work in advance. These changes made it possible for 140 men to do the work of 400 to 600 men. Taylor saved The Steel an average of $78,000 a year.

These two events—the invention of a machine and the invention of industrial management—are generally considered milestones of industrial progress. They are lauded in a culture that values above all else efficiency, speed, and profit.

It is a different world here, down on my knees. If I went any faster, I would miss the worm, and overlook the pea. I would lose touch with the ground I walk on, or crawl on, as the case may be.

—*Olivia Frey, Valley Creek Community Farm, Northfield, Minnesota. Used by permission.*

in order to make major changes in their lives through cleaner eating, who quit when they failed.

In 1998, David Trumble of Good Earth Farm in Weare, New Hampshire, calculated that over the previous six years his eighty-family CSA had a retention rate of 85 to 90 percent, which means that 50 percent of his customers were still with him, if you accept his math. David stressed the importance of keeping members:

> The issue of retention rate is key to the long-term economic sustainability of a CSA farm. For example, a 90 percent annual retention rate over five years yields a total 59 percent customer retention (90 percent times 90 percent times 90 percent times 90 percent times 90 percent = 59 percent). An 80 percent annual retention rate over five years yields a total 33 percent customer retention. A 70 percent annual retention rate over five years yields only a 17 percent total customer retention. At first, 80 percent sounds pretty good; but if in five years you only have one in three of your original customers, you could be facing long-term problems.

Our experience with the GVOCSA is more complex. Even though the number of households has grown to over three hundred, we have an easier time each year reaching our membership goal. The waiting list gets longer. At the same time, there is a large turnover rate, as high as 30 percent some years. But most of that turnover is among the newer members. We have a solid core of about one hundred households who have been with us for over ten years. I have also noticed that people will drop out because of complications in their lives—a long trip, a new baby—but then return a year or two later. Colleen Fogarty, quoted above, moved to Boston. After four or five years, she moved back to Rochester. Before she even had a house to live in, she had rejoined the GVOCSA. In our eighteenth season, Jamie Whitbeck, who had burned himself out on coordination our first year, joined again, with many apologies for the sixteen-year absence.

The more clearly people understand what they are getting into when they join a CSA, the greater the

likelihood that they will not be disappointed. In 1996, Deborah J. Kane and Luanne Lohr performed a study on shareholder retention, "Maximizing Shareholder Retention in Southeastern CSAs: A Step toward Long-Term Stability,"[1] which sheds some useful light on member expectations. The authors did telephone interviews with new sharers in seven different CSAs before the beginning of the season and again at the end. The interviewers found that "the tone of the fall conversations was considerably more subdued than the spring conversations." The contrast between the spring and fall remarks highlights the kind of misunderstandings that can occur:

Spring
"I'm looking forward to working with new vegetables that I've never eaten before."

"I'm assuming that these will be pretty much the only vegetables I buy."

Fall
"When I said I wanted variety, I really meant within the things I was used to eating."

"I never got a wide enough variety to really keep me from having to go to the grocery store."

The percentage of those interviewed who answered that they would pay more for the shares dropped from 66 percent in the spring to 39 percent in the fall, when 23 percent said they would only pay less. Yet in response to the Kane and Lohr questionnaire, of 196 respondents, 121 said they would rejoin, for an overall retention rate among the seven CSAs of 63 percent. Individual CSAs fared better, ranging up to 72.5 percent for New Town Farms, which members also rated highly for quality, quantity, freshness, and appropriate pricing of produce. Kane and Lohr conclude with a list of recommendations for increasing CSA retention rates:

- Offer rebates to members who refer new members

- Get sharers to do the recruiting
- Ask sharers to have friends or relatives pick up their shares when they are out of town
- Use T-shirts or tote bags with the CSA logo for free advertising
- Target local environmental groups, congregations, or businesses in recruiting
- Collaborate with other farms
- Offer U-pick or some form of choice in the produce
- Emphasize the basic crops, with exotics as the occasional treat
- Extend the season
- Offer value-added products
- Offer trial periods
- Set up a buddy system, pairing experienced members with new ones
- Communicate through meetings, resource booklets, crop lists, calendars, newsletters, cookbooks, and marketing materials
- Ask members what they want through voting sheets, surveys, suggestion boards, and boxes

I would add to their list:

- Offer members choices of as many different levels and ways to get involved with the CSA as possible: from once-a-year farm visits to core group membership to regular farmwork
- Try to involve members in setting long-term goals for the farm
- Include a good question-and-answer sheet about CSA in general and yours in particular
- Offer bulk orders of produce from the farm to freeze, can, or otherwise preserve with directions on how to do it
- Offer educational programs, farm tours, or on-farm activities for member children
- Welcome members to the farm for strolls, camping, skiing, and the like.
- Put up a decent Web site with all the information people need to be members of your CSA, and an e-mail List-serv for quick communications.

In a *New Farm* article, Linda Halley, who had farmed at Harmony Valley in Wisconsin for fourteen years, also suggested:

- every other week shares—much more convenient than two different share sizes
- flex plan—members pick the 15 weeks they want and you charge more because of the higher cost of keeping track
- pre-paid cards for use at farmers' market stand or on farm market—$100 worth of veggies for $95.[2]

The more sophisticated software available these days makes share flexibility easier for farmers who want to try it.

After sixteen years' experience chatting with members of the GVOCSA while overseeing distribution, Marianne Simmons has developed a theory about why some people just don't fit. She has observed a crucial distinction between what she calls deductive and inductive cooks. The deductive cook starts a meal by planning—looking up recipes, checking the kitchen for stocks on hand—and then goes out and buys ingredients. Unless these deductive cooks are willing to do their planning on the night they get their vegetables, Marianne explains, the uncertainties of CSA supply will make them very unhappy. They might last a season or two as members or even drop out after the first month. The inductive cook, by contrast, "flings open the refrigerator one hour before dinner, rejoices at the abundance, puts three-quarters of the meal on the table raw, and is happy that this is possible without stepping a foot in a supermarket." Marianne describes herself and her CSA friends as cooks of this kind and says they are content to let someone else give them a few surprises in their meals. Recruiting new members by word of mouth seems to work best, Marianne believes, because friends and colleagues have a chance to sort themselves out in advance, since they know what to expect. No advertising can be as effective. People who eat out often have trouble remaining in a CSA because they don't do enough cooking.

Marianne believes promoting return membership

Dancing around the Maypole at the annual May Day Celebration at Peacework Farm. PHOTO BY CRAIG DILGER.

starts with quality produce and evolves from there. The CSA made it easier for her to get the organic vegetables she wanted to feed her two sons. The local food co-op did not provide a reliable supply at predictable hours. Buying what is produced locally has become increasingly important to her. "If someone down the street made shoes, I would wear them, whatever they looked like," she explains. She has observed that other members, attracted initially by the food, develop an affinity for the farm and the farmers. The chance to work at the farm is, in her words, "a gift to the family." Creating an atmosphere that is friendly and tolerant is very important, she points out; a deeper understanding of what sustainability means takes time for people to grasp.

Although we have found word of mouth to be the most effective way of recruiting new members, at least once a year a few of the members and I make an appearance on a radio or TV talk show. The local public radio station has several sympathetic hosts who are willing to have us. Listening members can help by calling in with questions. During one of

these shows, I recognized the thinly disguised voice of core member Pat Mannix with a leading question that gave me the chance to plug our next orientation meeting. Lila Bluestone, another core member, volunteered to appear with me on a TV talk show. Before the interviews began, Lila was terribly nervous, repeating over and over the phrases we had agreed to emphasize. But in front of the camera, she blossomed. When the interviewer opened the show with a description of the CSA as a program for "needy" families, Lila corrected him ever so gently, saying, "Yes, the CSA provides the connection with the Earth and the clean, nutritious food of which so many of us in Rochester, of all income levels, are in desperate need." The wags at the next core meeting dubbed this the Liz and Lila Show.

The organization of the GVOCSA core group has been the main factor ensuring flexibility and endurance. Shared out among twenty-nine people, both the physical and the emotional work of maintaining the group do not become burdensome. (See chapter 6 for more details.) The stories of other

At a celebration at Brookfield Farm, farmer Dan Kaplan drives a tractor pulling a wagon full of member families. PHOTO COURTESY OF BROOKFIELD FARM.

CSAs are strikingly different. *Farms of Tomorrow Revisited*, by Trauger Groh and Steven McFadden, has some important lessons for understanding what enables CSAs to last. The authors returned to the CSAs profiled in the earlier edition for updates after seven more years. I interviewed some of these farmers or their sharers after another eight years.

Community Farms

SINCE THE PUBLICATION of *Farms of Tomorrow Revisited*, the Temple-Wilton Community Farm, Trauger Groh's home CSA and one of the two oldest CSAs in the United States, went through some lean years, only to re-blossom from the energy and attention attracted by the struggle for adequate secure land (see chapter 5). Development pressure and the lack of long-term tenure of high-quality land had led Trauger, Lincoln Geiger, and Anthony Graham to dissolve their joint farming association. After a few years on his own as "The Community Farm," Trauger retired from farming. Just when things looked blackest, deeply committed members found ways to secure land for Lincoln and Anthony. Their CSA membership has climbed back up to over one hundred families, and there is a long waiting list.[3]

MAHAIWE HARVEST CSA, which began at Indian Line Farm in 1985, thrives today because it has successfully gone through a transition from the cooperative effort of a group of part-time farmers to one full-time farmer and from more active member participation to less. Charlotte Zanecchia, a member throughout most of the farm's nineteen-year history, says that "the key thing is to have a farmer who is happy and keeping his customers happy." For a few years after the CSA's move to new farmland, several farmers continued to grow the food for the project. In 1993, David Inglis took on the job as head farmer and, as the other growers left, gradually took over the entire farm. At present, David and his family do all of the bookkeeping and share selling, and run whatever events occur at the farm, while he produces all the food for the shares. David has focused on building the soil and improving the quality of the produce, while offering members more choices, but dropping chicken and egg shares until he can do them properly. As a result, membership has grown and attrition from year to year has decreased. The core group has retired, and the board of Mahaiwe Harvest Land Trust has sold its protected Sunways Farm to another land trust that makes the land available to Project Native, a youth initiative that grows and sells native plants for landscaping, remediation and gardening, while David farms across the road on twelve acres that he has purchased himself. (See chapter 5 for more details.) He honors the role of the Indian Line core during the early years for creating the model of CSA and for ensuring the continuity of the group through the move. The present form of Mahaiwe Harvest CSA, however, suits his personality and satisfies the members. David is very emphatic on this question:

> It does not really matter to the integrity of your CSA whether you have an active core group, no core group, more festivals than you can shake a fist at, or no festivals. The essential point is that you come to your arrangement that you as the farmer and the members of the farm have come to by a process that is shared. . . . What do you do if you don't want to play that [organizer's] role? Maybe you want to grow the best vegetables that you know how to grow, for people you love to grow them for. I'm here to tell you that that's okay.

Charlotte Zannechia, who worked with Robyn Van

En on the very first CSA and played a key role in the continuing development of Mahaiwe Harvest, recalls "heart-wrenching" moments from the past, but she is pleased with the outcome. In answer to why she had stayed with the CSA so long, she replied:

> I always wanted to have organic produce. The CSA liberated me from having to grow it all myself, though I still garden. I became enslaved to the fundraising for the land. I have had to turn over the land for which I worked so hard to raise money to the farmer. Letting David have his space is the best gift I could give him. But when I eat those first spring crops—all the food, really—it is a spiritual experience for me. We are lucky to eat such diverse crops grown locally. And I see that this kind of farming is what more and more young people want to do. So Robyn and I were right, back at the beginning. The reality of it all has been like teaching a baby to crawl and walk. Finally we are in young adulthood. If David Inglis left, he could sell the business. He has a lot of equity . . . we would trust his choice of replacement should he ever leave.

BROOKFIELD, THE THIRD EARLY CSA profiled in *Farms of Tomorrow*, suffered through two very difficult years while the marriage of Nikki and Ian Robb, the initial farmers, disintegrated. Having known the Robbs as meticulous farmers, I realized there was serious trouble when I visited the farm on a NOFA summer tour in 1993 and could barely find some of the crops under the weeds. Under the Robbs' management, as at Temple-Wilton, members were asked to support the entire farm without regard to the precise amount of produce they received, at the cost of $450 per adult. With Dan Kaplan as farmer, beginning in 1995, Brookfield has revised this community farm model. Members purchase a share, as with most CSAs. The Board of Directors of the Biodynamic Farmland Conservation Trust, to which the former farm owners, the Fortiers, deeded the farm, has also introduced a business incentive into the farmer's pay scale. The board pays the farmer a base salary of $15,000, with incremental increases based on how many members he attracts and retains and how well he manages the business.

Dan manages Brookfield well. With the blessing of his board of directors, he lowered the share price to $300, increased membership from 150 to 530, and introduced the "Mix-and-Match" system from the Food Bank Farm, providing maximum choice of produce for the members who pick up at the farm. In the years since, the price has gradually risen to $440 for a full season share and $160 for a winter share. Dan also added drop-off sites in Boston. When I visited again during the summer of 1997, tractor cultivation had eliminated the weeds. In 2005, I had the pleasure of touring the new barn and CSA center; members and supporters have contributed over $150,000 toward the total budget of $250,000! Dan goes out of his way to make the members feel welcome to drop by for a walk or to bring their children to watch a calf being born. Brookfield holds regular festivals, but at the Northeast CSA conference, Dan said something very funny about them. He said that members like to know that festivals are going on, even though they don't attend them much. As Dan sums up the past eighteen years: "Things have changed. . . . We've lost the romantic illusion of community, but gained a real community. Brookfield Farm is not seen anymore as something special and 'apart from the real world,' but rather just as part of the community, the way the local churches or schools are part of the community. We are here. We are not the focus of the community, but we are a part of it."[4]

ROXBURY FARM SURVIVED AN UPHEAVAL similar to mine. When his marriage dissolved, Jean-Paul Courtens had to leave the land that belonged to his wife's family. Without missing a growing season, his farm team found new land and kept up production for six-hundred-plus shares, despite the fact that Jean-Paul could not work full-time through the transition because he was caring for his three children, helping them survive this painful dislocation. Roxbury members are contributing money to repay the bridge loan from Equity Trust to buy the new farm. Jean-Paul and his partner, Jody Bolluyt, have a ninety-nine-year lease with Equity Trust on 150

acres and a mortgage on 100 acres, a farmhouse, and farm buildings. Share numbers have risen to over one thousand in four communities—two pickup sites in New York City, one in the Kinderhook area of the farm, another in Westchester County, and a fourth in the Albany capital region. In 2004 the Sustainable Agriculture Research and Education (SARE) Program gave Jean-Paul the Patrick Madden Award for Sustainable Agriculture.

After a decade of steady improvements, the two other CSAs profiled in the original *Farms of Tomorrow* have gone through gentler, but fairly complex successions (see chapter 5). Financial viability allowing for respectable wages for the farmers with health insurance and even a pension fund made finding a new generation to farm Kimberton and Caretaker farms easier. Through U-pick or preorder systems, they have found ways to introduce more choices for their members, and they participate in Collaborative Regional Alliance for Farmer Training (CRAFT) and SAITA apprenticeship programs. (See chapters 7 and 19 for more details.)

Ironically, the smooth and easy service and convenience that marketing experts tell us are what U.S. consumers demand have not been what has made these community farms last. Quite the opposite! These farms have asked a lot of their members—advice, involvement, time, and money. They have shared their real stories including family hardships and economic realities. And they have given members genuine opportunities to exert their imaginations and become active participants in a great social experiment.

Subscription Farms

CSAS THAT FOLLOW THE SUBSCRIPTION, rather than the community farm model, also have found ways to solidify their existence. (Subscription CSAs supply members with a box of produce each week and most ask only for payment—not a work requirement—in return, and have other markets for a significant portion of their production.) In its fourteenth year in 2006, Full Belly Farm in Guinda, California, has reached maximum capacity at 800 shares, with a waiting list of 150. With year-round production, the farm offers an enviable selection of vegetables, fruit, and nuts as well as separate flower and lamb shares. Although most of their recruiting is by word of mouth, the turnover rate is high. The farm is working toward getting all members to make more of a commitment by paying quarterly, instead of monthly. Full Belly welcomes visits by members, but few come. To keep in touch, the farmers do occasional visits to the porches where members pick up shares, taking along a lamb or other attraction to ease conversation. As farmer Dru Rivers says, "The vegetables speak for themselves." A few members, though, say the weekly newsletter is even more eloquent than the vegetables.

> The more information a CSA can share with members, the better off it will be. Face-to-face interactions are, of course, the most effective. Farmers should take every convenient opportunity to meet with members; hear their ideas, suggestions, and complaints; and in turn tell them about the realities of the farm.

❧ **FOR MANY YEARS, SIX HUNDRED CSA SHARES** represented just a quarter of Be Wise Ranch's production. Bill Brammer added a forty-member CSA to his wholesaling as a community service in 1993. The growth to six hundred was almost entirely by word of mouth. After developers took 80 percent of his best land in 2005, Bill saw expanding the CSA and his on-farm market as the way to keep his farm alive. People regularly call to say they have eaten Be Wise food at a friend's house and that "it's the greatest they ever tasted." Bill invites them to sign up for a four-week trial period, after which 20 to 30 percent say it is too much food. They switch to buying organic produce at the local health food store—which is okay with Bill, since he supplies most of the produce there too. The only regular member involvement is through hosting drop-off points: Bill cuts the host's price in half for a ten-share site and gives a free share for twenty. The host makes reminder calls to those who forget to pick up. Instead of a frequent newsletter, Bill gives new members an elaborate packet of information when they join, posts announcements at pickup sites, and provides recipes on the farm Web site. After many years as president of the California Certified Organic Farmers (CCOF),

Bill is happy to make decisions without the benefit of group process. Nevertheless, he finds his relationship with members "very satisfying." He says it is "fun and a challenge to grow this much food," and he enjoys the many letters of appreciation and calls telling him the kids ate all the sugar snap peas before they even got the share home.

❧ **INSPIRED BY THE BOOK** *The E-Myth Revisited: Why Most Small Businesses Don't Work and What to Do about It*, by Michael E. Gerber, the farmers at Angelic Organics in Illinois made a careful study of each job on the farm in order to create a clear management structure that would provide each worker with a thorough understanding of his or her responsibilities. Although he is not writing about farms, Gerber is astute in analyzing how small businesses can completely eat up the people who start them. His book offers a systems approach to making your business franchisable: "The Entrepreneurial Perspective . . . views the business as a network of seamlessly integrated components, each contributing to some larger pattern that comes together in such a way as to produce a specifically planned result, a systematic way of doing business."[5]

Spreading franchised Old MacDonald's CSA farms around the country may not be our mission, but Gerber does have some very useful pointers on detaching the function of a job from the personality performing it. Angelic Organics' goal has been to free up the staff for other creative work besides farming. They were not able to do this fast enough to prevent one of the initial farmers, Kimberely Rector, from having to go elsewhere to pursue her calling as a graphic artist. By 2005, Farmer John had spearheaded the construction of adequate infrastructure, created effective systems, and upgraded equipment. He was able to take most of the year off to promote his film, *The Real Dirt on Farmer John*, and to write two more books. The CSA has generated enough revenue to begin making up for two decades of undercapitalization. Share numbers have grown to over twelve hundred, they have enlisted member help in purchasing more land, an increasingly active core group helped produce the handsome *Farmer John's Cookbook* and helps with member recruitment and farm financing, and they continue to publish an outstanding newsletter. One core member, Tom Spaulding, was so inspired by Angelic Organics that he gave up his city life, moved to the farm, and launched the CSA Learning Center at Angelic Organics.

∞ **FOR THE FIRST FEW YEARS** after Jim Kinsel took over the CSA that he renamed Honey Brook Organic Farm, he knew all of the members by name. As the CSA has grown into the largest in North America, with over twenty-one hundred shares, Jim has not been able to maintain one-on-one direct contacts with so many people. As often as he can, he hangs out at the on-farm market where three-quarters of the members pick up their shares each week. To keep the connection as vivid as possible, Honey Brook has increased the frequency of newsletters to weekly editions in the boxes they deliver and monthly editions on the Web site blog and has added a suggestion box at the farm market. Jim is trying to attract more member involvement in work parties with activities for which you don't move much but have a lot of time to talk, such as carrot thinning. The farm is also introducing "Kids' Blog": eight-year-old "Farmer" Kyle, the son of a member, who is convinced he wants to farm, will write monthly entries on his experiences at the farm. Jim finds working with enthusiastic youngsters like Kyle "good therapy for the soul."

Jered Lawson, a close observer of and technical adviser to CSAs, has expressed the concern that the larger subscription projects will crowd out smaller ones while not providing a meaningful connection between the sharers and the farms. Watching CSA membership on a few farms grow into the hundreds, Jered feared that CSA might fall into the clutches of agribusiness as usual, with the farmer behind a desk, labor reduced to the status of "input," and expansion and profit as the main goals. The evolution over just a few years of Full Belly, Honey Brook, and Angelic Organics suggests that size and the subscription model do not necessarily destroy the possibility of meaningful farmer-member partnerships. With time and creativity, the larger farms may yet figure out how to provide their membership with, in Jered's words, "the opportunity to keep moving toward greater conscious support of their farm's, and society's, general economic, social, ecological, and cultural health."[6]

Communications

John Peterson's writings and the Angelic Organics newsletter offer another important clue as to how to maintain a vital CSA: good communications. Steve McFadden, one of the authors of *Farms of Tomorrow Revisited*, goes so far as to say that "those [CSAs] are most successful that communicate most clearly." Whatever medium you use, the more information a CSA can share with its members, the better off it will be. Face-to-face interactions are, of course, the most effective. Farmers should take every convenient opportunity to meet with members; hear their ideas, suggestions, and complaints; and in turn tell them about the realities of the farm. Once you move into signs, flyers, e-mail, Web sites, and newsletters, you venture onto somewhat less predictable ground. Participants in a workshop on communications at the Northeast CSA conference, after a discussion of effective signage, concluded that there is no overestimating people's capacity to *not* read signs. Where you must use signs, fewer is better, and the lettering must be large and clear. Scott Chaskey described the system at Quail Hill, where all 150 members pick for themselves: Right out on the beds, they post on wooden stakes computer-generated signs with large, emphatic lettering stating, "Pick Only Outer Leaves!" "6 beets," "12 tomatoes."

For all CSAs, but especially for those where members have few personal contacts with farmers, a regular newsletter is an essential part of communications. Virginia CSA farmer John Wilson exclaimed, "CSA

Cheap Food and Flat Tires

Erik [an intern at Angelic Organics] gets up at 2:00 in the morning twice a week to get the veggies to you on time. He checks shareholder names against a flow chart as he loads a hundred boxes (1 1/2 tons) of produce and numerous buckets of bouquets from the cooler into our blue truck by flashlight. Coffee mug and cellular phone beside him, he guns out of the farmyard at 3:45 A.M.

The last time Erik headed the blue truck towards Chicago, the right front tire shredded. It was 4:30 A.M. Erik gulped coffee, then unloaded half your boxes to get at the spare. He bounced up and down on the breaker bar to remove the stubborn lug bolts. He changed the tire. Although heady with exhilaration from wresting those lugs to the ground, Erik did not forget to reload your boxes before again heading toward the Big City.

We've had a lot of flat tires this year. It's because the roofs are going. The winds zing shingles onto the yard and into the flowers. I find roofing nails in the driveways. Yesterday I found one on the kitchen floor. Someone must have tracked it in.

The last roofing we did here was in '74. We did all ten buildings. Then, I farmed big until the money ran out. In the mid-eighties the roofs started to leak—slow leaks at first. It took a blast of a storm and gusty winds for a meager drip to find its way inside. Now a mere drizzle can puddle the floors. When I pick up the phone after a rain, the receiver is sometimes cradled in a little pool of water.

Once the countryside was boisterous. Barns were red and straight. Livestock strutted in the farmyards. Children played with ducks. Now—take a drive from Chicago to our farm. Look for farmers walking in their fields: They are not there. Count the children playing in farmyards; where are the children? Look at the farmsteads. The paint goes. The roofs go. The ridgepoles go. The countryside is desolate, propped up by a mock luster of chemicalized corn and soybeans, and occasional gentrified hobby acreages.

I can't isolate the causes of this dilapidation of our countryside. I know about mechanization and computerized farming and the millions of tons of chemicals that seem to have stripped the people from the land, but what causes what, anyway? What caused the willingness to have the barrenness occur this way?

People work hard on this farm. There's barely enough money to pay them. There's barely enough money for the myriad other immediate expenses that accrue to this small but complex operation—the expenses that somehow get the operation into the next month. I know other farms have seen their roofs going and couldn't bear it and sold out. Farming our way into poverty amongst a hail of rotting shingles and roofing nails while affluent consumers stalk food on the cheap yet demand the most expensive "health care" is an insult. Getting a flat tire in the middle of the night because our roofs are blowing away just might have some relationship to the cheap food mindset that pervades this country.

Community Supported Agriculture is a step toward restoring this farm. It offers a deeper relationship than can be achieved through money spent at the organic counter for produce from "somewhere in California." But that California produce influences the price we charge for our memberships. We are often asked how our prices compare to store prices. Large, heavily capitalized organic operations in a remote temperate climate set the price standard for Midwestern organic farms. The true cost of sustainable farming—a farming operation that really takes care of itself, its machines, its buildings, its land, its workers, its future—is far from the cost of our shares.

—*John Peterson, Angelic Organics, Caledonia, Illinois. Used by permission.*

Schoolchildren tour the ALBA farm in Salinas, CA. FROM WEB SITE.

really does all the things people say—community connection to the farm. I love it. The newsletter creates as much community as picking up the produce." The newsletter does not have to be fancy. West Haven Farm posts weekly news to most of its members by e-mail. Andy Griffin at Mariquita Farm e-mails out delightful weekly essays blending reflections on vegetables and philosophical musings along with a list of the share contents. Many CSAs dash off weekly one-pagers, some of them handwritten. Willow Pond has a regular format with a list of the season's vegetables along the top. Farmer Jill checks off what's in the share that week, does a quick drawing of one vegetable with a note about its nutritional value, fills in a few lines on "Upcoming Events," and jots down a recipe in the box underneath. A tear-off section at the bottom invites member feedback. T and D Willey's weekly "What's Growing On" includes a short farm update from Denesse, a list of what's in the share boxes, and as many recipes as will fit. Rosie Creek Farm in Ester, Alaska, has a weekly one-pager with a short essay, the share list, and a recipe.

Hidden Valley Organic Farm has a two-pager, front and back. Like Willow Pond's, it includes a list of all the vegetables with checkmarks by the selections of the week. Most of the first page reports on

the state of the crops, the weather, things to come, and special requests from the farmers. Recipes cover the back page. Stephen and Gloria Decater publish a more elaborate, but less frequent newsletter for Live Power Community Farm, illustrated with photos of farm animals, children, and festivals. Cumulatively, the Decater's articles provide a narrative of their lives on the farm through the seasons.

I haven't read all CSA newsletters, so I may be judging unfairly, but surely the Pulitzer Prize must go to Angelic Organics. Edited by farm office manager Bob Bower, this newsletter absolutely sparkles with writing talent (both Farmer John's and others'), energy, good humor, and information about food.[7] Editing and mailing or e-mailing a regular newsletter is an excellent way for members to contribute to their CSA. If I am any example, it is not hard to find non-farmer members with better computer skills. The GVOCSA has had a series of talented editors. Since its second year, the GVOCSA has had a newsletter that appears six times a year: four issues during the growing season and two in the winter. I have the pleasure of contributing "Notes from the Farm," a continuing series of reflections on the crops, the weather, the people, and the critters at Rose Valley and then Peacework.[8] My farm partners take turns at analyzing the end-of-season surveys and veggie questionnaires. The editor and other members contribute messages from the core group; articles on topics such as homeschooling, how one person consumes an entire share, or the global significance of buying local; as well as recipes and notices of events. The May mailing to members is a fat packet of information that includes the membership list, a list of core members with their responsibilities and phone numbers, the work schedule, a map to the farm, maps of the farm and distribution sites, a page of suggestions on how to be ready for farmwork, and the bulk order price list. We post much of this on the Web site and print it in a Member Guide as well.

The GVOCSA also encourages members to purchase *The CSA FoodBook for a Sustainable Harvest*, which I wrote during the winter of 1993 (with some help from my partner David) as a guide to the selection of vegetables, herbs, and small fruit we pro-

Children run through field alongside of horses pulling a plow at Live Power Community Farm, Covelo, CA. PHOTO BY NANCY WARNER.

vided in the shares. Presented by season, the entries includes a little history of the crop, nutritional information, how to store for the short term and long term, anecdotes from growing at Rose Valley Farm, and recipes. Proceeds from sales of *The CSA FoodBook* go to the GVOCSA scholarship fund for low-income members. I added one hundred more recipes from GVOCSA members and other CSAs in 1996. (See the "CSA Resources" section.)

The CSA newsletter is blossoming into a new literary form, unveiling previously undiscovered talents. One member of Harmony Valley CSA commented, "I can't decide if the newsletter or the produce is the best thing you do—but since we can't eat the newsletter, I guess it comes in second. . . ." Mariquita Farm members feel the same way about Andy Griffin's writing. Annie Main of Good Humus Farm says that after twenty-two years of farming, the necessity of writing weekly essays has changed the way she experiences her life and work: "I live it looking for things to write in the newsletter. It has heightened my awareness of where I'm at." In antic-

ipation of a future collection of CSA essays, I have chosen a few of the best examples to whet the reader's appetite (see the sidebars sprinkled throughout this and other chapters).

In addition to a newsletter, Harmony Valley gives its CSA members a calendar packed full of useful information. Noted on the appropriate dates are farm events open to members; the farm's work schedule, including planting and harvesting days; payment reminders; holidays; the farm family vacation plans; and, at the end of the year, reminders to sign up for the next year. Some years, the illustrations are lovely nature scenes or vegetables; other years there are photos of the farm's workers on tractors; but both May and July have diagrams on how to flatten the box used to deliver members' vegetables. The cover bears a map to the farm and the words, "You are always welcome!" (A cautionary note on open invitations: *always* means seven days a week, twenty-four hours a day. You can be very welcoming to members while restricting visiting hours to the times convenient for the farm.)

Web Sites

At the pinnacle of up-to-date communications, CSAs are creating Web sites. When I wrote the first edition of this book, I was rather sarcastic about the value of harnessing the World Wide Web to sell local vegetables. I was wrong. People who live only two miles from our farm have discovered us via the Web. These days, almost every CSA has a Web site. There are hundreds of them, with more information than anyone but a total geek or a PhD candidate could possibly absorb. The Biodymanic Farming and Gardening Association, the Robyn Van En Center for CSA Resources, and Local Harvest all have links to Web sites of listed CSAs.

Angelic Organics has a Web site, of course, where the curious can view the map of pickup sites around the Windy City, archived newsletters with recipes, and farm photos. On the T and D Willey Farm Web site, you can sort back newsletters by date or by vegetable recipes, a great resource for anyone seeking food ideas. You can also take a photographic tour of the farm. Jean-Paul Courtens and Jody Bolluyt have posted their farm manuals on the Roxbury Farm site, sharing all the details of how they grow and harvest each vegetable and how they maintain fertility.

Michael Axelrod, a Web wizard in the GVOCSA, set up a simple site for us in 1996 where members can read where and when to get our vegetables and access the archive of my "Notes from the Farm." More recently, Mike added Tom Ruggieri's hand-tinted black-and-white photos of Peacework, and a recipe book where members can post favorite recipes.

Yet another dimension of CSAs is the role they can play in agricultural research. Harmony Valley offers its members a "unique opportunity to support some cutting edge research" in progress on the farm. Farmer Richard de Wilde has been working on demonstrations of the use of hedgerows for habitat for beneficial insects and of compost for controlling plant diseases. Several CSA farms in New York are breeding new varieties and trialing old ones as part of the Organic Seed Partnership. Since my farm did most of the trials on an early sweet pepper, the breeders have decided to name this new variety "Peacework." Both my farm and Roxbury are cooperating in the Cornell Soil Health Project to develop a simple and inexpensive test for biological life in a farm's soil. The Practical Farmers of Iowa, in cooperation with Iowa State University, have research projects on several CSA farms. Both the Sustainable Agriculture Research and Education (SARE) Program and the Organic Farm Research Foundation offer research grants to farmers for projects of this kind. (See "CSA Resources" for addresses.)

Educational Programs

CSA-linked educational programs have grown dramatically during the eight years since the first edition of this book. An entire book could be devoted to this topic. CSA farmers are finding creative ways to share the natural beauty of their farms and their agricultural knowledge with members and the general public. Heidi Mouillesseaux-Kunzman argues in her master's thesis that educating the public about the community benefits, rather than the individual health or taste aspects of CSA, is one of the best ways for farmers to overcome the obstacles to increasing CSA membership: "By including the concept of community . . . CSA educational strategies should have a greater chance of establishing the sense of social coherence, interdependency and mutual responsibility necessary to a CSA's long term success."[9]

Educational outreach is basic to many CSAs. Several biodynamic CSAs have close relationships with Waldorf schools and regularly host programs for the children. Live Earth Farm and Good Humus Farm in California run summer camp programs for children. Learning centers at faith community CSAs offer studies in earth literacy. The Center for Urban Agriculture at Fairview Gardens in Goleta, California, gives classes in cooking and gardening, workshops, tours, and apprenticeships. Winter Green Community Farm in Oregon, Live Power Community Farm in California, and Clark Farm in Connecticut provide tours and classes for local public school children. Caretaker Farm in Massachusetts, Willow Pond Farm in Maine, and Willow Bend Farm in Minnesota have special gardens by and for the children of members. Happy Heart Farm in Colorado welcomes student visitors from around the world. One day a week, three disabled adults help Richard Sisti at Catalpa Ridge with transplanting, seeding, weeding, and harvesting. Over the years, dozens of classes, from kindergarten to graduate students, have toured and helped with work at Silver Creek and Peacework farms.

Several CSAs extend their educational projects out into their local communities. The nonprofit Calypso Farm and Ecology Center near Fairbanks, Alaska, makes education its primary focus. Besides growing food for sixty shares, Susan Willsrud and Tom Zimmer, the Calypso farmers, and Megan Phillips, the farm's education director, provide ecological gardening classes, tours, and field trips on topics such as boreal forest ecology, natural dyeing, and spring planting. In 2005, eleven hundred students visited their thirty-acre farm. Their most ambitious project, however, is the Schoolyard CSA Garden Initiative. So far, they have organized summer CSA projects at three Fairbanks schools. At a workshop at the Community Food Security Conference in 2005, Megan described their approach. At each school, Calypso helps teachers, parents, and students form a Garden Committee to set up and fund the school garden. Calypso provides the school's teachers with curricular materials on how to use the garden in studies from kindergarten through sixth grade level. During the summer, high school students, employed by Calypso's EATing (Employing Alaska Teens in Gardening) project, participate in an eleven-week training program in which they grow food at their school garden and learn to run a CSA business. Eighteen shares cover the pay for two student gar-

> Marianne Simmons has developed a theory about why some people just don't fit. She has observed a crucial distinction between what she calls deductive and inductive cooks. The deductive cook starts a meal by planning—looking up recipes, checking the kitchen for stocks on hand—and then goes out and buys ingredients. . . . The inductive cook, by contrast, "flings open the refrigerator one hour before dinner, rejoices at the abundance, puts three-quarters of the meal on the table raw, and is happy that this is possible without stepping a foot in a supermarket."

deners, who also get to take home some of the food. The Calypso staff invites inquiries from others who would like to start school CSAs.

The CSA Learning Center at Angelic Organics has blossomed from an offshoot of the farm core group into a significant community food security nonprofit with seven full-time staff people and offices at the farm and in Chicago. Their mission statement declares, "The CSA Learning Center empowers people to create sustainable communities of soils, plants, animals and people through education, creative and experiential programs offered in partnership with Angelic Organics, a vibrant Biodynamic community supported farm." For five years, their Harvest Shares Program raised funds to give up to ninety subsidized shares in Angelic Organics to low-income people in the Chicago area. In 2006 the program shifted its funding to support a staff person whose job is to dialogue with congregations and inspire them to support shares and do their own farming or gardening. (See "Faith Community Supported Gardens" in chapter 21 for more detail.) The Learning Center runs educational programs of several kinds, reaching over thirty-five hundred people a year. For farmer development, the center facilitates the Midwest CRAFT program and organizes field days, farm tours, and one-day sessions on "Farm Dreams," for potential farmers. The center did a pilot project with three refugee farmers from Bosnia, Mexico, and Nigeria as a prelude to developing an urban incubator center for new migrant farmers. At eight garden sites around the city, the center provides technical assistance and leadership development for youth and adults. The center also took the lead in organizing a Food Security Summit in Illinois, which has led to the establishment of a governor-appointed Food Action Council for the state.

At Angelic Organics Farm in Caledonia, the center offers agro-ecology workshops in such homesteading skills as soap making, composting, and garden planning, and custom-designed workshops for all sorts of groups, such as Boy and Girl Scouts, community centers, and schools. Center staffers manage the farm's herd of twelve goats and participate in the Clean Harvest Guild, a soap-making cooperative employing Latina women who live near the farm. To remain grounded, Tom Spaulding and the other staffers spend a quarter of their time doing Angelic Organics farmwork. In 2005, World Hunger Year selected the center for the national Harry Chapin Self-Reliance Award.

Community Building

I really just wanted to write a few words about farm festivals, but I have heard so many farmers whose on-farm events have been poorly attended express disappointment about community building that I find myself compelled to address the much broader subject of the meaning of community. The number of people dancing around a maypole, roasting marinated vegetables over an open fire, feasting together on fresh strawberries, or taking turns on the crank of a cider press—is that the true indicator of community?

Certainly, some CSA farms hold fabulous festivals. Two thousand people attend the Hoes Down Harvest Festival at Full Belly Farm, run with the volunteer help of CSA members. Out of 800 families, 125 come to strawberry-picking day and 80 to 100 to garlic-braiding day at Harmony Valley Farm. In their first year, Seeking Common Ground CSA could not promise large amounts of produce, so they supplemented the shares with conviviality and celebrations. They held weekly courses and found a theme for a festival every single month. Peacework holds an annual Maypole dance, followed by a wildflower walk and potluck dinner on the Sunday closest to May 1. No matter how cold and rainy the weather, we always attract enough neighbors and CSA members to hold every ribbon on the pole. David Inglis,

who has set in abeyance some of the festivals once held at Mahaiwe Harvest CSA until members come forth to arrange them, still enjoys the "burning of the hat" ceremony at the end-of-the-year party he organizes himself. Children love to see him burn his hat.

Distribution time, when members converge on a pickup site, provides the occasion for many CSAs to foster a sense of community. The GVOCSA holds "pickup parties" once a month at distribution, where members are lured with name tags and snacks to linger and socialize. The Magic Beanstalk did its distribution in a church basement where other activities were also going on, such as the sale of other farm products, games for children, refreshments, and information tables, and where groups of chairs were arranged to invite members to sit down and interact. Farmers who make deliveries to CSA pickup points distant from their farms often stay for the distribution time to socialize with members.

> In surveys of CSA members in all regions of the country, the desire for "community" ranks very low among the reasons for joining. Most people are not looking for more busyness in their lives. . . . Perhaps we need to think about community differently, as a sharing of values and a commitment to act on those values, even in very modest ways.

Nevertheless, in surveys of CSA members in all regions of the country, the desire for "community" ranks very low among the reasons for joining.[10] Most people are not looking for more busyness in their lives. Squeezing in the time to pick up the food amidst all their other obligations is hard enough. As Cynthia Cone and Ann Kakaliouras stated in their article "The Quest for Purity: Stewardship of the Land and Nostalgia for Sociability," "'Community' for many [CSAers], if not most, seems to be an expression of longing, a nostalgia for imagined linked relationships of our rural past—a kind of community that is difficult for CSA members to realize given the demands and constraints of their lives"[11]

But a sense of community does not grow only out of group activities, nor is it an abstract concept. Rather, it is a complex web of interconnections among people. In his explanation for remaining with the CSA, Mark Pierce, the production editor of the GVOCSA newsletter, who also helps us put up our hoop houses every spring, describes how community affects his life:

More Than Good Eats

Some of you, perhaps, did not fully grasp what you were getting into when you signed up to receive a box of vegetables each week for twenty weeks. You did not realize that you were signing on to have your week's meal plan monopolized by fresh vegetables demanding to be cooked and eaten before they pass their prime. That you wouldn't have as much room for dessert after a big dish of Pasta Primavera made with your abundance of fresh organic veggies. That a grocery store tomato would never satisfy you once you had tasted our farm-grown, freshly picked, vine-ripened gems.

But did you stop to think about the further implications of belonging to a CSA? In supporting Angelic Organics, you are not just buying a commodity. You are supporting a dwindling way of life, the education of future farmers who learn here, a sense, even, of mission—which regards farmland as worth something more than its real estate value.

Sure, the money you pay goes towards the production costs of organic vegetable raising. Fuel, labor, seed, soil amendments, and plant protections such as row covers are provided for by your checks. But your money also enables relationships. It backs Farmer John's relationship with this, the land he grew up on. It allows him to continue labor negotiations with his machinery, coaxing it to help him out for one more year. It enables his relationships with those who work for him and those who buy from him.

You might come out of this year with new favorites, and familiarity with obscure vegetables with which you can impress your friends. ("Yes, the Rutabaga Bisque I made last week was absolutely delightful, darling.") You may have much stronger opinions about what constitutes "yucky" and "yummy" in the vegetable kingdom. You may pick up a few new recipes, or make up some of your own. But you can take your membership further than that if you wish.

Community is a word with potential. Community can simply mean the people who live in a selected geographical area. But it can also mean those with whom one has a more meaningful relationship than simple geographical proximity. You may find yourself swapping recipes with other shareholders when you pick up your box. You might wish to come out and volunteer for a day or more on the farm; get a taste of country/break from the big city. You might wish to join our new Core Group, which was founded this year to take over the farm's non-production tasks such as marketing, special events, and surplus vegetable distribution to the needy. Engaging yourself in one of these ventures, you are at risk of discovering within your heart the feeling of community.

Community in this sense is not a commodity that is purchased; it is built through actions. We at Angelic Organics propose to bring fresh healthy produce into your life. Nothing to sneeze at in and of itself. We also give you the option of taking your membership further; to make your involvement with Angelic Organics encompass not only what you pull out of your vegetable box, but what you put into your CSA.

—*Kristen the Cook, Angelic Organics Farm News, Caledonia, Illinois, July 1997. Used by permission.*

The number one reason I continue to be active, in all honesty, is the sense of community that the CSA provides. The members are people concerned with personal and civic health and environmental issues. They tend to be educated, kind, peace-loving, and warm individuals. That, more than anything else, has brought me joy in my association with the CSA. The apartment I rent, the pets I have, many friends I have, opportunities that have come my way have all come about through CSA associations. The bonuses are that I feel I can further goals that I believe in—local, quality, earth-friendly sources for food, a renewed connection for people to the food and land, and peace and understanding in general. The other bonus, not to be downplayed, is delicious, fresh produce that I can bite into without washing (if I want). That little thing seems so liberating sometimes . . . what moderately health-educated person dares to bite a factory farm strawberry without washing it?

Perhaps we need to think about community differently, as a sharing of values and a commitment to act on those values even in very modest ways. In his master's thesis, Tim Laird quotes a farmer who comes closer to this understanding of community as "the people's involvement, helping at the farm with promotion, making friends, creating a connection to the Earth and each other."[12] Of the farmers Tim surveyed, 63 percent believed they were achieving success at building community in this deeper sense and mentioned that this was the most gratifying aspect of CSA for them. One of the farmers wrote:

> Everyone feels a part of what is being done, what is going on. It is so much more satisfying than money for vegetables. Members have visited the farm in the off-season. They bring their children. They see where the fruit and produce come from and learn the process. We (farmers) meet concerned people who trust us enough to give us their money without contract or guarantees—we care about each other. We are into something mutually upbuilding.[13]

Jack Kittredge would agree, concluding his essay "Community Supported Agriculture: Rediscovering Community" by saying: "To the extent that alternative institutions recenter our lives in small, local groups of people with whom we have mutual obligations, they can rebuild human community."[14]

———————————•———————————

The determined work of a relatively small group of people keeps each CSA afloat. Local, fresh, tasteful food is the key. But for Community Supported Agriculture to flourish—and not merely survive—many farms and gardens must grow, cross-pollinating promiscuously with one another, every good idea lingering in the public sector for each of us to sniff suspiciously and snatch up if it suits us. There is no formula, except a little imagination and a lot of hard work.

CHAPTER 12 NOTES

1. Deborah J. Kane and Luanne Lohr. "Maximizing Shareholder Retention in Southeastern CSAs: A Step toward Long-term Stability" University of Georgia, 1997
2. Linda Halley, "CSA Notebook," *New Farm* (November 9, 2004).
3. For the story of Buschberghof, the farm in Fuhlenhagen, Germany, where Trauger farmed for fifteen years before coming to the United States—and the model for Temple-Wilton—see the book *Short Circuit: Strengthening Local Economies for Security in an Unstable World* (1996), by Richard Douthwaite.
4. Trauger Groh and Steven McFadden, *Farms of Tomorrow Revisited* 143.
5. Michael E. Gerber, *The E-Myth Revisited: Why Most Small Businesses Don't Work and What to Do about It* (Harper, 1995), 72.
6. McFadden and Groh, *Farms of Tomorrow Revisited*, 86.
7. Archived on the Angelic Organics Web site (www.angelic organics.com).
8. Many years are archived on the GVOCSA Web site (www.gvocsa.org).
9. Heidi Mouillesseaux-Kunzman, "Civic and Capitalist Food System Paradigms: A Framework for Understanding Community Supported Agriculture Impediments and Strategies for Success" (master's thesis, Cornell University, 2005), chap. 8, p. 1.
10. See studies by Gerry Cohn, Cone and Kakaliouras, Kane and Lohr, and Kolodinsky and Pelch in "CSA Resources."
11. Cynthia Cone and Ann Kakaliouras, "The Quest for Purity: Stewardship of the Land and Nostalgia for Sociability," *CSA Farm Network* (Vol II): 29.
12. Timothy J. Laird, "Community Supported Agriculture: A Study of an Emerging Agricultural Alternative" (master's thesis, University of Vermont, 1995), 94.
13. Ibid., 101.
14. Jack Kittredge, "Community Supported Agriculture: Rediscovering Community," in *Rooted in the Land: Essays on Community and Place*, ed. William Vitek and Wes Jackson (Yale University Press, 1996): 260.

PART FOUR

THE FOOD

GROWING THE FOOD

Care of the soil and care of the soul are not separate things.
—Satish Kumar, NOFA Summer Conference, August 2005

This we know: the Earth does not belong to man. Man belongs to the Earth. This we know: all things are connected. Whatever befalls the Earth, befalls the sons of the Earth. Man did not weave the web of life. He is merely a strand in it. Whatever he does to the web, he does to himself.
—Attributed to Chief Seattle

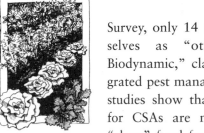

Eating a Community Supported Agriculture (CSA) share is like having your own garden, but with much less work. For the farmers who produce that share, however, the inverse is true: Growing a large variety of crops so as to have an appealing selection and combination every week over a six- to twelve-month season is a challenge. The mix required is similar to growing for a farm stand or a farmers' market booth. Depending on your previous marketing experience (or lack of it), growing for a CSA may require learning how to produce a much wider range of crops than you have done before. You will need to make careful rotation and crop succession plans to be successful.

Although in theory a conventional farm could start a CSA, in practice, almost all of the existing CSAs are committed to organic or biodynamic methods. Several farms, such as Pat and Dan LaPoint's Hill 'n' Dale in Pavilion, New York, have started CSAs while in transition out of chemical use. Bob Muth of Muth Farm in New Jersey said that he used CSA as a way into organic from conventional vegetable farming. Of the 312 farms that answered the question on cultural practices in the 1999 CSA Survey, only 14 (4.4 percent) listed themselves as "other than organic or Biodynamic," claiming low-input or integrated pest management practices. Market studies show that the potential members for CSAs are mainly people who seek "clean" food for themselves or their children. Most of the CSA brochures include descriptions of the farm cropping system. Here are three typical ones, gleaned from the hundreds of brochures I have read:

Common Ground Farm in Olympia, Washington: "We farm organically, though we are not 'certified organic.' We grow nutritious vegetables by attending to the health of the soil and plants. We use composted manures and cover crops to maintain fertility, and minimal tillage with light machinery to avoid compaction and erosion. Damage from pests or disease is minimized through our crop selection and rotations, and the judicious use of row covers. Increasing local food self-reliance is part of our work, and toward that end we trial vegetable varieties that extend the harvest season both early and late."

Farmer Steven Decater plows with horses at Live Power Community Farm, Covelo, CA. PHOTO BY NANCY WARNER.

Brian makes beds at Harmony Valley Community Farm in Viroqua, Wisconsin. PHOTO COURTESY OF HARMONY VALLEY FARM.

Grange sur Seine in La Broquerie, Manitoba:
"We try to run a traditional mixed farm. We have chickens, sheep, goats, pigs, two cows for milk, and horses for working the land. We use solar and wind energy for electricity and water heating. We have been largely self-sufficient by growing a huge garden, and we want to share this with others. We care for the Earth and aim to pass it on to our children in as good or better condition as when we received it."

Fiddler's Green Farm in Brooks, California:
"At Fiddler's Green we are committed to sustainable organic farming practices. Organic means no use of fertilizers, herbicides, nor pesticides that are synthetically produced. Sustainable means we want to continue farming without depleting our natural resources of land and water. Therefore, we rotate our crops, plant cover crops, apply organic compost, introduce beneficial insects, use mechanical and hand cultivation for weed control, select proper irrigation methods, consistently care for our soil, and grow seasonally."

In this chapter we will not attempt to teach how to grow a wide variety of vegetables and fruit. Use the many good resources available already. For an introduction to organic farming, *The Real Dirt:*

Farmers Tell about Organic and Low-Input Practices in the Northeast, which I helped Miranda Smith edit, is a good place to start. Eliot Coleman's books *The New Organic Grower* and *Four-Season Harvest* are essential reading. Vern Grubinger's *Sustainable Vegetable Production from Start Up to Market* is the best single volume for professional vegetable growers, organic or conventional, I have seen anywhere. With support from Sustainable Agriculture Research and Education (SARE), Northeast Organic Farming Association (NOFA) farmers have created a series of concise, practical manuals on such topics as *Organic Soil Fertility Management; Organic Weed Management; Vegetable Crop Health: Helping Nature Control Diseases and Pests Organically; Soil Resiliency and Health: Crop Rotation and Cover Cropping on the Organic Farm; Compost, Vermicompost and Compost Tea: Feeding the Soil on the Organic Farm;* and *The Wisdom of Plant Heritage: Organic Seed Production and Saving.*

If you are a beginning farmer, an excellent way to learn the trade is through an apprenticeship or internship with an experienced farmer or, better yet, with several farmers. You can also learn a lot by working on a variety of farms, even for short periods. Joining one of the many farmer networks that exist in most states and attending conferences and farm tours are other good ways to pick up useful information. These days, many conventional growers' conferences often include helpful sessions on biological controls and soil-building techniques. In some parts of the country, Extension agents are knowledgeable about organic farming, a big change from even five years ago.

I learned the hard way, by jumping from a large garden right into full-time farming, but I really do not recommend that approach. While it's true that you learn from your own mistakes, you can shorten the list and reduce the pain through hands-on training with someone who really knows the trade/art of farming. I did have the benefit of many trips to France, where I visited small farms and market gardens and observed some excellent examples of well-run professional operations. By the time I moved to Rose Valley Farm, I had completed the equivalent of an apprenticeship during eight years at Unadilla Farm in Gill, Massachusetts.

At Unadilla, we started with a single crop of leeks in 1981 and then increased to about ten crops on four acres of raised beds. At the same time, we grew a full range of garden crops for our own use. We marketed to restaurants, food co-ops, a farmers' market, and directly to neighbors. We also sold through two grower co-ops. Although we were the only organic growers in the Pioneer Valley Growers Association and took a lot of razzing, I learned a lot about quality, "packout" (the industry standards for packing produce), and how the vegetable business works. In 1985, I participated in the creation of Deep Root Organic Truck Farmers.

I had discovered CSA in France in 1977, but my partner in Gill was not interested in trying it. At Rose Valley, David was open to the idea, and his long-standing connection with Alison Clarke and the Peace and Justice Education Center (soon to be renamed the Politics of Food) in Rochester provided a perfect group with which to start. In terms of pro-

duction and marketing strategy, the CSA meant adding quite a few more crops. David had already established an acre of asparagus and half an acre of blueberries and was producing garlic, snow peas, assorted greens, broccoli, winter squash, and some root crops. We added shelling peas, herbs, flowers, cucumbers, onions, green beans, and a wider variety of greens. When arrangements with a neighboring farm to produce tomatoes, summer squash, melons, and potatoes did not work out, we began to grow those as well. Our friends at Black Walnut Farm were happy to grow these crops but did not want to take on a fair share of the organizational work of the CSA.

After nine years, David lost his appetite for juggling that many crops. By contrast, I still find fun in discovering new ones, especially when I can fit them into established cropping patterns. For example, planting mizuna along with bok choi, Chinese cabbage, and lettuce in the spring or throwing in a row of senposai next to the kale and collards for the fall involves little additional work. Growing for the farmers' market, we were used to planting greens sequentially every ten days so as to have a steady supply. It took a year or two to figure out that the best planting sequence for carrots was two three-hundred-foot beds every two weeks from mid-April through late July. Supplying a steady flow of broccoli all through the fall is trickier; shifting waves of cold and heat can cause broccoli plants started as much as a month apart to ripen at the same time.

At Peacework Farm, we grow in five-foot-wide permanent beds with grass strips in between them. A Celli spader allows us to turn in cover crops easily and to work parts of beds so that as soon as we have harvested a crop, we can spade in the residues. With four fields and close to three hundred beds, rotation planning has become an elaborate game, taking into consideration the previous two years of crops, residues that might interfere with planting, irrigation

> As a rule, farms producing for CSAs seem to be able to provide an average of twenty shares per acre. CSA farmers tend to reduce the amount of land needed per share as they become more experienced. Increased mechanization lowers the handwork per share, but it also lowers the number of shares per acre. The most intensive farms rely almost entirely on hand labor.

Community Supported Seeds?

By c. r. lawn

Monsanto's 2005 purchase of Seminis, the world's largest vegetable seed company, sent tremors through the seed community. Seemingly the climax of the unprecedented seed industry consolidation in the last thirty years, it left the ten largest multinational seed companies in control of half the world's commercial seed sales.

Monsanto has not limited its dominance to biotech traits (controlling 88 percent of the world's total area planted to genetically modified seeds in 2004) or to vegetable seeds. It has also bought out two of the world's largest cottonseed companies, including Delta and Pine Land, co-holder, along with the United States Department of Agriculture (USDA), of the original Terminator technology patent. (Dubbed "Terminator" by the Rural Advancement Foundation International, this genetic engineering technique prevents plants from setting viable seed so that farmers have to purchase new seed every year.)

Only they know whether they are planning to incorporate transgenic traits into their vegetable seed lines, but clearly their aim is nothing less than control of the world's agricultural varieties, the germ plasm, whether for cotton or tomatoes. As the chairman and chief executive officer of Seminis under Monsanto, Mexican billionaire Alphonso Romo, put it, "Seeds are software. And we have the seeds." As farmers and gardeners, we are dependent on an industry that views seeds not as the common heritage of ten thousand years of agriculture but instead as proprietary intellectual property for which we must pay each time we use them.

I hear you protesting, "I don't buy from Seminis, I buy from Johnny's, Fedco, Territorial and Nichols." Do you use Celebrity or Big Beef tomatoes, Fat & Sassy or Red Knight peppers, Gold Rush or Sundance squash? You sow Seminis. How about Silver Queen corn, Jade or Indy Gold beans, Sugarsnaps? You sow Syngenta, another multinational behemoth.

These are industry standard seed varieties that few commercial growers, whether conventional or organic, would want to be without. Even small independent seed companies are heavily dependent on huge wholesalers for some of their best varieties.

Few seed retailers produce many of their own seed crops. That carrot hybrid in the Fedco catalog was not grown in Maine, nor was seed for that luscious broccoli in Vesey's catalog grown on Prince Edward Island. For the most part, the best retailers run trials in plots located in their climatic region to assess the relative merits of varieties but contract with wholesalers for their seed production.

Yet there are signs of change on the horizon. Could we have imagined in 1970 that a fringe movement of "hippies and nutcases" would not only achieve recognition and respectability but also become the fastest-growing agricultural movement and market? Would we have believed that a handful of farmers' markets would turn into thousands? That a remarkable new form called CSA would take root and, in less than twenty years, sprout more than seventeen hundred branches in North America alone?

Today we stand in relation to the seed as we stood two generations ago to the soil and one generation ago to the creation of community. Seed is at once the beginning and the last frontier of sustainable agriculture as we start the twenty-first century. Imagine if in the next generation we put as much energy into our seed systems as we have into rebuilding our soils and our communities. What transformations might we achieve?

Actually, Karl Marx would be proud. The antithesis to industry consolidation and its concomitant loss of both genetic and social diversity is well under way. One of its pioneer species, the Seed Savers Exchange, arose to fill the void created by the threatened extinction of many of the open-pollinated varieties being dropped by the seed industry. SSE successfully revived interest in the heirloom varieties and helped spawn a whole generation of quirky alternative independent seed companies that are still adopting many of the old varieties for commercial production, ensuring their continued availability in the marketplace. In time old classics such as the Brandywine tomato, Jenny Lind melon, and Winter Luxury pumpkin have joined émigrés such as Cosmonaut Volkov tomato and Bulgarian Chile pepper to take their place alongside the Big Beefs in the growers' pantheon. New direct markets such as CSAs, restaurants, and farmers' markets build demand for these specialty varieties, which may not ship well but taste terrific.

The market for such niche varieties is too small to interest most large wholesalers, so idealistic alternative seed companies have stepped in. Many, including High Mowing, Turtle Tree, Seeds of Change, Southern Exposure, and Fedco, now have networks of small-scale farmer-growers to produce regional and heirloom specialties that might otherwise not be available. Alternative farmer-breeders such as Tim Peters and Frank Morton in Oregon and Brett Grohsgal in Maryland are creating new varieties with desirable colors and forms, increased disease resistance and cold tolerance. The SARE grants, a relatively small program within the USDA, have disproportionately influenced these trends. By funding such seminal gatherings as the Penn State biodiversity conference in 1997, and grassroots groups such as Restoring Our Seed (2002), Save Our Seed (2003), and several West Coast initiatives put together by the Organic Seed Alliance, SARE has brought together preservationists and seed companies, university plant breeders and organic growers, midwiving in the birth of participatory farmer-breeder networks to encourage farmer seed production, enhance seed quality, and reconnect farmers to their ancient heritage of crop selecting and breeding.

Cooperating with NOFA-NY in the Public Seed Initiative and Organic Seed Partnership (OSP), Cornell University is bringing the expertise of its plant breeders together with the needs of organic farmers for varieties specifically adapted to practices of sustainable agriculture in cold climate conditions. A superior melon, Hannah's Choice, has already been released as a result of this collaboration, and more melons, zucchinis, and peppers are in the works. Other land-grant universities have begun to follow Cornell's lead, rededicating themselves to their historic mission of breeding for the public sector, not just for large private seed companies. In meetings in 2003 and 2005 the Summit for Seeds and Breeds for 21st Century Agriculture set as its key goal invigorating public domain plant and animal breeding to meet the needs of a more sustainable agriculture.

Yet it would be wise for CSA not to overly romanticize these initiatives. Even the small seed companies are highly competitive and infused in a culture of secrecy. Their interests as they grow and face economic pressures may not entirely coincide with those of a growing CSA movement. It is not outrageous to believe that seeds are too important to be left entirely in the hands of seed companies, however benign. Just as a crucial contribution of CSA has been to put eaters in touch with the realities of farming, so CSA could go one step deeper in the next generation by putting them in touch with the process of selecting varieties and growing the seeds themselves.

Beyond sowing the old and new varieties coming from these programs and seeds people, why might the CSA movement want to participate in seed work? The benefits are potentially legion and include:

1. Preservation of outstanding regional varieties even where the demand is too small to interest seed companies

2. Varietal adaptation to specific local microclimates

3. Seed independence or, more accurately, regional seed interdependence as opposed to dependence on multinational corporations or on seed retailers

4. Some defense against genetically modified organism (GMO) pollution

5. Giving members an opportunity to learn seed production skills

6. Sharing seed stories

How might CSA farms participate in seed work? One way is what Elizabeth Henderson and her partners do on Peacework Farm: working with the OSP to assess some of their breeding lines. Could CSAs provide their own seeds for gardening members? While feasible for perhaps one or two varieties each year, this might prove difficult on a larger scale because CSA farms typically offer such a diverse choice of varieties in their boxes. Isolating varieties for cross-pollinating seed crops to ensure varietal purity will prove difficult under these circumstances. However, production of the largely self-pollinating crops such as beans, lettuce, and tomatoes might be more feasible.

A great deal of exchange of produce is already taking place among CSA farms. This cooperation could be extended to seeds. Seeds, in fact, have several advantages over produce in this regard, being light and easily shippable. As the density of CSAs grows in any region, there will be increased opportunities to exchange varieties that are well adapted to the climate of that area. Perhaps several CSAs could join forces to create a "CSA of CSAs," each adopting one or two or several varieties and trading them among one another and their members. Such decentralization would reduce CSA dependence on seed companies and bring CSA much closer to being a closed system for seed. Another, perhaps more realistic, scenario is for one or a few CSAs to become seed specialists, growing crops only or primarily for seed rather than food production, and marketing to other CSAs. Maybe in the next generation CSA will become a seed community as well as a food community.

c. r. lawn, founder of Fedco Seeds, has a good perch from which to inform us of what is happening in seed production and distribution.

needs, and how far we have to move equipment. Our rule of thumb is to allow three years before planting a crop of the same family in any of the beds. Thanks to our fields' scattered locations, we can move potatoes an entire mile to outfox the Colorado potato beetles. We use three row spacings and have cultivating equipment to match: one row for crops like winter and summer squash; two rows for broccoli, corn, and peppers; and three rows for lettuce, carrots, and spinach. Our system allows for a lot of flexibility in adding new varieties and experimenting with new crops.

I am still searching for the best sequence for crop rotations. After a lot of reading and twenty-five years of experience, I don't think we understand very much yet about how to optimize our rotations. Farmer and gardener observations over centuries confirm the value of rotations for soil fertility, pest and disease reduction, and yield improvement. We have some rough guidelines: Do not plant crops of the same family repeatedly in the same field; do not plant crops that share diseases in the same field; alternate deep- and shallow-rooted crops. Another approach is to progress from sod to large-seeded crops or transplants to smaller-seeded crops that require better weed control. In *The New Organic Grower*, Eliot Coleman suggests a sequence based on his observations, some of which coincide with mine, such as potatoes following corn. But the explanation for why rotations work remains a mystery, perhaps because the factors are too many and complex for the human mind to disentangle. With the help of mycologist Christine Gruhn, we began to explore the role of mycorrhizae in relation to rotations. It took three years to even know what questions to ask. After three years of sampling roots, Christine has data showing that some mycorrhizal crops, such as lettuce, benefit from following other mycorrhizal crops, while others, such as garlic, show no difference. The Northeast Organic Network (NEON) project conducted a study on rotations with twelve highly experienced organic vegetable farmers. The Dacum chart they produced shows the rotations these farms use. Scrutinizing this chart raises as many questions in my mind as it answers,

and the farmers all admit that intuition and flexibility in the face of changing conditions are essential. We have only scratched the surface of this fascinating puzzle.

Supplying a CSA gives a grower the opportunity to use varieties that commercial agriculture avoids because they are too delicate to handle, do not ship well, or are not adequately uniform in size and shape. Robyn wanted to include recommended varieties in this book, but I think the choice of variety is too region- or even farm-specific for any one list to be very helpful. Among CSA growers there is a lot of interest in heirloom varieties. I overheard two members of my CSA puzzling over what that term means. One asked, "What are heirlooms, anyway?" In a voice of authority, Jeff Mehr replied, "They are members of VAR. (Vegetables of the American Revolution)." We may need to organize a small army of seed producers in defense of VAR to ensure that open-pollinated and non–genetically engineered seed continues to be available. (See the essay by c. r. lawn on the politics of seeds and the seed industry in the sidebar on p. 168–169 in this chapter.)

As a rule, farms producing for CSAs seem to be able to provide an average of twenty shares per acre. Through his repeated surveys, Tim Laird noticed that CSA farms tended to reduce the amount of land needed per share as they became more experienced.[1] Since CSAs range in size from three to over two thousand members and the farms cover every level of mechanization, it is hard to generalize. Increased mechanization lowers the amount of handwork per share, but it also lowers the number of shares per acre. The most intensive farms rely almost entirely on hand labor. John Jeavons teaches that with biointensive techniques it is possible to feed one person a vegan diet for a year using only four thousand square feet. Utilizing a judicious mix of mechanical and hand methods, Sam Smith of Caretaker Farm in Williamstown, Massachusetts, grew enough food for 170 households for ten months of the year on six and one-half acres, of which two acres were always under cover crops—an average of forty shares per acre. The Food Bank Farm in Hadley, Massachusetts, supplies 500 families and gives an equal number of pounds of produce to the Food Bank on forty acres, for over twenty shares per acre. These farms have in common careful planning combined with the ability to be flexible and adjust to unpredictable weather conditions.

While not abandoning the concept of sharing the risk, many CSA farms purchase a few crops from other farmers. The summer Rose Valley was under water, we bought winter squash from our friend Rick Schmidt. When major crops fail to thrive, such as spring spinach, or winter squash, David Trumble, at Good Earth Farm in New Hampshire, informed his members and bought replacements from other organic farms. If less essential vegetables such as bok choi or kohlrabi were in short supply, his shares went without. Terry Carkner at Terry's Berries Farm, which is located in a cool-climate area in Washington State, regularly buys hot weather crops from other farms to fill out the selection in her shares and at her farm stand. Subscription CSAs fill out their boxes with supplements of crops they do not grow themselves. They pay the going wholesale price for the produce or trade with other growers and do not necessarily expect to make money on the deal. Some farms, such as Grindstone Farm in New York State, buy in produce they do not grow during the warm months and then continue supplying boxes with purchased fruit and vegetables through the winter as a moneymaking venture. Each CSA must find the right balance between member satisfaction and the financial value of risk sharing.

The team behind CSA Works—Linda Hildebrand, Michael Docter, Dan Kaplan, and Scott Reed—created a planning chart based on the preferences of their combined seven hundred sharers. They have graciously allowed me to reproduce the chart in this book, along with their explanation of how to make the best use of it.[2] (See p. 174.)

On the Roxbury Farm Web site, Jean-Paul Courtens and Jody Bolluyt have posted their production manuals (www.roxburyfarm.com). One provides the details for the planting of every crop, including varieties, spacing, dates, and fertility management. Another manual gives the same level of detail for harvesting and share size.[3] Share charts

from farms in Wisconsin and Maine provide a sense of the range of crops and the range of quantities provided by different farms. The chart from Jill Agnew at Willow Pond Farm in Sabattus, Maine, which gives both the projected amounts for three years and the actual amounts for two, is especially interesting. There appear to have been crop failures or serious shortfalls in a few places—kohlrabi and zucchini did not produce for Jill in 1994. And some spectacular bounties—5 pounds of peas planned for 1994 and 40 delivered, 40 pounds of tomatoes planned for 1995 but 103 pounds delivered. Overall, though, an experienced grower like Jill is able to predict production levels with a remarkable degree of accuracy while caring for sheep, horses, chickens, turkeys, two children, and a husband. (See page 176.)

Once you have calculated what to plant and how much, you need to figure out the timing. Drawing up a careful plan of where to plant each crop based on your rotations is a good winter activity. You also need approximate dates for starting in the greenhouse or direct seeding. Actual timing and placement will depend on real field conditions. I usually feel very successful if I rate 75 percent accurate. Another critical date to note is the outside limit. For example, I learned from hard experience that at Rose Valley if we planted carrots after the end of July, they would not have time to mature before the end of the growing season. Spinach planted after mid-June is a gamble, unless you have irrigation to prevent it frying in the heat. Brassicas planted after June 20 will not suffer attack by root maggot. These dates, of course, will vary according to your particular site.

Succession planting combined with variety selection makes it possible to spread out the ripening and harvesting of a crop. We usually do two or three plantings of corn at two-week intervals. For each planting, we combine a faster-maturing corn with a slower variety. That way we can usually provide a steady flow of corn over six or even eight weeks. We have also discovered that broccoli started in different ways comes to maturity at different times. We seed the same variety of broccoli in the ground, in Speedling flats, and in soil blocks on the same day and set them out at the same time. The field starts

On Successions and "Aesthetic Gluttony"

"Rocket" is another name for arugula. I planted it last Friday, just ahead of rain. On Sunday, it was poking through the dirt. On Monday, it was a row of little dots heading down the field.

The many crops here, with their diversity of rhythms, are a bit like a jazz performance: one crop of potatoes per season; five of arugula; beets—four; winter squash—one; cucumbers—two; carrots—four; mizuna—three; lettuce—five.

We brought a thousand pounds of melons a day out of our sow pasture. The melons are gone now, the field mowed, plowed, disced, seeded to oats and vetch. First planting of cucumbers and zucchini—gone. Onion field—plowed up and seeded, onions in storage. Fields are gingerly tended then pulverized and put to rest, or planted again.

These different rhythms and different shapes of crops offer a dazzling visual feast. Fuzzy rows of carrots streak to the west, flanking scalloped tufts of green and red lettuces.

Palm tree–shaped Brussels sprouts transform a service drive into The Grand Boulevard. Massive cabbage leaves flop in a savoy sprawl on the dirt and gradually hug themselves into a big ball. Blue-green broccoli leaves jut huge and rigid while, deep in the center, a tuft of yellow-green begins its fast journey toward broccoli destiny.

Enormous heirloom tomatoes hang voluptuously on avenues of trellising, blushing yellow, pink, purple. Gladiolas spire, then gush red, magenta, white, rose.

To gaze at the whole lush display of textures, forms, colors, to notice the daily changes (yes, there are daily changes), is a privilege of being a farmer. "Aesthetic gluttony"—AG for short—the team here has dubbed it. Pink morning mists and a long row of east-facing sunflowers ready to greet the day are casually acknowledged: "Really AG."

—John Peterson, Angelic Organics, Caledonia, Illinois. Used by permission.

mature first, the soil blocks second, and the Speedlings last. When weather cooperates, this strategy can give you a steady harvest flow over three weeks from the same planting date. Similar effects are probably possible by varying irrigation. Again, these subtleties may be site specific.

As a guide to succession planting, Scott Chaskey and Tim Laird at Quail Hill Farm developed a spreadsheet to record crop, variety, greenhouse seeding date, direct seeding date, area or linear feet planted, amount of seed needed, number of plants, number of plants or flats per bed or row, amount harvested per summer or winter share, and comments (see the sample on p. 176). Steve Decater created a similar spreadsheet with additional columns for dressing applied, biodynamic (BD) sprays 500 and 501, weather when sown, and field or bed number. Once set up on a computer or simply copied, charts like these are easy to use.

Biodynamic growers use a zodiac calendar, available from the Biodynamic Farming and Gardening Association under the title *Stella Natura: Inspiration and Practical Advice for Home Gardeners and Professional Gardeners in Working with Cosmic Rhythms* (see "CSA Resources" for how to acquire). It provides a guide to the optimum planting and harvesting times for leaf, flower, fruit, and root crops according to events in the solar system. The calendar includes an explanation of its use. Some growers swear by this method. Anne Mendenhall, a seasoned biodynamic farmer, points out that real field conditions do not always accommodate the cosmic rhythms, so growers should not feel too upset if they cannot follow this scheme to perfection.

While CSAs stress seasonal eating, most CSA farms try to extend the growing season in order to provide desirable variety in the shares and to increase the amount of food members purchase from the farm. The added cost of season extension needs to be factored into share prices. Farm notes in the CSA newsletter are a good place to educate members about the care and labor involved in babying along early crops or keeping late crops from freezing. A CSA farmer in northern Vermont uses

A field of potatoes at T and D Farms in CA. PHOTO COURTESY OF T AND D FARMS.

her skill at producing early tomatoes in a greenhouse as a way of attracting new members. She displays the tomatoes at her farm stand, but when people want to buy, she informs them that the quantity is limited so only members of her CSA enjoy the privilege. Signs them right up!

An alternative to growing all the crops for a CSA on one farm is for several farms to cooperate. Each farm can produce the crops it does best, and the production system can be simpler. On the other hand, a cooperative venture presents complexities of a different order: the need to agree on quality control, scheduling, fairness in allocating crops, and dividing the proceeds. (See chapter 19 for more on this topic.)

Animals
By Robyn Van En

There are several ways that a group of sharing members can obtain meat and poultry products. The first choice may be the primary farm site that is already providing the vegetables for the shares. The second may be to connect with another farm that is already

producing these items or is in a position to start livestock production. The *Resource Guide for Producers and Organizers* from Iowa State University Extension has a useful section on "CSA Livestock and Animal Products," which outlines some of the regulatory complexities of selling animal products in Iowa. Regulations may vary from state to state.

A primary goal of CSA is to develop each site to its most diverse and sustainable potential, so animals invariably come into that equation. Introducing animals can expand the product selection for members beyond vegetables, while providing the farm with an accountable and consistent source of manure for the compost piles. You do not have to use the animals for other products; they may, instead, be "farm pets," who help out by building up the biological activity on the grassland through their foraging and manuring.

Biodynamic farmers believe that the animals define the farm as an organism. Trauger Groh writes, "The exchange of the feeds from the farm organism to the herds, and from the herd's manure back to its soil, creates a process of mutual adaptation. . . . The aim and the result of this local interchange between

CSA Crop Planning Chart

Crop	Per share per year average	Yield per row foot	Annual goal per 100 shares in pounds	Yield per 100 row feet	Rows needed @ 100 feet per row	Percentage of total acreage	No. of weeks harvested
Arugula	3.74	0.17	374	17	21.99	1.48	20
Basil	0.25	0.05	28	5	5.60	0.38	9
Beans	3.20	0.15	320	15	21.33	4.30	10
Beets	16.28	1.00	1,628	100	16.28	1.64	19
Bok choi	5.00	0.50	500	50	10.00	0.60	14
Broccoli	15.00	0.25	1,500	25	60.00	12.11	11
Broccoli raab	1.25	0.50	125	50	2.50	0.25	10
Brussels sprouts	5.69	1.00	569	100	5.69	1.15	3
Cabbage	19.93	2.00	1,993	200	9.96	2.01	11
Cantaloupe	9.00	4.12	900	412	2.18	0.79	5
Carrots	66.05	1.49	6,605	149	44.35	4.47	18
Cauliflower	9.66	0.84	966	84	11.49	2.32	4
Celeriac	3.00	1.49	300	149	2.01	0.20	2
Celery	2.00	1.50	200	150	1.33	0.13	0
Chard	3.27	1.10	327	110	2.98	0.30	14
Cilantro	1.50	0.40	150	40	3.75	0.25	19
Collards	2.70	1.00	270	100	2.70	0.54	13
Corn	65.00	0.66	6,500	66	99.21	18.02	6
Cress	0.10	0.25	10	25	0.40	0.03	1
Cucumbers	12.00	4.51	1,200	451	2.66	0.97	11
Dill	0.45	0.30	43	30	1.43	0.10	11
Eggplant	10.00	1.50	1,000	150	6.67	1.35	6
Endive	1.20	0.20	120	20	6.00	0.40	10
Fennel	0.96	0.54	96	54	1.77	0.18	3
Han tsoi soi	0.50	0.45	50	45	1.11	0.11	8
Garlic	2.35	0.45	313	45	6.95	0.47	3
Garlic flowers	0.26		26		byproduct		2
Kale	7.08	1.50	708	150	6.18	1.24	16
Kohlrabi	1.91	0.51	191	51	3.76	0.38	5
Leeks	5.00	0.59	500	59	8.42	0.57	4
Lettuce	22.08	0.20	2,208	20	110.41	7.45	20

How to Read the Chart

The first column shows the average amount of each crop consumed per shareholder over the course of the entire season. The second column shows our yield per row foot for each crop. The third column is simply the first column multiplied by one hundred shares, which provides the total amount we need to grow for each one hundred shareholders.

On your farm, instead of one hundred, you would multiply by the number of shareholders you have. The fourth column, "Yield per 100 row feet" is simply the yield per foot multiplied by 100. To customize this for your farm, you would want to multiply yield per foot by your most common row length.

Crop	Per share per year average	Yield per row foot	Annual goal per 100 shares in pounds	Yield per 100 row feet	Rows needed @ 100 feet per row	Percentage of total acreage	No. of weeks harvested
Mizuna	3.37	0.57	337	57	5.92	0.40	20
Mustard greens	1.93	0.57	193	57	3.39	0.23	21
Onions	8.00	0.70	800	70	11.45	0.77	0
Parsley	1.10	0.50	110	50	2.20	0.22	17
Parsnips	3.00	0.48	300	48	6.21	0.63	1
Peas	2.00	0.20	200	20	10.00	2.02	2
Hot peppers	1.50	0.76	150	76	1.97	0.40	10
Sweet peppers	10.51	1.04	1,051	104	10.14	2.05	10
Potatoes	30.00	0.80	3,000	80	37.50	7.57	4
Pumpkins	9.88	1.40	988	140	7.06	2.56	2
Radishes	2.62	0.33	262	33	7.95	0.53	12
Rutabagas	4.16	1.40	416	140	2.97	0.30	4
Scallions	5.60	0.40	550	40	13.75	0.92	14
Acorn squash	6.44	1.20	644	120	5.37	1.95	5
Blue Hubbard squash	1.73	1.40	173	140	1.23	0.45	1
Buttercup squash	2.44	1.66	244	166	1.47	0.53	3
Butternut squash	13.14	2.00	1,314	200	6.57	2.39	4
Spaghetti squash	4.90	1.20	490	120	4.08	1.48	3
Summer squash	19.00	5.18	1,900	518	3.67		
Sweet Dumpling squash	1.00	0.50	100	50	2.00	0.73	1
Delicata squash	2.45	0.50	245	50	4.90	1.78	3
Tatsoi	3.00	0.20	300	20	15.02	1.01	18
Tomatoes	36.59	5.03	3,659	503	7.27	1.47	9
Tomatoes, plum	9.00	2.50	900	250	3.60	0.73	4
Turnips	1.83	0.33	183	33	5.50	0.56	6
Watermelon	24.95	7.09	2,495	709	3.52	1.28	4
TOTAL	405.30	51,273.96	686.00	100.00			

Planning for Contingency

As the Colorado potato beetle has proven a few times over, 660 row-feet of eggplant does not equal 1,000 pounds of baba ghanoush. The actual crop amounts listed on the chart assume an ideal world. You should add in a contingency factor of 5 to 25 percent, depending on how important the crop is to your members. You don't want to run out of staples like lettuce, corn, tomatoes, or carrots—so your contingency for these crops might be close to 25 percent. Smaller margins will work for the lesser crops. If you run out of fennel or kohlrabi, you are not likely to cause a riot.

Succession Planting Planning Chart

Crop	Variety	Greenhouse seeding date	Direct seeding date	Area/Lin Ft	Seeds needed	#Plts (PL/SD)	Plant (#Bed) #Trays/type	Harv/Share	Harv/WS
Arugula			4-19, 5-6, 5-23, 6-11, 7-10, 8-2	2 beds/1,200	12 oz.		6 (1 row)	24 bunch	
Beans	Jumbo		4-30, 5-17, 6-6, 7-1, 7-29	10 beds/4,000	25 lb.		5 (2 rows)	20 lb.	
Beets	Windsor Fava		4-1	2 beds/800	5 lb.		1 (2 beds)	2 lb.	
	Chioggia		4-14	4 beds/2,400	3 lb.		1 (4 beds)	25 lb.	
	Formanova		5-11	2 beds/1,200			1 (2 beds)		
	Lutz Green Leaf		6-4	4 beds/2,400			600	1 (4 beds)	25 lb.
Bok choy	Mei Qing Choy		8-2		1 bed/600	1/4 oz.	5/128		5 heads
Broccoli	Green Valiant	6-30		4 beds/1,600	3 oz.	2,700	13/128	10 heads	
	Emperor	7-20		6 beds/2,400			19/128		
Brussels sprouts	Prince Marvel	5-16		1 bed/600	1/4 oz.	600	5/128	10 lb.	10 lb.
Burdock	Takinogawa		4-13		1/4 lb.		1 (1 bed)	10 lb.	
Cabbage	Early Jersey Wake	6-12	4-3, 5-16, 6-12	2 beds/800	9 beds/3,600	1/2 lb.	2,400 (800)	7/128	12 heads
	Storage #4	6-12			1,000 seeds	500	4/128		10 heads
Chinese cabbage	Blues	8-2		1 bed/600	1/4 oz.	600	5/128	3 heads	
Carrots	Thumbelina		4-4	6 beds/3,600	2 lb.		1 (6 beds)	50 lb.	50 lb.
	Narova		5-1	4 beds/2,400			1 (4 beds)		
	Bolero		6-13	6 beds/3,600			1 (6 beds)		
Cauliflower	Violet Queen	5-16, 6-12		8 beds/3,200	12 gr.	2,000 (1,000)	8/128	10 heads	3 heads
	Amazing								
Celery	Ventura	3-29		2 beds/1,200	1/4 oz.	900	7/128	6 stalks	
Celeriac	Diamante	3-29		1 bed/600	1/4 oz.	450	4/128		10 lb.
Corn	Sugar Buns	4-30	5-17	10 beds/400	20 lb.	2,000	1(5b)+16/128	50 ears	
	Kiss 'n Tell		6-5, 22	20 beds/8,000			2 (10 beds)		
Collards	Champion	5-23		1/2 bed/200		150	2/128	8 bunch	7 bunch
Cucumbers	Marketmore 86		5-26, 6-22		6 beds/1,200	7 oz.		2 (3 beds)	20 lb.
Eggplant	Rosa Bianca	4-3		2 beds/800	8 gr.	600	9/72	10 lb.	
Fennel	Zefa Fino		5-3, 6-2	1 bed/600	2 oz.		2 (1/2 bed)	8 stalks	
Garlic	French Red		Mid-October	5 beds/3000	90 lb.		(5 beds)	30 bulbs	15 bulbs
Greens	Spring Raab		4-19, 5-6, 5-23, 6-11, 7-10, 8-2	12 beds/7200	25 lb.		6 (2 beds)	24 bunch	
Herbs	Sacred Basil	4-3			1/4 oz.	400	4/128	18 bunch	
	Borage	4-7				150	2/72		
	Cilantro		5-3, 6-2	2/3 bed/400	1/2 oz.		2 (1/3 bed)	10 bunch	
	Fern Leaf Dill		5-3, 6-2	2/3 bed/400	8 gr.		2 (1/3 bed)	10 bunch	
	Garlic Chives	4-3			pkt.	72	1/72	10 bunch	
	Tarragon	4-3			pkt.	72	1/72		
	Sorrel	4-3			pkt.	72	1/72	10 bunch	
Kohlrabi	Kolibri		6-19	1/2 bed/300	1 oz.		1(1/2 bed)	10 lb.	

Source: Quail Hill Farm

CSA Distribution

CROP	1	2	3	4	5	6	7	8	9	10	11	12	13	14	15	16	17	18	19	20	21	22	23	24	total	unit	pyo	on farm	del
BEAN, SNAP						1	2	2	2	2	2	1	1	2	2	2	1								20	#	yes	20	0
BEAN, EDAMAME											1	2	2	2	2	2	3	3							17	#	yes	17	0
BEET				1	1			1	1					2		1.5			1.5		2		2		13	#	no	13	13
broccoli										1			1	1	1					2	1				7	#	no	7	7
BRUSSELS' SPROUTS																							1		1	pc	no	1	1
cabbage						1			1		1					1		1			1		1	1	7	hd	no	28	28
cabbage, chinese		1	1			1																			3	hd	no	3	3
canteloupe								1		1															2	pc	no	4	4
carrot				1	1	1	1	1	1	1	1					1	1	1	1	1	1	1	1	1	16	#	no	16	16
cauliflower																	1								1	pc	no	2	2
celeriac																			1		1				2	#	no	2	2
celery													1		1		1		1		1				5	pc	no	5	5
chard			1				1				1	1		1				1							6	#	no	6	6
collards																									0	#	no	0	0
corn, sweet								0.5	0.5	0.5															1.5	dz	no	7.5	7.5
cucumber						1.5	3	3	2		1	1	1	1											13.5	#	no	13.5	13.5
daikon												1				1		1			1				4	pc	no	4	4
eggplant									1				1	1	2	2	3	2	2						14	#	no	14	14
escarole																	1		1						2	#	no	2	2
fennel						1	1				1														3	pc	no	1.5	1.5
garlic														0.5											0.5	#	no	0.5	0.5
garlic (scapes)		1		1																					2	#	no	2	2
greens	1	2	2	1	1	1			1	1	1	0.5	1	1	1	1		1	1					1	19.5	#	no	19.5	19.5
kale								1					1	1		1	1	1		1		1	1	2	11	#	no	11	11
kohlrabi																									0	pc	no	0	0
leek														2		1		2		2			1		8	pc	no	8	8
lettuce	2	2	2	1	1	2	2	1	1			1	1	1				1					1		19	hd	no	19	19
onion											1	1	1		1		1		1		1				7	#	no	7	7
parsnip																					1.5		1.5	1.5	4.5	#	no	4.5	4.5
pea, shelling				2																					2	#	yes	2	0
pea, snap			2	2																					4	#	yes	4	0
pea, snow		2	1	1																					4	#	yes	4	0
pepper, hot											0.3	0.3	0.2	0.2	0.2	0.5	0.5	0.5	0.5						3.1	#	no	3.1	3.1
pepper, sweet											1	1	1	1	2	2	1	1	2						12	#	no	12	12
potato													2	2	2		2		2		2		2.5		14.5	#	no	14.5	14.5
potato, sweet																									0	#	no	0	0
radicchio				1						1															2	pc	no	1	1
radish		1		1	1									1		1									5	bu	no	2.5	2.5
rutabaga																		1.5		2		2			5.5	#	no	5.5	5.5
scallion			1		1	1	1		1	1	1														7	bu	no	7	7
spinach	1																								1	#	no	1	1
squash, summer			1	1.5	1.5	1	1.5	1	1	2	3	3	1	1	1										19.5	#	no	19.5	19.5
squash, winter														3		3		3		3				6	18	#	no	18	18
tomatillo											1	1	1	1	1	1	1	1	1	1					10	#	yes	10	0
tomato, cherry									1		3	3	3	3	3	2	1	1							20	#	yes	20	0
tomato, plum											2	4	4	4	4	4	3	1							26	#	yes	26	0
tomato, slicing							1	2	2		2	5	6	6	6	3	1	1							35	#	no	35	35
turnip	1		1											1							2		2	2	9	#	no	9	9
watermelon											1														1	pc	no	3	3
y - basil						1	1	1	1	1	1	1	1	1	1	1	1	1							13	bu	no	1.625	1.625
y - cilantro				1	1	1	1	1	1	1	1	1	1	1	1	1	1	1							15	bu	no	1.875	0
y - dill				1	1	1	1	1	1	1	1	1	1	1	1	1	1	1							15	bu	no	1.875	0
y - parsley					1	1	1	1	1	1	1	1	1	1	1	1	1	1	1	1					15	bu	yes	1.875	0
zf-blueberries																									0	pt	yes	0	0
zf-raspberries																									0	pt	yes	0	0
zf-strawberry	1	2	2	1																					6	qt	yes	12	0

	on farm	del
tot #/share	454.4	333.7
avg.#/wk	18.9	13.9
$/# share	0.8	1.1
tot#farm	163566.0	53729.7
tot#farm (not including herbs and flowers)	217295.7	

From Brookfield Farm, Dan Kaplan.

animal and forage is permanent fertility."[4] Whether you adhere to biodynamic teachings or not, developing a site to its most sustainable potential requires recycling nutrients. The ideal situation is to create a complete system on site where you do not have to import any of the necessary inputs. This entails the production of all feed for the animals at the site and then use of all of their waste products in the compost to reincorporate the nutrients derived from the fields back into the soil and the crops once again.

While this has been traditional practice at farms and gardens around the world for thousands of years, some recent writings on the subject provide a theoretical understanding of the processes involved. I highly recommend Alan Savory's work on Holistic Resource Management, Robert Rodale's writings on Regenerative Agriculture, and Rudolf Steiner's essays on agriculture.

If a CSA farm does not have animals already, the farmers or the core group need to discuss the various options and pros and cons to decide which animals are appropriate and to determine the member demand for animal products. I recommend extensive research, especially if animal husbandry is a first-time venture for the production crew. Since all members may not be interested in shares of meat or eggs, the core may want to create a separate budget and share price for these items. Some CSAs include eggs and meat in the regular shares, while others do not. Silver Creek Farm offered shares of the lamb, chicken, and eggs they raised themselves, as well as beef and goat cheese, which neighboring farms delivered. Supplying other products is another way to get the larger community involved with your CSA. Your community may have lots of home gardeners who do not want a vegetable share but who might be interested in eggs or meat from free-range animals that have been raised organically, treated humanely, and not dosed with antibiotics or hormones.

The presence of animals at a CSA site may be a moral, religious, or philosophical issue for some of your members. While only 2 percent of the population are vegetarian, the percentage among CSA sharers is probably a lot higher. Members of one CSA took exception to the use of animal byproducts as soil amendments and wildlife deterrents, arguing that their use helps to perpetuate the livestock industry. The core group acknowledged their concern but decided to overrule their objection. The majority felt that the needs of the soil and crop protection were more important and that integrating a sustainable number of animals into the farm and raising them in a humane way could not be equated with supporting the industrialized livestock industry. The discussion provided a good occasion for the education of all members on the difference between humane animal husbandry and industrial meat production.

Even if a CSA is based on a corner lot in an urban center, some animal waste products can be incorporated into the compost piles to achieve optimum microbiological activity and to build up nutrient levels in the soil. At an urban site, you might not be able to keep animals, but you can buy bags of dehydrated manure from garden supply stores. In a few cities, urban gardeners have been able to gain the cooperation of the mounted police, who were happy to have a good use for the cleanout from their stables. A small amount of manure goes a long way when mixed with the plant wastes from the garden as the basis for the compost, and, ideally, members could return their kitchen and yard scraps to add to the piles. Quail Hill Farm explored community supported composting for a few years, supplying members with biodegradable bags, designed by Woods End Research Laboratory, for their food scraps—though they dropped the project when the price of the bags got too steep. Rabbits are the most suitable animals for an urban site. They provide excellent manure for a contained nutrient cycle. The site would have to be large enough to grow their food and to give them room to run around. Security also could be a problem.

Animals are a lot of work and a daily responsibility. Even to board or allow animals at a CSA site for their company and pasture management reflects on the whole project. The CSA group must be sure that the animal owner complies with local health and animal maintenance codes and that insurance coverage acknowledges the animals' presence and insures accordingly, for everyone's protection.

What is a Share?

The chart below shows what we estimate to constitute a share. Shareholders share the risks and the bounties inherent in growing. As you can see here, yields fluctuated according to yearly weather variations. Weights are in pounds.

Vegetable	Est. '94	Actual '94	Est. '95	Actual '95	Est. '96
Apples	—	—	10.0	10.0	10.0
Asparagus	2.0	2.1	2.0	3.0	2.0
Basil	1.0	4.5	2.0	3.0	2.0
Beans	20.0	21.0	20.0	18.0	20.0
Beans, dry	2.0	2.0	2.0	—	2.0
Beets and greens	15.0	17.0	15.0	6.8	15.0
Broccoli	10.0	11.8	10.0	19.0	10.0
Brussels sprouts	2.0	3.0	2.0	—	2.0
Cabbage	15.0	9.0	15.0	6.6	15.0
Celeriac	—	—	—	0.5	0.5
Chinese cabbage	4.0	5.0	4.0	5.7	4.0
Carrots	30.0	24.0	30.0	30.0	30.0
Cauliflower	5.0	10.5	5.0	4.5	5.0
Cucumber	15.0	25.5	15.0	13.0	15.0
Daikon	2.0	0.4	2.0	2.0	2.0
Eggplant	3.0	0.0	3.0	2.2	3.0
Flowers	pick your own		pick your own		
Garlic	—	—	—	0.25	0.25
Herbs	pick your own		pick your own		
Kale	5.0	2.5	2.0	2.4	2.0
Kohlrabi	2.0	—	2.0	—	2.0
Leeks	1.0	2.0	1.0	3.3	1.0
Lettuce	20.0	12.0	20.0	23.0	20.0
Melon	20.0	15.0	20.0	29.0	20.0
Onions	15.0	4.3	15.0	15.5	15.0
Parsley	1.0	1.5	1.0	1.5	1.0
Parsnips	—	—	—	1.0	1.0
Peas	5.0	40.0	5.0	7.5	5.0
Snap peas	2.0	2.3	2.0	1.0	2.0
Snow peas	2.0	0.5	2.0	2.0	2.0
Peppers	2.0	0.8	3.0	2.0	3.0
Potatoes	50.0	14.5	50.0	36.0	50.0
Pumpkin	30.0	25.0	25.0	10.0	25.0
Radish	5.0	1.5	5.0	2.0	5.0
Rhubarb	—	—	—	2.6	2.5
Rutabaga	10.0	0.0	10.0	10.0	10.0
Scallions	4.0	2.9	4.0	1.5	4.0
Spinach	2.0	3.4	2.0	1.5	2.0
Winter squash	30.0	15.0	25.0	5.7	25.0
Summer squash	15.0	31.0	20.0	8.0	20.0
Swiss chard	5.0	2.5	5.0	3.0	5.0
Tomatoes	40.0	21.6	40.0	103.0	40.0
Turnip	4.0	3.3	4.0	9.0	4.0
Watermelon	4.0	5.0	4.0	14.0	4.0
Zucchini	10.0	—	—	20.0	20.0

Source: Willow Pond Farm CSA.

Most CSAs have members who would be interested in shares of eggs and stewing chickens, so hens may be the first critters a group will consider. Acquiring and outfitting a flock of chickens is still a major step. It is difficult to set up a small-scale chicken operation that combines fair regard for the birds' well-being and relative happiness with cost-effectiveness. The benefits, however, are many: You can't beat the eggs for flavor, freshness, and nutrients; and the litter and waste from the chicken house and yard are excellent in the compost pile. You can expect 112 pounds of manure per year per bird. Covering the costs is another matter. Chicken tractors are a worthwhile contraption, but you must consider the work of construction and the time it takes to move them around the field. Chickens are vulnerable to wild predators and can also be temperamental.

I had always had a flock of mixed-breed chickens, ten to twenty at a time. With the CSA project going and demand for eggs increasing dramatically, I thought I'd try a bigger flock. Friends leaving town offered me two hundred hens, so we moved them to Indian Line Farm. Once the hens and roosters settled into their new "digs," free-ranging with organic grains, they all seemed happy and healthy laying eleven dozen eggs a day. Until, that is, I went away for two days, during which time they missed some component of their feed ration and stopped laying for an entire month! Of course, they continued eating, but we got less than a dozen eggs from two hundred hens. We could not supply the accounts we had established for eight-dozen eggs a day. I'm sure that part of the problem was that the birds were mostly hybrids and temperamental. Our plan was to build up to mixed breeds over time, but along with the coon and coyote problems and the rate at which the flock ate down their ranging area, the chicken business proved to be way more trouble than it was worth to us. Since our experience, I've read similar stories, so I can't stress enough the need to research and talk to experienced folks before you begin. You do not want to compromise your own energy reserves or the existing CSA during your learning process. (See *Humane and Healthy Poultry Production: A Manual for Organic Growers*, by Karma Glos [NOFA, 2004], for more information.)

Preliminary research may determine that your production crew or CSA site is not prepared or appropriate for having animals. Holistic Resource Management training in ecological decision-making is very helpful in figuring out what steps you can take to have animals in the future. In the meantime, you can pursue a relationship with local animal farmers, who can supply your members with an expanded product selection and your compost piles

with manure. When CSA projects start networking with other farms, the possibilities are myriad. CSA has the capacity to involve a wide variety of farm operations in supplying consumers with a broad selection of regionally produced products. Transcontinental and intercontinental relationships of product exchange among sustainable producers are within the realm of possibility.

———————————●———————————

In a world of growing population and shrinking fertile land, CSAs, even those with relatively unskilled farmers, have proven their capacity to produce enough food for twenty families or more on each acre. As CSA farms mature, their production becomes more intensive, whether on one farm or several associated farms. Where industrialized agriculture seems to have passed its peak of productivity, and more chemicals no longer means greater output, biological farms with community support offer long-term prospects of unlimited promise.

CHAPTER 13 NOTES

1. Timothy J. Laird, "Community Supported Agriculture: A Study of an Emerging Agricultural Alternative" (master's thesis, University of Vermont, 1995), 29.
2. CSA Works has disbanded, but Michael Docter is willing to answer questions.
3. www.roxburyfarm.com.
4. Steven McFadden and Trauger Groh, *Farms of Tomorrow Revisited: Community Supported Farms—Farm Supported Communities* (Kimberton, PA: BD Association, 1997), 24.

HANDLING THE HARVEST

Suitable temperature and moisture, and the separate storage of incompatible types of produce
are the keys to keeping fruits and vegetables fresh and attractive as long as possible.
—TRACY FRISCH, *"How to Keep Fresh Fruits and Vegetables Longer with Less Spoilage"*

Good-quality produce starts with good harvesting practices. Skilled produce farmers recognize the peak of crop maturity and schedule harvesting at the best time of day to prevent loss of nutritional value. With sharers or without, the harvest crew needs certain basic tools: sharp knives; clippers; rubber bands for bunching; containers for harvesting; a cart, wagon, or truck to haul the food in from the fields; a setup for washing the produce in clean, cold water; and a cool place to store it once picked. Harvesting and post-harvest handling take up 50 percent or more of the time and energy involved in the production of fruits and vegetables. This is the underside of vegetable production, which small-scale farmers need to make more efficient. (For a look at how far you can push efficiency, see *CSA Works Harvest Video*, made by Michael Docter and Linda Hildebrand to demonstrate their techniques for harvesting six thousand pounds of produce with three people, a flatbed truck, and a tractor.[1])

Plant Respiration

The energy that plants store through photosynthesis is released by cellular respiration, both while a plant is living and after it has been harvested. Of course, energy that is lost through respiration is no longer available to us when we consume the food. Therefore, the greater the rate of respiration post-harvest, the faster a fruit or vegetable loses nutritional quality. The following chart will help you understand what happens to your crops at various temperatures. For example, at any given temperature, bunched carrots respire more rapidly than do topped carrots. Heat is more damaging to some crops than others. (See p. 184–185.)

CSA harvesters also need to understand a lot more about how to maintain quality after produce has been picked. At Roxbury Farm, Jean-Paul and Jody divide crops into three harvest groups. Greens are crack-of-dawn crops to be picked with the cool of night upon them. The harvest crew passes lettuces and bunched greens on as quickly as possible to the packing shed crew to submerse in cool water and store in a cooler. At Peacework, we harvest lettuces early on the morning before distribution so that we can cool them for a full twenty-four hours before putting them in shares. The longer cooling period preserves their vitamins and extends the shelf life in our members' refrigerators. Next come crops that wilt less readily, such as cucumbers, broccoli, summer squash, and tomatoes—mid- to late-morning harvest will do for them. With the exception of the tomatoes, these also get the cool water wash and cooler treatment. Last come the bulk crops—potatoes, sweet potatoes, onions, garlic, winter squash—which you

Members help farmer Katie Lavin wash lettuces at Peacework Farm. PHOTO BY KATE LATTANZIO.

can harvest at any time and which require curing appropriate to each crop.

Different crops have differing storage requirements. Tracy Frisch's little booklet "How to Keep Fresh Fruits and Vegetables Longer with Less Spoilage" is an excellent guide to optimum storage temperatures for maintaining produce quality and nutritional value. If kept from obvious wilting, many vegetables will still look all right after several hours of storage at the wrong temperature, but they will have lost a significant portion of their nutritional value and storage life. Watershed Organic Farm CSA provides members with a handbook that summarizes briefly how to store the vegetables to retain freshness and nutritive content. The Genesee Valley Organic Community Supported Agriculture (GVOCSA) and Madison Area CSA Coalition (MACSAC) food/cookbooks contain this information too. (See "CSA Resources" for a selection of publications.)

An Extension newsletter on direct marketing gives this sensible advice on postharvest handling:

> In general, a fruit or vegetable is at its prime eating quality when it is harvested (the few exceptions to this are fruits or melons that may be ripened after harvest). This does not necessarily mean that a vegetable would not be better if it were left on the plant longer, but simply that it does not continue to improve after it is removed from the mother plant. In fact, the quality of that item a few hours, days, or weeks later will depend on how well the rate of deterioration is controlled. From a practical standpoint, this means controlling the storage or display environment, especially temperature and humidity, and thus reducing deterioration due to respiration and transpiration (water loss).[2]

There is also a lot of information on crop standards

Washing potatoes at Fair Share Farm in Kearney, Missouri.
PHOTO BY TOM RUGGIERI.

Farmer Ammie Chickering welcomes members and explains the morning's work at Peacework Farm. PHOTO BY ELIZABETH HENDERSON.

and post-harvest handling on the Internet, accessible through www.agnic.org/agbd/, the National Agricultural Library's listing of agricultural resources.

The goal of this chapter is not to teach you how to harvest crops (only hands-on experience can really do that), but rather to raise some important considerations.

All the studies of CSA members show that access to high-quality, organic, fresh, and locally grown produce is at the top of the list of reasons for joining. (See master's theses by Dorothy Suput [1992] and Tim Laird [1995]; and Rochelle Kelvin's Rodale survey [1994].) Gerry Cohn at University of California at Davis surveyed the members of ten CSAs in California in 1995. He concluded that "the three most important aspects in the decision to join a CSA are: (1) fresh, seasonal produce; (2) environmental concerns; and (3) supporting a local farmer. The three least important aspects are: (1) only source of organic produce; (2) education about farming; and (3) convenience. It is also notable that price is not considered particularly important."[3] In Vermont, however, Jane Kolodinsky found that price was a more important factor.[4]

Since freshness and quality are so important to CSA sharers, harvesting must be designed accordingly. Greens must be cut while the day is still cool

(never left in the sun), washed, and stored in a cool place as quickly as possible. Most vegetables left sitting in a warm place drop drastically in nutritional value within minutes. (See the chart on p. 184–185 for information on plant respiration, an indicator of quality loss during storage.) You may not need expensive hydrocooling, but you must pay attention to detail. As soon as you can afford refrigeration, make the investment. It will repay you in savings of nutritional values, in convenience, and in improvement to the storage life of your vegetables. The walk-in cooler at Rose Valley freed us to pick some things the day before distribution and lowered my blood pressure. Setting up at Peacework, a walk-in was one of the first things we built. At the distribution (pickup) location we built a second cooler so that in warm weather the vegetables spend as little time as possible without refrigeration.

You must make your own decisions about cosmetic standards. As Robyn put it:

Many of the current CSA farmers come from an extensive farmers' market or restaurant supply background, where providing a good-looking product was integral to their success but required a lot of prep and groom time. Often these farmers still strive to maintain high standards of appearance, even though the extra work can impact the

Respiration Rates of Fruits and Vegetables

Commodity	0°C (32°F)	4–5°C (40–41°F)	10°C (50°F)	15–16°C (59–60°F)	20–21°C (68–70°F)	25–27°C (77–80°F)
Apples, summer	3–6	5–11	14–20	18–31	20–41	—
Apples, fall	2–4	5–7	7–10	9–20	15–25	—
Apricots	5–6	6–8	11–19	21–34	29–52	—
Artichokes, globe	14–45	26–60	55–98	76–145	135–233	145–300
Asparagus	27–80	55–136	90–304	160–327	275–500	500–600
Avocados	—	20–30	—	62–157	74–347	118–428
Bananas, green	—	—	—	21–23	33–35	—
Bananas, ripening	—	—	21–39	25–75	33–142	50–245
Beans, lima	10–30	20–36	—	100–125	133–179	—
Beans, snap	20	35	58	93	130	193
Bean sprouts	21–25	42	93–99	—	—	—
Beets, topped	5–7	9–10	12–14	17–23	—	—
Beets, with leaves	11	14	22	25	40	—
Berries						
Blackberries	18–20	31–41	62	75	155	—
Blueberries	2–10	9–12	23–35	34–62	52–87	78–124
Cranberries	—	4–5	—	—	11–18	—
Gooseberries	5–7	8–16	12–32	27–69	41–105	—
Raspberries	18–25	31–39	28–55	82–101	—	—
Strawberries	12–18	16–23	49–95	71–92	102–196	169–211
Broccoli	19–21	32–37	75–87	161–186	278–320	—
Brussels sprouts	10–30	22–48	63–84	64–136	86–190	—
Cabbage	4–6	9–12	17–19	20–32	28–49	49–63
Carrots, topped	10–20	13–26	20–42	26–54	46–95	—
Carrots, bunched	18–35	25–51	32–62	55–106	87–121	—
Cauliflower	16–19	19–22	32–36	43–49	75–86	84–140
Celery	5–7	9–11	24	30–37	64	—
Celeriac	7	15	25	39	50	—
Cherries, sweet	4–5	10–14	—	25–45	28–32	—
Cherries, sour	6–13	13	—	27–40	39–50	53–71
Citrus						
Grapefruit	—	—	7–9	10–18	13–26	19
Lemons	—	—	11	10–23	19–25	20–28
Limes, Tahiti	—	—	—	6–10	7–19	15–45
Oranges	2–5	4–7	6–9	13–24	22–34	25–40
Cucumbers	—	—	23–29	24–33	14–48	19–55
Endive	45	52	73	100	133	200
Figs, fresh	—	11–13	22–23	49–63	57–95	85–106
Garlic	4–14	9–33	9–10	14–29	13–25	—
Grapes, American	3	5	8	16	33	39
Kale	16–27	34–47	72–84	120–155	186–265	—
Kohlrabi	10	16	31	49	—	—
Kiwifruit	3	6	12	—	16–22	—
Leeks	10–20	20–29	50–70	75–117	110	107–119
Lettuce, head	6–17	13–20	21–40	32–45	51–60	73–91
Lettuce, leaf	19–27	24–35	32–46	51–74	82–119	120–173
Lychees	—	—	—	—	—	75–128

Commodity	Temperature 0°C (32°F)	4–5°C (40–41°F)	10°C (50°F)	15–16°C (59–60°F)	20–21°C (68–70°F)	25–27°C (77–80°F)
Mangoes	—	10–22	—	45	75–151	120
Melons						
Cantaloupes	5–6	9–10	14–16	34–39	45–65	62–71
Honey dew	—	3–5	7–9	12–16	20–27	26–35
Watermelons	—	3–4	6–9	—	17–25	—
Mushrooms	28–44	71	100	—	264–316	—
Onions, dry	3	3–4	7–8	10–11	14–19	27–29
Onions, green	10–32	17–39	36–62	66–115	79–178	98–210
Okra	—	52–59	86–95	138–153	248–274	328–362
Olives	—	—	—	27–66	40–105	56–128
Papayas	4–6	—	—	15–22	—	39–88
Parsley	30–40	53–76	85–164	144–184	196–225	291–324
Parsnips	8–15	9–18	20–26	32–46	—	—
Peaches	4–6	6–9	16	33–42	59–102	81–122
Pears, Bartlett	3–7	5–10	8–21	15–60	30–70	—
Pears, Kieffer	2	—	—	11–24	15–28	20–29
Peas, unshelled	30–47	55–76	68–117	179–202	245–361	343–377
Peas, shelled	47–75	79–97	—	—	349–556	—
Peppers, sweet	—	10	14	23	44	55
Persimmon, Japanese	—	6	—	12–14	20–24	29–40
Pineapples, mature-green	—	2	4–7	10–16	19–29	28–43
Plums, Wickson	2–3	4–9	7–11	12	18–26	28–71
Potatoes, immature	—	12	114–121	14–31	18–45	—
Potatoes, mature	—	3–9	7–10	6–12	8–16	—
Radishes, with tops	14–17	19–21	31–36	70–78	124–136	158–193
Radishes, topped	3–9	6–13	15–16	22–42	44–58	60–89
Rhubarb, stalk	9–13	11–18	25	31–48	40–57	—
Romaine	—	18–23	31–40	39–50	60–77	95–121
Rutabagas	2–6	5–10	15	11–28	41	—
Spinach	19–22	35–58	82–138	134–223	172–287	—
Squash, butternut	—	—	—	—	—	66–121
Squash, summer	12–13	14–19	34–36	75–90	85–97	—
Sweet corn, with husks	30–51	43–83	104–120	151–175	268–311	282–435
Sweet potatoes, uncured	—	—	—	29	—	54–73
Sweet potatoes, cured	—	—	14	20–24	—	—
Tomatoes, mature-green	—	5–8	12–18	16–28	28–41	35–51
Tomatoes, ripening	—	—	13–16	24–29	24–44	30–52
Turnips, topped	6–9	10	13–19	21–24	24–25	—
Watercress	15–26	44–49	91–121	136–205	302–348	348–438

Note: Respiration rate often shows as a range. To compute heat evaluation rates at harvest time, use either the highest figure or median one. To convert to Btu/ton (2,000 lb.)/24 hour day, multiply respiration rate by 220. To convert to kcal per 1,000 kg/24 hour, multiply by 61.2. Some data included for low temperatures that cause injury to certain commodities or cultivars, such as avocado, mango, okra, papaya, pepper, pineapple, tomato, and zucchini squash; these low temperatures are potentially dangerous and should be avoided.

Source: From R. E. Hardenburg, A. E. Watada, and C. Y. Wang, The Commercial Storage of Fruits, Vegetables, and Florist and Nursery Stocks, *Agriculture Handbook No. 66, USDA, 1986. With permission.*

Crew of members pack shares in packing shed at Vermont Valley Farm in Blue Mounds, Wisconsin. PHOTO COURTESY OF VERMONT VALLEY FARM.

share price. Other farmers do the absolute minimum, removing dead leaves and weeds, sloshing the carrots and radishes in a bucket of clean water to rinse off the dirt clods, spritzing the lettuce with water and storing it in the shade or a cool place to deter wilting, and removing the obvious slug. The results may not look like supermarket produce, but the quality, variety, and nutritional value cannot be beat.

CSAs can certainly work to set themselves apart by the aesthetically perfect appearance of their produce, but their budget will rarely cover the cost of the labor required. Pound for pound, CSA members get a tremendous return on their investment, even if they have to do an extra bit of washing or separate the radishes from the string beans in the bag. If this is a real issue with the members, discuss it at your annual meeting or ask for comments in the newsletter or a year-end survey. Better yet, get

members involved in distribution, and see what standards they can maintain in the rush to make the shares look as nice as possible in the shortest amount of time.

CSA is supposed to be socially responsible, meaning that all hours of labor should be paid for. Little measures such as not scrubbing the carrots, rubber-banding the herbs, or putting several items in one bag will save an amazing amount of time and materials. Understandably, this goes totally against the training or standards of some growers.

Members who participate in the harvesting are less likely to complain about dirt or bugs. We had a member who loved CSA and had an infant son who was allergic to almost everything, other than vegetables and fruit, so each week she would cook and purée almost an entire share for him. At one point, though, she couldn't refrain from telling us that she was annoyed with the mud puddle she'd get in her kitchen each month when she dealt with her winter share. We explained that crops stored in a root cellar must be left with the soil on to retain their moisture content, and invited her to help with the winter distribution.

She did show up to help. Eight hours at 35 degrees, handling 1,500 pounds of vegetables, wearing as many clothes as she could while still being able to move, taught her a lot. It was significantly colder outside the root cellar, and the thought of how many more hours it would take to wash and dry everything in those conditions made her realize how much she had taken for granted.

Finally, though, your standards are up to you and your members. Robyn and I argued about the name for this book. She wanted to call it *Everybody Gets Some Crooked Carrots*. At Peacework we do not include crooked carrots in the regular shares, but we do sell them in bulk for juicing. Our CSA members are mainly city people, described by core member Pat Mannix as "recovering Wegman's addicts"

(Wegman's is the local super-supermarket). We try to meet their expectations more than halfway. To get my partner Greg Palmer, who can't throw anything away, and a few of the members who are the same way, to adhere to our standards, I urge them to carry a special bag while harvesting. They get to keep the misshapen vegetables they just *have* to pick, although they know Ammie would reject them. Our own kitchen, the local soup kitchens, and the food bank are happy to take the culls.

———————•———————

Picking and packing are the forgotten end of vegetable production. Low-paid migrant farmworkers do most of this work. If we ever hope to have a just and sustainable food system, we must face this reality and figure out how to change it. Inviting the members of CSAs to help harvest their own food is a good beginning.

CHAPTER 14 NOTES
1. Michael Docter and Linda Hildebrand, *CSA Works Harvest Video* (Hadley, MA: CSA Works, 1997), video.
2. Cornell, Department of Agricultural Economics, no. 10 (1979).
3. Gerry Cohn, "Community Supported Agriculture: Survey and Analysis of Consumer Motivations" (research paper, University of California at Davis, 1996). Available from Gerry Cohn, 127-B Hillside St., Asheville, NC 28801.
4. Jane Kolodinsky, and Leslie Pelch. "Factors Influencing the Decision to Join a Community Supported Agriculture (CSA) Farm," *Journal of Sustainable Agriculture* 10, no. 2/3 (1997): 129–41; and Jane Kolodinsky, "An Economic Analysis of Community Supported Agriculture Consumers" (SARE grant #5-24544, 1996).

DISTRIBUTING THE HARVEST

As raw materials production is moved around the world to where land and labor
are cheapest, local economics and ecologies are destroyed for economic reasons.
—JOAN DYE GUSSOW, *Chicken Little, Tomato Sauce and Agriculture:*
Who Will Produce Tomorrow's Food?

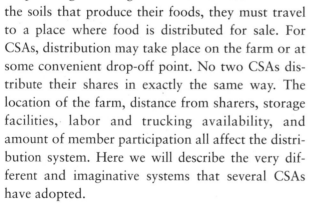

The easiest way to get your vegetables is to walk out into your garden and pick them. At the end of a workday on the farm, I take great pleasure in picking up a knife and a bag and "going shopping." Sadly, most people never experience this simple delight. Living at a distance from the soils that produce their foods, they must travel to a place where food is distributed for sale. For CSAs, distribution may take place on the farm or at some convenient drop-off point. No two CSAs distribute their shares in exactly the same way. The location of the farm, distance from sharers, storage facilities, labor and trucking availability, and amount of member participation all affect the distribution system. Here we will describe the very different and imaginative systems that several CSAs have adopted.

Farm Pickup

The simplest way to distribute the shares is to have members come to the farm to pick them up, assuming they live nearby. At Indian Line Farm (ILF), about half of the members picked up their shares at the farm. The farm crew, comprising the head gardener, his assistant, and member volunteers, picked the vegetables that morning and packed them

for the members. The distribution area was the cool, shaded downstairs room in a big old barn. A few jars of flowers decorated the four-by-ten-foot table built from salvaged lumber that held the shares. A blackboard served as communication center, with a list of what was in the harvest, notices, announcements, and space for members to write comments.

The other half of the sharers picked up their bags at the Berkshire Co-op Market, six miles away in Great Barrington. In Robyn Van En's words, "The Co-op had bought shares in the Apple Project and was very willing to offer a corner in the store for our town drop site, allowing regular Co-op shoppers to coordinate their shopping and introducing other community members to the Co-op for the first time. Early on we discovered that we had a pocket of members all living around the mountain in Sheffield. To cut down on their travel (and traffic at the farm), we provided these members with a list of their names and numbers, and let them work out their own rotation of pickups from the farm."

ILF also sent some shares to Lexington, Massachusetts, located about two and a half hours away. As Robyn described it:

The first year our delivery cost was factored into their share price. The second year, they worked out

Robyn Van En packs shares at Indian Line Farm.
PHOTO BY CLEMENS KALISCHER.

a rotation among themselves, and each share-holding member came out to the farm once per season (twenty-two weeks) and each month through the winter. The third year, we made arrangements for the local Berkshire Mountain Bakery to take the shares along with their weekly bread delivery to the Boston area. The Lexington households paid a portion of the bakery's transport costs and filled their own weekly bread orders at the same time.

Piggy-backing with the bakery truck was a real breakthrough, but the ultimate would have been a backhauling connection. That is, to identify a truck that delivers something to your community each week, and have that same, previously empty, returning truck haul shares back to your drop point in the urban center. You would need a site manager to help unload and tend the site, but it would be the most responsible way to go, short of bicycle trailers (like those used at the Topinambur CSA near Zurich, Switzerland), to avoid paying for auto insurance and burning more fossil fuels.

Hidden Valley Farm in Alna, Maine, actually did deliver some of its shares by bicycle! The farm collaborated with Bicycle Works, which transported the farm's Portland shares by bicycles towing covered trailers. Similarly, Jim Gregory's Fresh Aire Delivery and Recycling Service delivered Magic Beanstalk

shares in Ames, Iowa, for $2.80 a week during that CSA's first year of operation. Amy Courtney of Freewheelin' CSA near Santa Cruz personally hauls all twenty of her shares by bicycle with trailer.

The Indian Line group tried packing the shares into half-bushel baskets, switched to brown paper bags, then to large plastic bags, and finally ended up back with paper. When they realized they were using three hundred small plastic bags a week for carrots or greens that needed to stay moist, they found two members who were willing to contribute muslin and sewed six twelve-by-fourteen-inch bags for each share. "This allowed for two bags to be at the farm for the next harvest, two filled up with the current week's harvest, and two in the laundry or lost in space." They started by writing each member's name on the bag, then found they saved time by keeping a check-off list.

Quail Hill Farm on Long Island has the ultimate system: Members pick their own shares. From 9:00 A.M. till noon on harvest mornings, hundreds of people with knives and bags wander among the beds cutting spinach leaves, bunching greens, and carefully selecting ripe tomatoes. A board near the parking area provides a guide to the harvest, and signs with large, clear lettering designate the beds that are ready for harvesting. Each spring the farm staff gives an orientation walk for new members, who also get pointers from their fellow pickers. Farmer Scott Chaskey says that about 80 percent of the members turn out to pick during the summer, but numbers drop to 50 percent in the fall. Although members requested a wash area, few of them use it, taking bits of the farm soil home with the crops.

All the sharers in the Food Bank Farm near Amherst, Massachusetts, pick up at the farm. After the first two years of low membership retention, the farm made a concerted effort to learn what its members wanted. Michael Docter reports:

> During our third year, we implemented a mix-and-match greens table. We put all the collards, kale, lettuce, mizuna, arugula, radishes, etc., on one table and gave shareholders a plastic bag of a certain size and let them fill their bag with whatever they wanted. If someone did not like greens and

Basil Pickup: GVOCSA member and webmaster Mike Axelrod picks up his share at the Abundance Cooperative Market. PHOTO BY TOM RUGGIERI.

Amy Courtney of Freewheelin' CSA delivers shares to Santa Cruz, CA, on her bicycle. PHOTO BY CAROLINE NICOLA.

just wanted to fill their bag with lettuce, great. Let them eat lettuce.

Members were thrilled with the system. Finally, they had some real choices. . . . Not only were our members happy, but also the farm saved thousands of dollars in harvest labor under the new system. We no longer had to spend time counting heads of lettuce and bunching and tying greens. Distributing by volume meant we no longer needed to produce uniform head sizes. This enabled us to switch to a much less costly lettuce production system.

In August 1995 we broadened this method of distribution to include the majority of crops we raise on the farm. We place the carrots, beets, broccoli, eggplant, peppers, etc., on another table and give members a bag, which they can fill to a certain point with the vegetables they like.

At first, we were afraid that everybody would just take carrots. Our fears were unfounded. While carrot consumption did go up about 20 percent, we distributed just about the same amount of other crops as we had under the old system. Some folks took lots of carrots, others took lots of eggplant—but it all averaged out the same in the end. Consumer response was overwhelmingly positive.[1]

Members not only pick up their vegetables at Caretaker Farm in Williamstown, Massachusetts,

but they wash them too. Conveniently located near the compost pile is a sink with a supply of rubber gloves, scrub brushes, and aprons. After depositing their vegetable wastes from the previous week, and covering them with dry material from the nearby pile of leaves, they can step up to the sink and rinse this week's takings. This arrangement saves work for the farmers and keeps the kitchen floors of members clean.

Farm Stand and Farmers' Market Pickups

Farms that run farm stands or sell at farmers' markets have found ways to combine choice and convenience for their sharers. After leaving the boxes for his sharers under a tree near his farm stand and watching the produce wilt by late afternoon, Skip Paul had a brainstorm to combine the two enterprises. During its second year, the Wishing Stone Farm CSA in Little Compton, Rhode Island, started offering its members a line of credit at the farm stand. Rather than purchasing a uniform box, members could choose what they wanted from the stand at special CSA-only prices, with extra savings on crops that were abundant that week. Skip reduced their discount on regular prices from 20 to 10 percent. Although he sold the stand, the forty-five Wishing Stone members who pick up at its farmers'

Member families dig potatoes for their own share at Quail Hill Farm in Amagansett, Long Island. PHOTO BY KATHRYN SZOKA.

season, money sometimes remains in member accounts. For the first few years, the Goransons returned the unspent money. These days, they send a reminder card and if there is no response keep the change. They also offer winter shares; for $150 families pick up a box a month for five months.

Earthcraft Farm in Indiana sold a cluster of 180 units to the CSA members who picked up at their farmers' market stand, to apportion at their convenience. Members did not seem to worry about leftover units. Earthcraft's 1997 CSA handbook says, "If you only used half your units, but you ate wonderful meals, tried new recipes, stayed healthy, and enjoyed life with the help of Earthcraft Farm produce, then you got from the subscription what we tried to make available through it." The high retention rate suggests that members agreed.

In-Town Pickup

Farms that are located farther from their members need a more elaborate distribution strategy. Most Genesee Valley Organic Community Supported Agriculture (GVOCSA) members live in Rochester, New York, a thirty-five-mile, one-hour drive from Peacework Farm. Despite the distance, each share comes with the commitment to work three four-hour shifts at the farm helping to pick the vegetables. At the end of their shift, that week's working members transport the produce to the pickup point at the Abundance Cooperative Market, a member-owned natural food store. They store the vegetables, packed in large boxes and bags, in a refrigerated box cooled by an air conditioner on a timer. Peacework farmers write out detailed directions to the distributors so they know what to put in the shares. Late in the afternoon, the distribution crew, five or six members supervised by a core group member, arrange the produce for pickup. They weigh out and bag such items as spinach and carrots. Weighing produce is a job many children enjoy, although combining childcare and distribution does not always go smoothly. A distribution captain told me about a mother who came with a fussy baby in a backpack.

market stand in Providence can purchase a debit card for $350 and get the same deal. They agree to spend the entire amount within the twenty-week season. The farm has a certified commercial kitchen in its barn where a special staff cook and bake an assortment of products for sale at the market and to stores.

The Goranson Farm in Dresden, Maine, invites its customers to pay in advance and receive a discount at its farm stand. Anyone can buy at the stand, but the 275 or so families who pay in advance are making a special commitment. Jan Goranson asks, "Why give interest to a bank, when we can give a discount to our customers?" For $100, members receive $102 worth of food, for $200 they get $207, and for $400 they get $430. As at Wishing Stone, members pay up front and then take whatever they want at the stand until they use up their credit. At the end of the

CSA member bags her share according to the weekly list at Caretaker Farm. PHOTO BY CLEMENS KALISCHER.

The mother took the job of weighing onions. Every time she bent over to pick up more onions to place in the scale, the baby threw the ones already in the scale basket on the floor.

Members have an hour and a half to collect their shares, which they assemble according to the amounts posted. They check off their names on a list. Lines of various colors through their names signal that they have bulk to collect, that they should take boxes back to the farm, or that their payments are overdue. If there is something in the share they don't like, they can leave it or exchange it for something in the share box. They also pick up and pay for any bulk orders from the previous week, and make orders for the next week, a system that sometimes leads to confusion. Marianne Simmons recounted this bulk tale:

One day, member John Garlock came for his share and found 40 pounds of red cabbage waiting for him that his wife Marilyn had not mentioned ordering. Perplexed, he paid for the order and lugged it home. When I got home from distribution that evening, Marilyn called in a panic. "Marianne, why do I have fourteen red cabbages and what will I do with them?" "Sauer-kraut?" I suggested lamely, and explained that she had ordered them. The bulk order sheet indeed had her name next to "red cabbage 40 lb." The price for the cabbage was 40 cents a pound. Marilyn, it turned out, had left off the decimal point. The next morning, I was able to sell thirteen heads to St. Joseph's House of Hospitality for their noon meal.

A special order coordinator oversees sign-ups and collects payments for products from other farms, such as organic maple syrup, wine, and grape juice. (See chapter 18 for more details). When pickup time is over, a staff person from the Southeast Ecumenical Ministry takes unclaimed shares to distribute to families in the neighborhood who need the food.

Like the GVOCSA, the farms that deliver to New York City leave the details of distribution in the hands of their core groups. In most cases, the farmers themselves make the deliveries and then hang out during pickup hours for the chance to make contact with members. Negotiating New York City bridges, tunnels, streets, and traffic gives the farmers many headaches. The Just Food staff have become expert in dealing with would-be arresting officers, and its Tool Kit includes the phone numbers of all NYC police precincts. For the Roxbury CSA, core members take turns meeting the farm truck and helping to unload the vegetables for the shares each week. Then they oversee the pickup hours. Members make up their own packets, with freedom to select from the produce up to the posted amounts. Signs alert them to the relative scarcity or abundance of the various items. Just Food reports that people are pretty good about not hogging early-season tomatoes, and that they appreciate the allowance for choice. The core group also handles sales of items such as locally baked bread to members.

In San Francisco, members of Live Power Community Farm sort vegetables into share baskets. PHOTO BY NANCY WARNER.

Roxbury Farm delivers shares to New York City pickup site. PHOTO BY PETE LOWY.

Live Power Farm is located even farther away from its San Francisco members than these farms are from New York City. The farm crew packs the vegetables in bulk boxes and the Decaters and their sons take turns driving their unrefrigerated van the 3 1/2 hours to the pickup site, where there is refrigeration. The core group oversees the breakdown into shares at the Native Plant Nursery in the Presidio. Members take turns picking up for established groupings by neighborhood. A veritable farmers' market takes place each Saturday with flower and fruit shares from Good Humus Farm, grain shares from Jennifer Green, organic butter in bulk from Straus Dairy, and chicken and egg shares from two other farms. In the winter, Good Humus takes over the vegetables from Live Power. Coordinating this variety of shares requires a lot of organization; the farmers try to alternate share deliveries. Live Power also has drop-off points in towns nearer the farm.

Not all distribution schemes require such a high level of member involvement. Many CSA farm crews do all the bagging or boxing of shares and all the deliveries. T and D Willey Farm in California; Angelic Organics in Caledonia, Illinois; Harmony Valley Farm in Viroqua, Wisconsin; and Full Belly

Farm in Guinda, California, deliver by truck to distant cities. CSA members volunteer their porches or garages as drop-off points. In exchange, they get a reduction in their own share price or even a free share. Harmony Valley members take turns accompanying the farmers on the city end of the deliveries to help with unloading. Tony Ricci, who drives the produce from Green Heron, his Pennsylvania farm, 115 miles to sites around Washington, D.C., says he could be delivering anything. He becomes a delivery person. His main concern is to organize the process well using a computerized system, to time his trips to avoid traffic, and to keep from getting tickets. Two of his drop-off sites are in office buildings where he had contacts on the inside, one with the staff of the World Wildlife Fund.

With his characteristic relish of detail and good humor, Bob Bower of Angelic Organics described their delivery system at the 1997 Northeast CSA conference. They have twenty drop-off sites in a loop around the city of Chicago and its suburbs. Each site has a host who, for a $100 reduction in share price, is willing to allow the use of a porch or to leave a garage door open, doesn't mind signs being posted, and is congenial with other members.

Farmer Katie Lavin packs boxes into members' cars for the trip back to the pickup point at Abundance Cooperative Market.
PHOTO BY KATE LATTANZIO.

The farm crew selects sites that are shaded and cool, with protection from the elements. A month ahead of the first delivery of the season, the farm sends out a letter to members with clear, concise instructions on the process. Since the farm is two hours away and the farmers few, they have set up a buddy system among the members, "a tree-like structure for building community." The advantages of this distribution system, according to Bob, are the ease it affords for adding more sites, its flexibility, and the fact that no one has to staff the pickup hours. It also allows delivery close to the homes of the members, within a fifteen-minute drive. On the downside, it limits share choice possibilities, though they provide a swap box. (Encouraging members to come in early for a swap hour for trading vegetables with one another did not fly.) A survey of Angelic Organics' members convinced the farm to give up plans for changing to a bulk system at a central distribution site. Only 3 percent of the members favored that option, and many wrote answers like this one: "We like the present system because we eat things that we might not choose otherwise and end up enjoying."

Bob talks about the box-packing ritual at the farm as the "highlight of the week. It's really fun! We do it with a conveyor belt. We pack over three hundred boxes in less than an hour. We put bins full of each vegetable along the side of the belt. Heavy vegetables—carrots, squash—go on the bottom of the box. We'll have seven or eight people: one person making

boxes, five putting vegetables in them, and one person closing them. Then the box goes up the conveyor onto another conveyor right into the truck. The music's playing. It's really festive. And now the boxes are on the truck and we clean up. It's the culmination of the whole efforts of the week." The driver comes to the farm at six on Tuesday and Saturday mornings to make the eight- to nine-hour round-trip with the refrigerated truck. Before leaving, the driver picks up an instruction sheet on what to do at each site, target arrival times, and a check-off list to remind him to collect recycled bags and boxes and to tidy up. Bob credits the book *The E-Myth Revisited* by Michael E. Gerber (see p. 153) as the impetus for devising their packing and delivery procedures.

Bill Brammer of Be Wise Ranch in California uses a similar system for the six hundred shares he divides among sixteen delivery sites in and around San Diego. With the help of a computer expert, Bill designed a computer program that tracks billing and deliveries. The program makes it easy to allow members to select their own delivery schedule: weekly, biweekly, or whatever, with skips for vacations. It also prints out the harvest list for the four picking days a week. Every member call for a cancellation is double-logged, noted on the phone log, and then recorded onto the share program so that billing is accurate. Bill has thoughts of redesigning his system so that he can also offer home deliveries of individually chosen orders.

All the produce need not come from one farm. Several farms can coordinate deliveries to a central pickup point. In Lawrence, Kansas, six farms have joined together in the Rolling Prairie Farmers' Alliance. (See chapter 19 for more details.) On Mondays, they converge on the Merc, the local food co-op, to assemble their packets.

Home Delivery

The ultimate in member non-involvement is the home delivery, called in England a "box scheme." The farm crew picks, sorts, boxes, and delivers to the doorsteps of the members. Katy Sweeney of Malven Farm in

First veggie pickup on Monday, May 17th.

Deliveries will continue weekly for 20 weeks. Pick up your veggies Mondays 3 – 7pm.

On this page we hope to provide answers to all your questions. We won't repeat the driving directions here, so if you need driving directions, contact Erin at erintjohnson@comcast.net or 301 438-3927.

How our CSA pickup works (in 4 easy steps):

1. One box equals one share. Unless you're splitting a share, take one box.
2. Please put a checkmark beside your name on the signup sheet.
3. Some weeks we'll get supplemental produce, in addition to what's in your box. These supplemental items will be displayed in bulk. The quantities you are to take will be posted.
4. Please fold up and return your box next week.

What if you are splitting a share?

If you are part of a share group, it is your responsibility to work out with your partner(s) how to handle pickup.
The options are:

- You pick up and consume the share one week; your partner picks up and consumes the share the next week, and so on.
- One partner picks up and then all partners rendezvous later to split the bounty.
- You agree with your partners on a system to divide the shares at pickup location so that you separate and mark the shares for partners arriving after you. Please label the portion you leave for your share partner.

What if you are away?

When you are unable to pick up, it is your responsibility to arrange for someone else to pick up. You may arrange for your partner(s), a friend or prospective CSA member to pick it up. Any shares not picked up by 7 pm will be donated to the Friends retirement house.

CSA Etiquette

Except for the farmers, this is an all-volunteer operation. The farmers will not be present at the pickup location. You are likely to meet Ed Carlson. He is the person who transports the produce to our location and then closes up at night. Ed is a CSA member himself and does not work for the farmers. Please be nice to Ed!

If you have customer service issues, please contact Erin at 301-438-3927 or erintjohnson@comcast.net. If you are new to CSA, please remember that this is not like going to the grocery store. Things will look like they recently came from the field. That's a good thing.

On Tuesdays, you will receive an email listing everything that was contained in your share. The same info will be posted on this page.

FROM SANDY SPRINGS CSA, SANDY SPRING, MARYLAND

New Truck

I bought a new truck. Refrigerated. From now on, your boxes will be chilled when they arrive.

We were renting trucks to deliver your boxes. Hundreds of dollars a week for trucks that we have to pick up, return—trucks that didn't cool your boxes.

I spent months looking for the right truck—the right size, the right cooling unit, the right condition, the right price. I ended up buying the approximate truck. I'm already in love with it. Bob and I took it to Rockford last night to show it off. We went into restaurants and bars, asking people if they wanted to see our new delivery truck.

"That reefer unit sitting out in the parking lot?" they'd say. They already knew. There's something about a new truck. It just sits there and shines, like a new baby. People can't help but notice.

It's so seldom that I can go around saying, "I bought a new truck today"—twice so far. The other one I bought in '74, just before the floods that year. It was headed for Pakistan, even had a metric speedometer. The salesman told me it had been sitting on the dock, waiting for the ship, and the deal fell through. I used to wonder who in Pakistan was waiting for that truck. Every Pakistani I imagined waiting for it was so beautiful.

The truck hauled thousands of tons of corn, wheat, beans, thousands of hogs. Tons of building stone. Massive wooden beams.

An artist on the farm went to Egypt. She came back and made things black and white. (Did you check out our barn kitchen?) She painted black and white checks on the bed of my truck.

She made shrines: shrines to memories, shrines to souvenirs, shrines to shoes, to toothbrushes, to teeth—intricate assemblages of Scotch tape, cellophane, string, candles, glue. Paper mâché planets bobbed on wire armatures in her three-foot model of the universe. Extruded teeth gleamed.

One fall, she loaded up the truck with her shrines, her universe model. She called it the Traveling Curio Museum. She toured Wisconsin in my grain truck, pulled into small towns, staked triangular black and white flags from the four corners of the truck to the ground, lowered the catwalk. She set the tooth lady mannequin on the sidewalk. "Tell the tooth lady your dreams," the shrine builder implored the small townspeople. "Turn on the little tape recorder. She wants to know your dreams. Then come in." People ascended the ramp. Often looking for paintings of Wisconsin wildlife, which weren't there.

Twenty-three years later, the second new truck stalked me. I called all over Wisconsin and Illinois seeking a reliable used truck. Sometimes, especially in Chicago, the salesman would say, "Nope, had one a few weeks ago (or even a few years ago!). Seems there's one like that out your way. Some deal fell through, and there's one out there in Rockford somewhere." Refrigerated trucks can become legends.

Smyrna, New York, has been running her CSA this way since 1990. She says her members are too busy to participate in any way, and she prefers the control this gives her over her farm operation. Flying Dog Gardens in Avon, New York, also did home deliveries in the Rochester area for several years for an extra fee. The Organic Kentucky Producers Association (OKPA), a farmer-owned cooperative, similarly charged a fee for home delivery to the members of its CSA in the Lexington area. Tanyard Farm in West Hartford, Vermont, solicits members' help with deliveries to other members as a way to reduce share payments, as does Tracie's Farm in New Hampshire. Farmer Tracie Smith explains the value of this arrangement for rural CSAs: "This has been key in this area, where I live out of the way, and people want service being so busy themselves."

Entrepreneurial schemes are springing up around the country that offer home or office box deliveries. Some of these are quasi-CSAs in that they purchase all their produce from a set group of farms, which they list and describe in their promotional literature to give the flavor, if not much of the reality, of a farm connection. Others purchase vegetables and fruit for the boxes at a terminal market. To call this service a CSA would stretch the definition beyond credulity, since these operations lack any connections to specific farms.

The Rockford salesman called, offering a viewing of the new truck.

I said, "You can bring it out, but I'm not going to buy it. We're poor. We can't afford a new truck. We couldn't get the credit, anyway. In fact, don't even bring it. It's just wasting your time."

The salesman brought out the truck. He had grown up right down the road from us on a hard-scrabble farm. I hadn't seen him in thirty-five years. He had the soft manner that I remembered his whole family having.

I was disturbed at how right the truck looked, sitting in our yard. Sometimes things show up, and I know they're just the dense version of a space that's already there. If I look out of the corner of my eyes, I see that thing just plunking itself into an etheric outline, an Akashic unfolding. It's a hard thing to stave off, because it feels so much like destiny.

I bought a four-wheel-drive tractor like that this spring. I noticed it out of the corner of my eye as I was searching for other iron on a three-thousand-acre farm that was phasing out of cabbages. I wouldn't look at it directly. I wouldn't get near it. Bob mentioned it later—"The only thing worth thinking about that we saw today was that four-wheel-drive tractor, and you wouldn't even look at it."

"That thing scared me, Bob. It's too perfect. We shouldn't even think about it. We can't afford it." I bought it, because Angelic Organics already hosted an etheric space for it. That tractor's one of the main reasons we're on top of the work this year.

Farms fail for many reasons. One reason is too much junk propping up the operation—thwarted plans due to breakdowns, thwarted planting schedules due to chasing parts, thwarted deliveries. Another reason is too much debt. I try to balance the junk with the debt. This time I chose the debt; I bought the truck—five years to pay.

I had to sell my Pakistani truck at auction in '83. I tucked a little terracotta angel under its seat the morning of the sale. I knew the truck was supposed to stay on my farm, but I had to bring in every dollar just to keep the farmstead. A burly farmer bought it. He roared off in it through the spring slush.

I was driving by his farm a few years later; there was an auction at his farm that morning—the bank was selling him out. I stopped and checked for the angel under the seat; it was gone. I wasn't farming then, but I bid on the truck, because I knew it belonged on my farm. I was outbid.

A year later, I was leading a workshop on the farm crisis in a local high school. A hundred scared farmers made up the audience, wondering when they would be sold out, or needing to talk about their recent demise. Through a bank of windows behind the farms, I could see a highway. My old truck glided into view and disappeared.

—*John Peterson, Farm News, July 21–26, 1997, Angelic Organics, Caledonia, Illinois. Used by permission.*

Especially for CSA farms that are at some distance from members, getting distribution to work well for everyone involved while maintaining the high nutritional value of the food seems to present as much of a challenge as growing the food. Even for on-farm pickup, there may be tradeoffs between convenience (extended hours of share availability) and food quality (maximum freshness and appropriate refrigeration). CSA sharers and farmers around the country need to keep putting our heads together to solve this one!

CHAPTER 15 NOTE
1. Michael Docter, *Growing for Market* (February 1996): 4–5.

THE WEEKLY SHARE

We live in a world that has practiced violence for generations—violence to other creatures, violence to the planet, violence to ourselves. Yet in my garden, where I have nurtured a healthy soil-plant community, I see a model of a highly successful, non-violent system where I participate in gentle biological diplomacy rather than war. The garden has more to teach us than just how to grow food.

—ELIOT COLEMAN, *Four-Season Harvest*

How does it feel to have a garden of your very own? You await with anticipation the first tender greens of spring: dandelion greens, lettuce, asparagus, spinach, baby beets. Later come such delicacies as fresh peas, strawberries, baby carrots, and the earliest tomato. As summer reaches its peak, you enjoy the wonderful abundance of fresh corn, hot-weather crops such as tomatoes, eggplant, peppers, summer squash, cucumbers, and beans. Then comes the glut—tomatoes till you are forced by their excess to can them. As the nights cool, fall broccoli and cauliflower, sweeter carrots, onions, leeks, potatoes, and winter squash warm your tummy. And finally come the crops of late fall: rutabagas and turnips, Brussels sprouts, kale, parsnips, and pumpkins. Time to squirrel food away for the winter. This is what the Community Supported Agriculture (CSA) experience provides, without all the work.

Modern supermarkets boast of their ability to supply any food you want all year-round, if you can afford it. As a result, the post–World War II generations of consumers have completely lost touch with eating fresh foods in season. When recruiting new members to a CSA, you have to be very clear in explaining what the concept of "eating within the seasons" really means. We try hard to do this at the orientation sessions of the Genesee Valley Organic CSA, and yet every few years someone joins only to quit before the end of June because there are no tomatoes in the share.

Biodynamic teachings offer good health reasons for eating what is seasonal for your locality. The brochure for Union Agricultural Institute CSA, in Blairsville, Georgia, presents this biodynamic wisdom succinctly and refers members to *Louise's Leaves* for recipes (see the "CSA Resources" section):

Seasonally, spring greens such as spinach, lettuce, nettles, and poke cleanse and get you going for the summer. Eat them one way or another every day when you have them. It needn't be monotonous. . . . Almost too soon you'll move along to something else.

Summer fruits such as cucumbers, tomatoes, sweet corn, and peppers cool and give strength. Eat them like there's no tomorrow.

Fall greens like mustards and rape taper you back off, and roots like turnips and radishes give that inner heat and nerve power for loving winter. Stuff yourself to health.

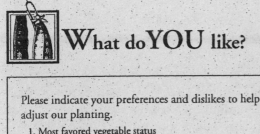

What do YOU like?

Please indicate your preferences and dislikes to help us adjust our planting.

1. Most favored vegetable status
2. A favorite, use large quantities
3. Use moderate amounts
4. A little goes a long way
5. I wouldn't eat it when I was a kid, and I won't eat it now.

Mesclun	1	2	3	4	Sweet Corn	1	2	3	4
Lettuces	1	2	3	4	Tomatoes	1	2	3	4
Radishes	1	2	3	4	Low acid	1	2	3	4
Salsify	1	2	3	4	Cherry	1	2	3	4
Kale	1	2	3	4	Paste, vine	1	2	3	4
Spinach	1	2	3	4	Zucchini	1	2	3	4
Baby Beets	1	2	3	4	Summer Squash	1	2	3	4
Baby Carrots	1	2	3	4	Acorn Squash	1	2	3	4
Broccoli	1	2	3	4	Butternut Squash	1	2	3	4
Diakon Radish	1	2	3	4	Cabbage	1	2	3	4
Pole Beans	1	2	3	4	Tatsoi	1	2	3	4
Green Beans	1	2	3	4	Leeks	1	2	3	4
Basil	1	2	3	4	Onions	1	2	3	4
Bell Peppers	1	2	3	4	Asparagus	1	2	3	4
Hot Peppers	1	2	3	4	Potatoes	1	2	3	4
Cucumbers	1	2	3	4	Pumpkins	1	2	3	4
Eggplant	1	2	3	4	Turnips	1	2	3	4
Shell Peas	1	2	3	4	Radicchio	1	2	3	4
Snap Beans	1	2	3	4	Winter Carrots	1	2	3	4
Watercress	1	2	3	4	Fennel Bulbs	1	2	3	4

Vegetable preference questionnaires, like this one from Wishing Stone Farm in Rhode Island, help CSAs plan.

A share from Peacework Farm. PHOTO BY ELIZABETH HENDERSON.

What and how much should you put in the weekly share? No one has found the magic answer to this most basic question. After surveying CSAs all over the country, though, Tim Laird came up with some practical recommendations. First, before you even begin growing the food, discuss with your members what and how much they want. Spend at least a winter planning. Next, do not overload the shares. More is not necessarily better. Especially when beginning, you need to plant extra to be sure to have enough, but you do not have to force all your food on your sharers. Excessive amounts of produce lead sharers to split shares or to drop out in embarrassment over the waste composting in their refrigerators. Finally, organize to offer some choices, if

possible. This is easier when sharers pick up at the farm or at a farm stand. (See chapter 15 for ideas on how CSAs arrange for choices and chapter 14 for a discussion of quality.)

Getting the right quantity of food can take some adjustment. Robyn described her experience at Indian Line Farm: "We started out intending a share to feed two to four people with a mixed diet or one to two vegetarians. We determined the 10- to 15-pound weekly average from the total weight we figured the household would need over the entire year. This turned out to be way too much for most households, especially as we distributed the shares twice a week. The next year, people found somebody to take the other half, so in the third year we cut back to one

199

distribution a week. Even at the reduced size, some people wanted a smaller portion. So we provided half-shares at a little more than half the price."

Farmers, who are used to their own capacious appetites, sometimes have a hard time believing city people can eat so little. The best way to establish the share size is to learn from your sharers how much they want. Take an annual poll of the members. Boise Food Connection and Wishing Stone Farm have good membership questionnaires. In the Genesee Valley Organic Community Supported Agriculture (GVOCSA) vegetable questionnaire, we explain that members will be receiving seven to eleven items a week. Given that, we ask how much they would prefer to receive of each vegetable or fruit. Surprisingly, most of them want similar quantities of most items. Bunched greens and root crops, such as turnips and rutabagas, arouse the widest deviations. We inquire whether the season's quantities satisfied them, whether they want items more or less often, and what vegetables they want in bulk. We borrowed Wishing Stone's category—"I wouldn't eat it as a kid and I won't eat it now"—to eliminate a few selections but found that for every member who hates Brussels sprouts, there is another for whom they are a favorite. So we give those items as a choice. Some CSAs provide as many as fifteen or twenty items a week.

Michael Docter of the Food Bank Farm in Massachusetts claims that the reluctant greens eaters among his sharers came to view his well-meaning attempts at increasing their greens intake as a "colossal collard conspiracy." The average between "none" and "big bunches" may be a small bunch, which won't really satisfy either group, so the GVOCSA found an alternative solution. Members who find that their shares have too many greens or roots can put them in the "Share box" for others to take. Those who get too few can order extra through the weekly bulk order. Magic Beanstalk CSA offered a greens share of an extra bunch a week. CSAs that have pickup at the farm can solve this more simply by allowing sharers to take what they want up to a given limit—the "Mix-and-Match" system.

To get the highest level of member response,

Angelic Organics includes a postcard in its share boxes every week, giving members the chance to comment immediately on quality and quantity. Of course, farmers who see their sharers face to face on a regular basis can get this information directly.

David Trumble at Good Earth Farm in Weare, New Hampshire, offered his local members a novel choice: They could sign up for either the "Dave's Mix" or the "Member Choice" option in either family (full-share) or single-sized (half-share) boxes. Those with Member Choice picked what they wanted from a computer-generated list in the previous week's box, up to a certain number of units, and called the farm by the cutoff time to place their order. The farm kept a second copy of the lists by the phone: Dave or his wife, Linda, checked off the orders as members called them in. If members forgot, they simply received a "Dave's Mix" box for that week. This system also gave the farm a better grasp of how often to put vegetables in all the shares. Close to half of the local sharers took the Member Choice option, a total of about twenty a year, which Linda reported was manageable. (Trying to do this for over one hundred shares drove Cass Peterson and Ward Sinclair to give up running a CSA.) The Member Choice option cost $80 extra for the single size and $100 extra for the family size. Ben Watson, the editor for this book, belonged to Good Earth Farm, and when he did not see this system mentioned in the draft version, he wrote:

> When I first joined Dave's CSA back in 1992, he didn't have the option of making your own selection, and my then-girlfriend and I didn't know what to make of Swiss chard week in and week out. I find the Member Choice option works better for me, particularly since I have my own garden, and thus don't need many peas, beans, tomatoes, and other crops. Instead, I supplement my own garden's production with Dave's (he grows better salad greens, potatoes, and carrots than I do, plus space-hogging crops like winter squash). In fact, I dropped out of his CSA in 1993, and then re-upped when he instituted the Member Choice and single-share size. Without these two options, I

Share Information

	May	June	July	Aug	Sept	Oct	Nov +
Asparagus	▓▓						
Beans			▓▓▓	▓▓▓	▓		
Beet Greens		▓	▓▓▓	▓▓	▓	▓	
Beets			▓▓	▓▓	▓▓	▓▓	▓
Broccoli		▓▓	▓▓	▓▓	▓▓	▓▓	▓
Cabbage		▓		▓▓	▓▓	▓▓	
Cauliflower			▓	▓▓	▓▓	▓	
Carrots			▓	▓▓	▓▓	▓▓	
Corn			▓	▓▓	▓		
Cucumbers			▓▓	▓▓	▓		
Eggplant				▓▓	▓		
Kale			▓▓▓	▓▓	▓▓	▓▓	
Lettuce	▓▓▓	▓▓	▓▓	▓▓	▓▓	▓▓	
Melon				▓▓	▓		
Onion				▓▓	▓▓	▓	
Peas			▓				
Peppers				▓▓	▓▓	▓	
Potatoes			▓▓	▓▓	▓▓	▓▓	
Pumpkin					▓	▓	
Radish	▓	▓▓					
Rhubarb	▓	▓					
Spinach	▓	▓			▓▓	▓	
Squash, summ.		▓	▓▓	▓▓	▓		
Squash, wint.				▓	▓	▓▓	▓
Sweet Potatoes							
Swiss Chard							
Tomatoes			▓	▓▓	▓▓	▓	
Watermelon				▓	▓		
Herbs		▓	▓▓	▓▓	▓▓	▓	
Flowers, PYO		▓	▓▓	▓▓	▓▓	▓	
Strawberries		▓▓	▓				

Crop	Amt/share	Units
Basil	0.56	lbs
Beans, green	9.76	lbs
Beans, purple	1.88	lbs
Beans, yellow	3.21	lbs
Beet Greens	0.97	lbs
Beets	14.6	lbs
Broccoli	4.88	lbs
Cabbage	3.83	lbs
Carrot, 2nd	1.24	lbs
Carrots	34.2	lbs
Celery	5.09	lbs
Chard	1.43	lbs
Corn	2.79	lbs
Eggplant	4.3	lbs
Kale	1.47	lbs
Leeks	0.41	lbs
Lettuce	33.9	heads
Melons	19.2	lbs
Mesclun greens	1.45	lbs
Onions	12.8	lbs
Parsley	0.24	lbs
Peas	4.9	lbs
Peppers, grn	5	lbs
Peppers, hot	0.82	lbs
Pickling cukes	6.47	lbs
Potatoes	34.2	lbs
Radish	2.28	lbs
Rhubarb	7.72	lbs
Summer squash	32.7	lbs
Slicing cukes	8.66	lbs
Snap peas	1.92	lbs
Spinach	2.9	lbs
Strawberries	3.86	qts
Tomatoes	36	lbs
Tomatoes, cherry	2.12	lbs
Tomatoes, green	0.66	lbs
Tomatoes, plum	4.43	lbs
Watermelon	14.3	lbs
Winter squash	15.9	lbs

SOURCE: MILL VALLEY FARM CSA.

Weekly Shares through the Seasons

Vegetables	Herbs	Fruit		Vegetables	Herbs	Fruit
Spring Crops				cabbage		
asparagus	oregano	strawberries		okra		
dandelion greens	dill			sweet corn		
mizuna	cilantro			tomatillos		
arugula	tarragon			broccoli		
radishes	chives			cauliflower		
garlic greens	garlic chives			lima beans		
spinach	thyme			butter beans		
corn salad (mâché)				(green soybeans)		
turnip greens				fava beans		
bok choi						
lettuce				***Fall Crops***		
mustard greens				broccoli	garlic	apples
radicchio				kohlrabi	parsley	pears
Chinese cabbage				onions	cilantro	
tatsoi				shallots	dill	
broccoli raab				potatoes	sage	
Hon tsai tai				celery		
senposai				daikon radishes		
cress				leeks		
snow peas				carrots		
sugar snaps				beets		
beets—root and greens				spinach		
scallions				chickweed		
shelling peas				bok choi		
escarole				tatsoi		
				radicchio		
Summer Crops				Chinese cabbage		
lettuce	dill	blueberries		pumpkins		
Swiss chard	basil	raspberries		winter squash		
green onions	summer savory	gooseberries		collards		
carrots	marjoram	cantaloupe		mizuna		
kale		watermelon		arugula		
purslane		blackberries		cauliflower		
vegetable amaranth		cherries		cabbage—green, red, savoy		
summer squash		peaches		burdock root (gobo)		
cucumbers		plums		long-keeping tomatoes		
beans—green, purple,		nectarines		lettuce		
Romano, yellow		apricots		Jerusalem artichokes		
sorrel				Brussels sprouts		
chicory				celeriac		
garlic tops (scapes)				scorzonera/salsify		
peppers						
eggplant				***Late Fall and Winter Storage Crops***		
tomatoes				broccoli	garlic	apples
early potatoes				potatoes	parsley	pears
				Vegetables	Herbs	Fruit

Vegetables	Herbs	Fruit
onions	cilantro	
carrots		
leeks		
turnips		
rutabagas		
Brussels sprouts		
kale		
collards		
spinach		
parsnips		
celeriac		
winter squash		
cabbage		
Chinese cabbage		

Hispanic Shares

carrots	garlic	[pineapple]
lettuce	wild mustard	[bananas]
onions	with seed	watermelon
potatoes	purslane	cantaloupe
cucumbers	sweet corn	cherries
tomatillos	cilantro	apples
tomatoes (Roma)	radishes	peaches
	beets	spinach apricots
broccoli	avocado	mango
cauliflower	jicama	
summer squash &		
squash flowers		
green beans		
cabbage		
peppers (sweet green, frying, and hot)		
nopales		

Macrobiotic Shares

macrobiotic diets usually do not include:
tomatoes
potatoes
eggplant
peppers
spinach
Swiss chard
some also exclude:
corn
zucchini
lettuce, except for romaine

would not have remained a member, despite my commitment to local agriculture.

In 2002, having become a stay-at-home dad, Dave cut back to twenty-five shares, and the next year joined with eight other farms in the Local Harvest CSA (see chapter 19).

On the other hand, Bette Lacina, who farmed on 1 1/2 acres at Under the Willow Produce in Sag Harbor on Long Island, presents a compelling argument against choice:

> Our concept was to tell people we are experimenting with growing on a small piece of land in an urban area where land is extremely expensive. If you want to support what little open land is left, you need to be willing to take what we can produce and what nature gives us. If it's a dry season or a wet season, we'll have different things. We can't guarantee anything. And the response I've gotten has been tremendous. I'm learning to eat things I never ate before. A lot of people say to me at the end of the season, they like not having a choice. It's a challenge to creative people.

To help you design your shares, we are including a list of vegetables, herbs, and fruit by season, which includes some fairly unusual vegetables along with the old garden standards. To keep CSAers in the North from turning green with envy, we have left out exotics such as figs, persimmons, avocados, and artichokes. If you can grow these, by all means put them in your shares. Farms with nearby Hispanic or other ethnic communities might try designing special shares. We have also listed choices for a macrobiotic share: by general agreement, solanaceous vegetables (potatoes, tomatoes, eggplant, peppers) are out, while controversy continues over such items as corn, zucchini, and lettuce. Check with your members, if they are interested in this option. In addition, we have chosen a few sample share descriptions from Winter Green Community Farm in Noti, Oregon; Caretaker Farm in Williamstown, Massachusetts; and Mill Valley in Stratham, New Hampshire.

Let your imagination devise creative forms of

COMMUNITY FARM PROJECTED CROPS PER LARGE SHARE

| 1995 GROWING SEASON | | | JUNE | | | JULY | | | | AUGUST | | | | | SEPT | | | | OCT | | | |
CROP	TOTAL	WEEKLY	12	19	26	4	11	18	25	1	8	15	22	29	5	12	19	26	3	10	17	24
basil	1 lb+								●	●	●	●	●	●								
beets	4 lbs	1 lb			●		●														●	●
blueberries	16 pts	2 pts							●	●	●	●	●	●	●							
broccoli	20 lbs	1 lb	●	●	●	●	●	●	●	●	●	●	●	●	●	●	●	●	●	●	●	●
burdock	1.5 lbs	.5 lb														●		●		●		
cabbage	4 heads	1 head			●											●		●		●		
carrots	30 lbs	1.75 lbs						●	●	●	●	●	●	●	●	●	●	●	●	●	●	●
cauliflower	7 heads	1 head					●		●		●		●		●		●		●			
chinese cabbage	3 heads	1 head				●											●		●			
corn	48 ears	8 ears										●	●	●	●	●	●					
cucumbers	30 fruit	3 fruit							●	●	●	●	●	●	●	●	●					
eggplant	8 fruit	1 fruit									●	●	●	●	●	●	●	●				
endive	3 heads	1 head	●																●			
flowers	occasionally																					
french sorrel	6 bun.	1 bun.	●	●		●										●		●				
garlic	40 bulbs	40 bulbs									●											
green beans	4 lbs	2 lbs						●	●													
green onions	8 bun.	1 bun.	●	●	●	●	●	●	●													
kale	5 bun.	1 bun.														●	●	●		●	●	
leeks	15 stalks	3 stalks														●	●	●		●	●	
lettuce	40 heads	2 heads	●	●	●	●	●	●	●	●	●	●	●	●	●	●	●	●	●			
melons	15 fruit	3 fruit												●	●	●	●	●				
red onions	6 lbs	1 lb									●	●	●	●	●	●						
storage onions	18 lbs	9 lbs												●					●			
pak choi	2 heads	1 head															●		●			
parsley	11 bun.	1 bun.	●			●		●		●		●		●		●		●		●	●	●
peas	2.5 lbs	1.25 lbs	●	●																		
new potatoes	20 lbs	10 lbs					●		●													
storage potatoes	40 lbs	20 lbs														●						●
pie pumpkins	3 fruit	3 fruit																				●
jack-o-lanterns	2 fruit	2 fruit																				●
spinach	8 bun.	1 bun.	●			●		●		●						●	●		●			
summer squash	25 fruit	2.5 fruit							●	●	●	●	●	●	●	●	●	●				
sweet peppers	24 fruit	2.5 fruit									●	●	●	●	●	●	●	●				
tomatoes	50 lbs	4 lbs							●	●	●	●	●	●	●	●	●	●				
winter squash	20 fruit	2/2wks+16																			●	●

SOURCE: WINTER GREEN COMMUNITY FARM.

shares. One little CSA grows ingredients for spaghetti sauce. The members gather in late summer to cook and can their sauce together. Many farms offer flower shares, a certain number of bouquets for a set price paid in advance. Separating shares by season seems to work for some farms that provide summer and winter shares. Tuscaloosa CSA divides its produce into spring/fall shares and summer shares. Seven Springs Farm in Virginia grows the basic share and contracts with other farms for a few crops, such as sweet corn, and for "sub-shares." In addition to the regular share, members can purchase a "gourmet sampler," twenty-five pounds of French filet beans, baby vegetables, shiitake mushrooms, and the like. In Portland, Oregon, John Martinson and Beverly Koch offer "The Birds and Bees Option," a mouthwatering array of seventeen fruits and berries, nuts, honey, and eggs, as an add-on to Urban Bounty CSA shares.

The Student Organic Farm at Michigan State University divides its shares into sixteen-week segments to coincide with the university calendar. A CSA called Own-a-Goat in Virginia allows goat producers to distribute raw goat's milk to people who want to buy it. The law prohibits raw milk sale but not drinking milk from a goat you "own," even though a farmer cares for it and milks it.[1] Thorpe's Organic Family Farm in East Aurora, New York, sells raw cow's milk this way. Silver Creek Farm in Hiram, Ohio, offered "Preserver shares" and received Sustainable Agriculture Research and Education (SARE) funding to set up a canning shed at the farm and to provide lessons in canning. It also sold beer shares for homebrewers, which included hops, malt, sugar, and yeast. Each year a different brewmaster set the recipe; everyone who signed up shared in the cost and received three six-packs of beer. (For ideas on connections with other farms, see Chapter 18, "Regional Networking for Farm Products.") Dan Guenthner told me about yet another kind of share. Back in 1990 he was excitedly giving a talk on CSA as a new alternative when a man interrupted him to say, "That's nothing new. In the seventies I belonged to a pot CSA. You paid your money once a year, and you got a share of the harvest."

CSAs can offer meat, eggs, and milk as part of the regular share or create separate shares for vegetarians and meat eaters. Several farms can associate to supply a wider array of products. Silver Creek Farm sold shares in its own lamb, chicken, and eggs but also goat's milk cheese and beef shares from neighboring farms. Rather than recruiting more members, CSAs can grow by diversifying product offerings to provide more of the food needs of current members. Wayback Farm in Belmont, Maine, took this path, gradually adding milk, meat, dry beans, and grains to its vegetable shares. In 2003, Essex Farm in upstate New York charged right into a CSA with the goal of providing all its members' food needs—vegetables,

The Production of Food in the Service of Land and Community

EATING WITH THE SEASONS

Spring
Arugula
Beets/greens
Lettuce
Radish
Rhubarb
Scallions
Shell peas
Snap peas
Spinach
Strawberries

Summer
Arugula
Green beans
Beets/greens
Broccoli
Carrots
Chard
Cherry tomatoes
Corn
Cucumbers
Garlic
Kohlrabi
Lettuce
Melons
New potatoes
Peppers
Raspberries
Strawberries
Scallions
Summer squash
Tomatoes

Fall
Arugula
Green beans
Beets/greens
Broccoli
Carrots
Celery
Garlic
Kale
Leeks
Lettuce
Onions
Potatoes
Pumpkins
Scallions
Spinach
Tomatoes
Winter squash

Winter
Beets
Brussels sprouts
Cabbage
Carrots
Garlic
Kale
Onions
Parsnips
Potatoes
Rutabagas
Winter squash

LINOLEUM CUT BY ELIZABETH SMITH, CARETAKER FARM, 1997.

fruit, raw milk, yogurt, maple syrup, eggs, meat, and grains—year-round. With twenty-five shares and a share price of $2,400 per person per year, Essex Farm could be economically viable—if the farmers, Mark and Kristin Kimball, can keep up with the workload. Native Offerings Farm sells shares of fruit, most of which it purchases from other farms, asking $10 a week for twenty weeks. The list includes both low-spray and organic strawberries, cherries, blueberries, raspberries, cherries, apricots, plums, peaches, apples, and pears. The markup on this fruit adds a significant boost to the farm income, with a profit margin of 44 percent. Stew Ritchie explains their strategy: "We want to increase the dollar value of each share by capturing a greater percentage of our members' total food purchases. The more staples they can pick up at the farm means less running around for them."[2]

Sharers' abilities to handle the flow of food vary greatly. By supplying recipes and tips on storage, the CSA newsletter can help members improve their skills. The GVOCSA newsletter has featured stories on members who have systems for blanching and freezing extra greens or have found ways to turn their porch into a root cellar. The days are gone when everyone grew up snapping beans and stocking the family root cellar. Broadcasting from Muncie, Indiana, Garrison Keillor, on his *A Prairie Home Companion*, once referred reverentially to the town as the "Mother Church" of canning, since it was the home of Ball canning jars. With his inimitable logic, Keillor blamed the failure of modern parents to can their own tomatoes as the cause of the destruction of their children's education. According to Keillor, in the days when parents canned, they were able to afford to send their children to Ivy League colleges.

Modern homes and apartments have very limited food storage space. Many CSAs have tried to organize "processing parties" to teach the basics of canning, freezing, or dehydrating, but most have not been successful; members are simply too busy to attend workshops. In some states, the Extension Service offers workshops and advice on food storage skills. The CSA newsletter could carry a schedule of such local offerings.

Winter Shares

With the exception of some of the California subscription farms, few CSAs are able to provide shares year-round. Mariquita Farm in Watsonville, California, cut back to thirty-six weeks to avoid boring members with week after week of kale. Other Californians supplement their farm's production with vegetables and fruit from other farms. Nigel Walker of Eatwell Farm in Davis claims that going year-round helps retain members; one of his slogans is "sustainability through continuity." The coordinator for the Eatwell CSA makes up the weekly box from the farm's production, supplemented by purchases from other farms or exchanges with other area CSAs. Like a retail store, Eatwell puts a 50 percent markup on the corn, citrus, and green beans it buys to fill out share boxes.

In more northerly parts of the country, some CSAs offer winter shares for an additional payment. When Peacework shares end in November, some of the members of the GVOCSA purchase winter shares from Blue Heron Farm in Lodi. Over many years of improving farm infrastructure, Lou Johns and Robin Ostfeld of Blue Heron have constructed high-quality storage areas, recently adding cement floors with quarter-inch steel mesh extending up the walls to keep out rodents. For storing root crops, they have two twenty-by-forty-foot coolers and one ten-

by-ten-foot cooler, which they keep at 36 to 38°F. Onions and garlic go in a six-by-eight-foot storage room at 40 degrees, and winter squash is in another six-by-eight-foot room at 50 degrees. Washing takes place in a "relatively" warm room that they are tightening up with insulation. They sell through the winter to restaurants and a few stores and provide one hundred winter shares, half to members in Rochester and half in Ithaca. The shares come in two sizes and six biweekly deliveries from January through March. Besides the standard root crops, onions, garlic, and winter squash, Native Offerings adds homemade kimchi, sauerkraut, and other lacto-fermented vegetables to its winter shares.

The Student Organic Farm at Michigan State University has replicated Eliot Coleman's winter hoop house system, enabling it to sell forty-eight weeks of shares in drafty East Lansing. The Simmons Farm in Rhode Island offers two share choices during the warm months: vegetables or vegetables with beef, eggs, and goat cheese. After the twenty-week vegetable season, members can continue through the winter with a weekly dozen of eggs and pound of ground beef.

Seasonal Eating Guides

After a few years of answering members' questions about odd vegetables, I wrote *FoodBook for a Sustainable Harvest*. Lovely illustrations by Karen Kerney enable readers to identify mystery crops. Entries for seventy crops include history, nutritional information, short- and long-term storage requirements, anecdotes from growing at Rose Valley Farm, and recipes, many contributed by members. Inspired by our example, the Madison Area Community Supported Agriculture Coalition (MACSAC) created *From Asparagus to Zucchini: A Guide to Farm-Fresh, Seasonal Produce*, beautifully illustrated by Bill Redinger. It contains essays on eating locally, write-ups with recipes for forty-six vegetables, profiles of the MACSAC member farms, and other resources. Joanne Lamb Hayes and Lori Stein contacted farms around the country and compiled them in the charmingly designed cookbook *Recipes from America's Small Farms: Fresh Ideas for the Season's Bounty*. Just Food has created a booklet of one-page "Food Tips," which is easy to copy for distribution to members. To accompany his film, John Peterson and the Angelic Organics crew have come out with *Farmer John's Cookbook: The Real Dirt on Vegetables. Seasonal Recipes and Stories from a Community Supported Farm*, which includes good recipes, beautiful photos and more of John's fine writing. (You will find information on how to acquire these books in "CSA Resources.")

In "Eating Seasonally: A Personal Reflection," David Bruce paints with bold strokes the broad cultural, historical, anthropological, and cosmic significance of this practice:

> Those who are stepping into seasonal eating by participating in a CSA deserve recognition for their extra efforts. It is revolutionary eating, refusing to be a part of the environmental degra-

dation that characterizes the current food system. It is charitable eating, wanting those who produce your food to be earning a decent standard of living, and so loving others as yourself. It is Zen eating, requiring mindfulness, a simplifying of and a concentration on what we consume. It is environmental eating, for what better action than to provide an example to those around you on living more harmoniously in your world? With revolutionary spirit, charitable hearts, thoughtful practice, and sensitive action, we can learn to eat seasonally. As we do so, we learn about our region and, with creativity and caring responsibility to place, we can develop our own regional epicurean fare.[3]

The most fitting conclusion for this chapter on the food of CSA is "An Ode to Community Supported Agriculture," by Anna Barker.

CHAPTER 16 NOTES
1. *FARM* (Summer/Fall, 1996): 1.
2. Stewart Ritchie, *Growing for Market* (March 2002): 12.
3. David Bruce, "Eating Seasonally: A Personal Reflection," *From Asparagus to Zucchini* (MACSAC, 1996, 2003, 2004, Madison, WI): 13.

Beauty and the Feast: An Ode to Community Supported Agriculture

In December air
Indoors growing together
Snow mist frosts the fields

We are cooking.
Recipes for diversity
must include hair-brained schemes and land pur-
chases,
awareness of trends, open houses with salsa
(home-canned,
of course), sizzle, regional/seasonal ingredients,
and Trust.

New Pioneers in the MERF* Movement (and else-
where, where ever
there is Common Ground) who receive grants and
give advice
along with healthy food, know about efficiency,
keep their knives
sharp, and enjoy the flavor of the batter in their
stories.

We "walk the talk," reach out to network and
touch 1-800-516-7797
while "cultivating education mechanisms with con-
sumer." Really.

Trust, grown from self-delivered knowledge
(as in midwifery) and
experience-based education result in believing and
form the basis
for growing the caring/carrying capacity. (Money
counts.)

See us: we are not a fad or a social echo; we are
The Land.
We are a community of sustainers, co-creators of a
Way of Life,
setting loose a system of nutrition based on the
soil, nurtured
by the craft of farming, by elders grounded in
integrity, giving
witness through words meeting actions.

Join us as we regain the heritage that lies deep in
the Earth,
our Mother, much-covered but alive beneath layers
of concrete,
asphalt, and square buildings.

Chew works, spit out ideas. Let the organic
process spread,
dynamically influencing source-labels that increase
Trust.
Through facts and knowledge gained from
experience
stimulate growth-through-the-seasons;
grow as individuals, as unique as your footprints
on soil after rain.
Participate in peeling back the "price per pound"
mentality and
share in linking lives with The Land in ways that
cultivate
community along with food security.

Eat close to home, and invite many guests to share
the feast.

In the cool spring air
Outdoors growing together
December warms fields

This ode (framed by two haiku) is a synthesis of words chosen by participants in "Eating Close to Home," the Upper Midwest CSA Conference held December 2–4, 1994, at the University of Wisconsin-River Falls, to summarize their experience in Saturday's workshop. Anna Barker, conference attendee and poet, created the synthesis and haiku.

*MERF—Madison Eaters Revolutionary Front

Reprinted with permission from the Minnesota Food Association Digest (December 1994): 2.

COMBINING CSA WITH OTHER MARKETS

The goal must be to raise all food in wholesome, sustainable ways while eliminating the poverty and inequality that deprive many of the ability to buy an adequate diet at any price. Food produced on prosperous and sustainable family farms should be the affordable food of choice for ordinary people everywhere.

—MARTY STRANGE, "Peace with the Land, Justice Among Ourselves"

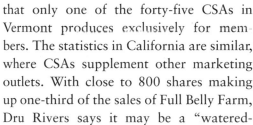

Although ideally Community Supported Agriculture (CSA) farms might be totally devoted to their members, in reality most farms and gardens must sell to other markets as well. As Steve Decater of Live Power Community Farm puts it, the market offers consumers the luxury of no commitment and total choice. Joining a CSA farm means commitment and requires flexibility. Some farms retain other established markets when initiating a CSA. Others, begun as dedicated CSAs, find that they need supplemental income and have excess production that their sharers cannot comfortably absorb. Too much produce in the weekly share has driven off more CSA members than too little.

Tim Laird's survey in 1993 found that 74 percent of the CSA farms were selling to outside markets. The more recent surveys of CSAs show an even smaller percentage of farms that are 100 percent CSA. Some farmers told Laird that maintaining diverse markets increased their security; others planned to give up outside sales as soon as they found enough sharers to ensure financial survival. Outside markets included restaurants, farmers' markets, food co-ops, local stores, and natural foods supermarkets.

I learned at the Northeast Organic Farming Association (NOFA) of Vermont conference in 1998 that only one of the forty-five CSAs in Vermont produces exclusively for members. The statistics in California are similar, where CSAs supplement other marketing outlets. With close to 800 shares making up one-third of the sales of Full Belly Farm, Dru Rivers says it may be a "watered-down" version of CSA, but she feels very good about it. Through careful record keeping, Richard de Wilde at Harmony Valley calculated in 1997 that the specialty crops, such as celery root and greens, which they grow for distributors, netted more income than their 335 shares at that time (since increased to 800). He admits, though, that the CSA is valuable for reasons other than the money. John Hendrickson's 2003–2005 study, "Grower to Grower: Creating Livelihood on a Fresh Market Vegetable Farm," concluded that CSA in the marketing mix helps to stabilize income: "Other marketing strategies are subject to the vagaries of the marketplace and weather."[1]

Since the early 1980s, food co-ops in many parts of the country have been the most faithful markets for local organic farms. Some, like the Wedge Co-op in Minneapolis, Minnesota, make preseason contracts with growers for future crops. At winter meetings, the co-op and the growers review the previous season and agree on crop projections. They also set

Signe Waller and a helper staff the Earthcraft Farm table at the Lafayette, Indiana, Farmers' Market. PHOTO BY ERNEST MCDANIEL/COURTESY EARTHCRAFT FARM.

prices for the year to come at a rate they consider to be the "Grower Sustainable Price," a premium that accounts for the higher cost of producing food in Minnesota and Wisconsin. The Wedge wants to buy as much as possible from local farmers and is willing to educate its customers about local farming conditions. The produce section also lines up second-string growers from whom it buys if the first string has production difficulties, so the co-op does not share the risk with the farmers, as many CSAs do. GreenStar Co-op in Ithaca, New York, similarly signs contracts with growers for specific crops.

Quail Hill Farm in Amagansett, New York, markets to some restaurants. One restaurant purchases bulk quantities of a few crops, while another takes two shares and features Quail Hill produce on its menu. Following the lead of Alice Waters, of Chez Panisse fame, chefs in major cities around the country have increased their purchases of fresh, in-season produce from local farms. A few restaurants and schools for chefs, such as the New England Culinary Institute in northern Vermont, purchase

regularly from local farmers and meet with them every winter to discuss specific crops and quantities. This incipient Restaurant Supported Agriculture (RSA) has yet to blossom into widespread advance contracting and payments to farms.

How do you decide what produce goes to your sharers and what to sell to other markets? Farms accustomed to selling to outside markets can easily calculate this by setting a dollar limit on the weekly share. For Peacework in 2007, that limit was $17 a week based on our wholesale prices. Each week, we plan a share that consists of at least seven or eight items that would have sold at the farmers' market for $17. If the share is short one week, we make up the difference another week. The consumer-run Farm Direct Co-op in Massachusetts aims to pay the farmers who supply their produce 80 percent of the highest retail price they can find for an item in local stores. If you are not familiar with market prices, a set number of items and a set number of pounds could be your guide. Any produce that remains after the shares are filled is available for sale elsewhere.

When we began the Genesee Valley Organic Community Supported Agriculture (GVOCSA), the thirty-one shares consumed a small percentage of our production at Rose Valley. Gradually, the number of shares has grown to three hundred, which accounts for over 90 percent of that year's production. We do not devote a particular section of our ground to the CSA. Instead, we divide production into crops grown exclusively for the CSA and crops for both the CSA and outside markets. The Veggie Questionnaire establishes the amount of these crops that members want per week. In periods of great abundance, we give them up to one and a half times the optimum share. For example, most of our sharers agree that one pound of beans per week is a good supply, so we never give them more than one and one-half pounds of beans at a time. Any beans beyond that go to other markets. A heavy crop of broccoli in 2006 brought our members a few weeks with two heads of broccoli instead of one. (See chapter 13 for the quantities several CSAs plan to supply.)

In his survey, Tim Laird found growers who make a much clearer distinction between CSA and market crops. One farm had "separate crews do the planning and work, very independent of each other." Other farms designated particular crops, such as white eggplants, yellow tomatoes, garlic, or fall crops, for outside markets. Martha Jacobs, at Slack Hollow Farm in Argyle, New York, who ran a CSA of thirty shares along with her main production for other markets, found that trying to grow a few rows of CSA crops along with her wholesale crops led to confusion. Trying to save a row of kale amidst many beds of harvested lettuce can lead to a weedy mess. So Martha set off one acre for CSA crops and treated it like a big garden, with many crops grown on a small scale rotating within that acre.

In addition to running their own CSA, Slack Hollow grew the carrots and onions for the Hudson-Mohawk CSA, an important connection for both mutual support and friendship. Janet Britt and her crew enjoyed helping with planting and harvesting and paid Slack Hollow for the onions in advance. The year when Slack Hollow lost all their onions to disease, the two CSAs suffered together, though Slack Hollow was able to soften the loss by substituting leeks and garlic. The moral Martha draws is the importance of being flexible. For a few years, Slack Hollow expanded carrot production to supply two other CSAs.

The real crunch can come during times of shortage. Drought, flood, disease—any of the plagues the Sovereign of the Universe inflicted upon the Egyptians—can descend upon a farm. Do you give as much as possible to your sharers, or do you supply the other markets for money that the farm may need badly? For the GVOCSA, the CSA always comes first. GVOCSA members sign a contract for the entire season and pay in advance of receiving our produce. None of our other markets gives us this level of commitment. Many other CSA farmers feel the same way. When Tom Meyers ran the market garden at Hawthorne Valley, he observed that if you short other markets, you feel it in the pocketbook immediately; if you short the CSA, you'll feel it next year, when the membership dwindles.

In 1992, the year of the big rains, we cut out all other markets, bought some extra boots and members slogged with us through the mud. Share packets were heavy on survivor greens such as kale and collards, but no one complained. The warm and steady support from our members was wonderful for our morale and helped us through what could have been a devastating time. Quality problems can also give you pause in deciding where or whether to sell a crop. Good communications with your CSA members may enable you to give them vegetables that you cannot sell elsewhere for aesthetic reasons. Late sweet corn that was delicious but wormy threw Martha Daughdrill, who ran Newburg Vegetable

> In 1993, 74 percent of the CSA farms surveyed were selling to outside markets. Some farmers felt that maintaining diverse markets increased their security; others planned to give up outside sales as soon as they found enough sharers to ensure financial survival.

Farm with her husband in Maryland, into a quandary:

> It tasted so fantastic, but it was wormy up the kazoo! We took a little bit to market, but we sold hardly any. So we decided to give it to the CSA folks with a disclaimer, and by and large, people were thrilled. It was the best corn of the season. . . . When we got our postcards back, one of the coolest things was one card that said, "More corn! More wormy corn next year! My kids have a worm habitat set."

(See chapter 14 for more thoughts on quality.)

Offering a weekly bulk order, in addition to the prepaid share, will encourage your members to purchase more from you so that you have less need of outside markets. Harmony Valley calls this system "Shareplus." A bulk-order system will also help you avoid dumping more produce on your members than they really need. You can include in your bulk list produce rated as seconds, such as crooked carrots or nicked potatoes, for a lower price. For the GVOCSA bulk order, each week at distribution we provide a list of items that will be available for the following week, with the price for each item. Members sign up for the amount they want and calculate what they will owe. The next week, those bulk items appear at distribution and members pay the coordinator. Since we do not supply winter shares, we try to stimulate bulk purchases for canning, freezing, and root cellaring by providing information on how best to store the food and offering "Squirrel Bulk" order forms in October. Some members buy just a little extra through the season; others stock up energetically for the whole winter.

The Food Bank Farm has invented another way to get members to buy more from them. It offers six different "value-added" packets with recipes. In Michael Docter's words, "We sell a 'Pickle Package,' which consists of all the fresh ingredients needed to make pickles: cukes, garlic, onion, hot pepper, a bunch of dill, all prepackaged in a plastic bag with a recipe included. The package looks beautiful and people eat them up. We do the same thing with salsa, tomato sauce, pesto, and a number of other recipes."[2] You can purchase these "value-added recipe kits" from them.

If you do not have experience with other markets beyond your CSA, inform yourself thoroughly about market standards for quality and packaging. The nearest Extension office should be able to provide you with federal and state standards for produce. With some markets, you can jeopardize all future sales by delivering your produce in the wrong kind of box. A restaurant will have different requirements from a retail store. Produce that might be acceptable to your CSA members or at a farmers' market stand may not look polished enough for the bright lights of a supermarket, where perceived quality is based on cosmetic perfection. The produce manager at my local co-op complained to me about a new farmer who delivered his vegetables to the store unwashed, expecting that store staff would do all the prep work.

If you plan to sell your produce as "organic," you may need to have your farm certified. The National Organic Program, administered by the United States Department of Labor (USDA), requires certification for all farms with sales over $5,000 a year. Your state may have its own organic standards and enforcement mechanism. (See chapter 11 for more information.)

———————————•———————————

On a farm run with a high level of ecological consciousness, nothing need go to waste. Top-quality crops will go for sale to primary markets, CSA or other. Processors or neighbors may provide outlets for culls. Crop that is not of salable quality but is still edible can go to the emergency food supply system. Whatever is left can feed livestock, the compost pile, or the vast "microherd" in the soil.

CHAPTER 17 NOTES

1. John Hendrickson, "Grower to Grower: Creating a Livelihood on a Fresh Market Vegetable Farm," 2004 (online article, www.cias.wisc.edu).

2. Michael Docter, *Growing for Market* (July 1996): 6.

REGIONAL NETWORKING FOR
FARM-BASED DEVELOPMENT ECONOMICS

As physical resources are everywhere limited, people satisfying their needs by means of a modest use of resources are obviously less likely to be at each other's throats than people depending upon a high rate of use. Equally, people who live in highly self-sufficient local communities are less likely to get involved in large-scale violence than people whose existence depends on world-wide systems of trade.
—E. F. SCHUMACHER, *Small Is Beautiful*

In her keynote speech at the Pennsylvania Association for Sustainable Agriculture (PASA) conference in 1996, Sarah Vogel, then commissioner of agriculture in North Dakota, made a very significant distinction between *economic development* and *development economics*. Economic development is the familiar phenomenon: Towns, counties, states, and even nations compete in offering incentives to entice enterprises from outside to select their community. Development economics works differently: It begins with an assessment of the existing resources of a community and creates an economic strategy to strengthen and build on those resources. In North Dakota the department of agriculture funded a feasibility study for a pasta cooperative proposed by the state's wheat growers. The cooperative pasta factory has become a thriving enterprise, adding value to the raw commodities produced on grain farms of the region. When CSAs network with other area farms for supplemental products, they are applying development economics to smaller, more diversified farms.

The sharers assembled in a CSA make a perfect group for bulk purchases of all kinds of sustainably produced items from other farms or enterprises.

Berries, fruit, cheeses, wines, maple syrup, bread, eggs—even nonfood products and services are possibilities. Linking up with CSAs can give a boost to the marketing of specialty farms, while members benefit from access to hard-to-find local products.

For these purchases beyond the regular share payment to go smoothly, however, some questions need to be answered. First, who is going to be responsible for choosing the products and making arrangements for buying and delivering them? If the CSA farmer handles other people's products, how should the farmer be remunerated? Who sets the quality standards for outside purchases? What is the procedure if a product is not acceptable? Does payment go through the CSA treasury or directly to the outside producer? Who handles these payments and makes sure no one is shortchanged?

The Genesee Valley Organic CSA (GVOCSA) has done outside purchasing for many years and has tried out a variety of approaches. Our experience highlights some of the problems that can arise. In 1993, I suggested that members might like to try getting chicken from Backbone Hill Farm. A member of the core group agreed to be chicken coordinator.

Backbone Hill's Beth Rose supplied us with her price list, and the coordinator took orders. (A few vegetarian members were mildly put off by a meat deal.) Beth then delivered the chickens to Rose Valley, where we kept them in the walk-in overnight for members to deliver to Rochester the next day together with the week's shares. Rose Valley's part in the deal was small, but Beth gave us a few chickens for our help in connecting her with a new market.

The first year, Beth brought the chickens in large boxes on which she wrote the bulk weight. The chicken coordinator then weighed out the chickens one by one for the orders. As one might have predicted, the two scales did not agree, or the smaller quantities did not round out exactly to the bulk weight. Whatever the cause, there was some trouble over exactly how much the members owed Beth for their chicken, and I was the apologetic intermediary. The next year, Beth switched to weighing each chicken herself and making deliveries directly to Rochester. The GVOCSA chicken eaters and I regret Beth's passing. Her surviving partner, Kim McKnight, keeps the chicken business going, though he no longer markets directly to the GVOCSA.

With another neighboring farm, we arranged for bulk purchases of organic strawberries, with the farmer delivering the berries to our walk-in for transshipment. His first berries were beautiful, top-quality. In the second batch, however, I noticed bruised berries, yet I did not want to tell my neighbor to take them home. No one had put me in charge of his quality control. The distribution coordinators ended up selling some of these berries for less than the agreed price, and the farmer returned the money to members who paid the full price and complained. When we discussed the incident at the core meeting, we decided that in the future the quality decision would be up to the distribution coordinator.

Since the strawberry disappointment, the GVOCSA core has handled all outside purchases directly. A member volunteers to be coordinator for a given product and takes charge of sign-ups, communication with the producer, and collection of payments. GVOCSA has purchased organic wine and grape juice, maple syrup, sheep cheese, strawberries, blueberries, and low-spray and organic apples. Our members get the price advantage of bulk orders, and we farmers get to mind our own business!

In 1998 the GVOCSA invited Rick and Deb Austin, who raise free-range chickens up the road from us, to offer chicken and egg shares. Instead of offering shares, every other week Rick sets up a table at distribution to deliver preordered eggs and frozen chicken. The Austins also offer pork, lamb, and Thanksgiving turkeys. This system works smoothly for everyone. Besides including the Austins' flyer in a regular mailing, the core has little involvement.

Rather than sharers making bulk purchases as a group, in some CSAs the farm does the buying and sells to the members. Michael Docter's Food Bank Farm came up with a scrip system to simplify the bookkeeping. For $10 the farm sells a "Scrip Card" with five $2 punches on it. Farm staff divide the produce into $2 packets: three pounds of pears, four pounds of apples, or five pounds of potatoes for $2 each. According to Michael, they purchase field-run produce, unsorted and ungraded, at a discount and resell it to members for a good price while maintaining a profit margin of 20 to 40 percent. Each scrip card is numbered, and a perforated section is torn off at each sale to provide the farm with a receipt.

Dan Kaplan of Brookfield Farm, near Amherst, Massachusetts, buys products from other farms and resells them to his CSA members, who pick up at the farm. He adds a small markup to cover his crew's labor, and they literally eat or drink the leftovers. Financially it is a wash, but Dan considers the additional products a useful service both to members and to the other farms.

Local Marketing Networks

In more and more localities, groups of farmers are clustering together to build local markets. Since 1998, Grown Locally, "a community farming cooperative" in Northeast Iowa, has been adding more farms and more product offerings every year. Initially, Michael Nash and Solveig Hanson of Sunflower Fields Farm ran a CSA and started inviting neighboring farmers to offer their products to the members. Michael and Solveig grew the vegetables, while neighbors provided optional shares of fruit, baked goods, eggs, and honey. The informal cooperation proved so encouraging that the group of farmers decided to form a limited member farming co-op. Membership has increased from nine to fourteen farms, and grants from Sustainable Agriculture Research and Education (SARE) and other funders have enabled the co-op to develop an online ordering system and home delivery service. Most recently, they have constructed a processing facility at Sunflower Fields Farm where they can wash, chop, and pack produce for sales to local institutions. Over the first winter of 2006–2007, they processed excess winter squash into frozen squash and "scrumptious squash casseroles" and made apple pie kits from slightly damaged apples. By shredding cabbage and peeling or dicing potatoes, the co-op is able to supply a hospital in Decorah. Solveig and Michael turned their CSA over to the co-op but continue to provide most of the vegetables for the shares. Grown Locally offers customers the choice of shares of vegetables, fruit, baked goods, eggs, and honey or the option of buying these crops and a growing list of other products, such as pork, soaps, herbs, and so forth, as individual items. In 2006 they experimented with selling their products on consignment at the Lady Bug Landscapers shop in Decorah.

Michael and Solveig say that the co-op is growing slowly but steadily as the members take more ownership and as more and more people in the region grasp the value of eating locally grown food. Sunflower gets a significant portion of its farm income from the co-op. Most of the other farmers are part-timers, who are free to participate in the co-op at any scale. The co-op allows them to make a little money from their farms while keeping full-time jobs. In his previous life, Michael says he had felt "displaced." In rural, economically depressed Northeast Iowa, however, he and Solveig sound like

they have found their place and also a deep satisfaction in helping to build the local community-based economy.

At Research Triangle Park, a complex with fifteen hundred employees in Raleigh-Durham, North Carolina, a committee of the employees teamed up with the Extension Service to create connections with local farms to "enhance the quality of life" for all involved. Once a week, six farms deliver shares of their produce to a parking lot at the research complex. People who work there, as well as employees of other area companies, have a choice of vegetable, flower, fruit, and meat shares. The employee committee does outreach, organizes sign-up days, and maintains a Web site for the project. Celia Eicheldinger, co-chair of the employee committee, said that in the first year, 2002, there were too many farms and not enough buyers. Two farms dropped out mid-season, creating a bad experience for all. In 2006 they limited the choice to six farms so that each one gets enough orders to make the trip worthwhile. Celia herself buys multiple shares: vegetables from Hannah Creek, fruit from Mystic Farm, beef and poultry from Nu-Horizon, and flowers from Highland Creek. When I asked her why she goes to all the trouble, she answered, "I feel more comfortable eating food from folks I know. Local farms are far better than distant factory farms. I believe in sustainable agriculture very strongly. It's the future. It will be a tough sell, but we will get there."

Agricultural Development

In several parts of the country, organizations that formed to reconstruct the rural economy have chosen CSA as one of their tools. Closely related to (or even overlapping with) community food security efforts, these projects combine support for local farms with the creation of job opportunities and access to improved nutrition for low-income residents. In Story County, Iowa, where 990 farms cover 90 percent of the land and gross sales of agricultural products topped $15 million in 1994, the organizers of

Magic Beanstalk CSA had a hard time finding enough skilled vegetable growers. As farms have gotten bigger, Iowa's rural communities have declined. The poverty level, though not as severe as in Appalachia, is a distressing 9.2 percent. The Magic Beanstalk was one of a number of projects stimulated by the Kellogg Foundation–funded Shared Visions program to rebuild the local economy.

The story of the Magic Beanstalk goes back to 1993, when Shelly Gradwell got snowed in at the Northeast Organic Farming Association of New York (NOFA-NY) Conference, where Katy Sweeney gave a workshop on Malven Farm CSA and many of us who attended spent the extra day in an impromptu discussion on breaking new ground. Shelly brought the CSA concept to her graduate studies and Extension work in Iowa. In 1995, Mark Harris grew one acre of vegetables, supplying the thirty shares for Magic Beanstalk's first season. That year, Iowa saw its first three CSAs. In 2006 there were forty-two CSAs in Iowa. Shelly Gradwell has since moved to Alaska, where she married Jerry Brennamen, a fisherman. They sell sockeye salmon and fresh and smoked halibut to Farm To Folk (the renamed Magic Beanstalk) members through Nick's Wild Fish.

The Magic Beanstalk never reached its early goal of becoming an intentional community on its own land, but it did serve as a market for six farms for over a decade. When the Kellogg funding ended, the project continued as a cooperative CSA, but disagreements among the producers mounted, with communication among growers a particularly bitter issue. In 2005 the survivors decided on a new strategy: instead of a multifarm CSA, Farm To Folk offers the choice of vegetable, fruit, or meat shares from several farms, bread shares from a bakery, and a range of other products "à la carte" from both farms and craftspeople. Share prices are higher than Magic Beanstalk's.

In their promotion, Farm To Folk stresses their commitment to "providing our producers with fair compensation for their labor." One of the members is Wholesome Harvest, a coalition of forty small family farms that produce organic and pasture-based livestock and are "committed to revitalizing decimated rural communities and Fair Trade principles." A 10 percent fee on sales from the farms remunerates Marilyn Andersen and Deb Edmonton, both nonfarmers, as coordinators. Deb says she will be satisfied if her salary covers the price of a share to feed her family. Marilyn, who has stayed with the project from the beginning despite minimal compensation and who sells her weavings through Farm To Folk, hopes to grow the network from its current gross sales of $30,000 to $40,000 to a level where it can provide a decent salary for a coordinator. With so many university people in the area, Marilyn calculates that asking for an up-front payment that members can use up as their schedules allow will be more viable.

The most successful effort at sustainable development based on CSA that I have been able to find is taking place north of the U.S. border in the province of Quebec. Perhaps even more than in the United States, since the early twentieth century, Quebec has lost farms to development and witnessed the consolidation and industrialization of those that remain. The number of farms has dropped from two hundred thousand to twenty-eight thousand, leaving many rural communities depopulated and depressed. CSA appeared as a way to reweave the torn rural social fabric. Since 1995, Équiterre has built a network of one hundred farms that provide organic vegetables and meats to 8700 households. A small but incredibly lively nonprofit, Équiterre offers sustainable and socially equitable choices from a "perspective integrating social justice, economic solidarity and defense of the environment." Its motto is "Change the world, one step at a time." Besides creating CSAs, the organization tries to lure people out of their cars and onto feet, bicycles, or public transportation; promotes Fair Trade products; and provides energy audits to make homes more efficient. Like a fresh northern breeze on an oppressively hot summer day, Équiterre pursues practical projects without cant or jargon.

Of the hundred farms in the Équiterre CSA network, eighty-two sell shares and eighteen supply complementary products—honey, apple products,

Farmer Michel Massuard of the Équiterre network hefts a mighty cabbage at Le Vallon des Sources, Quebec, Canada. PHOTO COURTESY OF LEVALLEN DES SOURCES.

cheeses—to the members of the CSAs. In founding the network, Équiterre had two main objectives: "to boost the development of Quebec organic agriculture by creating consumer demand, and to make organic and local products more available to citizens by allowing them to support farms directly.

In the early nineties, inspired by news from across the border, a few farms in Quebec started doing Community Supported Agriculture. Some members of Équiterre decided to try a CSA with Cadet-Roussel (the name of the farm, taken from a children's song about a mischievous boy, is a charming French in-joke), a farm in Mont St-Gregoire with a drop-off point in Montreal. The success of the first season led to the idea of establishing a network of CSA farms. Elizabeth Hunter, a staff member of

Équiterre, took on the job of coordination. The members of the network agreed on four criteria for participation:

1. Members make a commitment to share the risks with the farmers by paying in advance without guarantee of quality, diversity, or quantity, and the farmers make CSA production their highest priority.

2. The farms "work in respect for the environment and biodiversity," using no synthetic pesticides or fertilizers, and are certified organic.

3. All products in the shares are local: during the summer, a minimum of 75 percent comes from the main farm and the rest

from other regional farms. In winter, 50 percent can come from other regional farms.

4. To encourage healthy communication, transparency, and trust between the farm and its members, each project includes a social dimension—the creation of a core group, member gatherings, work days at the farm, etc. Either the farmer or the members can initiate these activities.

Équiterre has done a remarkably effective job of coordinating this network, providing technical assistance to the farmers, promoting the CSA concept with the public, and linking citizens with farms.

A mentoring program connects first year CSA farmers with an experienced farmer who helps them plan their season and provides support during that first critical year. To satisfy the growing popularity of CSA membership, Équiterre does outreach at agricultural schools, farmers' events and through the media to find farmers who could benefit from the direct link with people to develop their farms.

In 2000, Équiterre held a well-attended regional CSA conference, where I got to dust off my French as a keynote speaker, sharing the podium with noted French farm and food activist José Bové. The conference celebrated the release of Elizabeth Hunter's book *I Grow, You Eat, We Share: Guide to CSA* (*Je cultive, tu manges, nous partageons: guide de l'agriculture soutenue par la communauté*, Équiterre, 2000), the first handbook on CSA in French. *I Grow* is an excellent, practical guide to every aspect of CSA, from finding land to distribution, with good examples of farm brochures, press releases, newsletters, surveys, share contents, prices, and farm budgets. It also includes Équiterre's study comparing prices between Quebec stores and CSAs, which concludes that members save from 10 to 50 percent on organic food. Farms for which Équiterre recruits members pay 2

percent of their CSA income, with a maximum of $350 a year, to support its services. They also collect $15 from each member as a "gesture of solidarity" toward the life of the network. Associate farms pay $10. These payments fully fund Équiterre's CSA project and pay for 2 1/3 staff people.

Farmer Stephen Homer tells me that he does not have to worry about recruiting new members, since Équiterre does such a good job of promotion with its Web site, map of CSAs, brochures, newsletter, and press releases. When Elizabeth Hunter left for Lebanon, Frédéric Paré took over as coordinator with Isabelle Joncas. He has done two economic studies of Quebec CSAs (cited in chapter 9). Together, they plan a study to assess the potential market for future CSAs in the province. The growth of the Équiterre network, from one farm to one hundred in ten years, shows what is possible when talented nonprofit staffers and farmers work together.

———————————•———————————

Whether a CSA sells other farms' products as supplements to its own production, additional shares, on-farm purchases, or through bulk or cooperative sales depends on the personalities and individual needs of everyone involved. In developing these markets, it is important for CSAs to be aware of other struggling local sustainable businesses, such as food co-ops, so as not to create more competition where cooperation might be possible. The Abundance Cooperative Market in Rochester, New York, finds that, whatever it loses in vegetable sales to the GVOCSA during the summer, it more than makes up in additional sales of fruit and other products on CSA pickup evenings and in highly food-savvy members in the long run. The progress of efforts such as Grown Locally in Iowa and Équiterre in Quebec offers hope to all of us who want local food systems to replace the global supermarket.

PART FIVE

MANY MODELS

MULTIFARM CSAs

Farmers are pitted against one another in the prevailing system of commodity production,
in ways that work to the detriment of all. . . . A new production ethic will emphasize
solidarity, mutual support, and interdependence among farmers.
—Vision Statement/Call to Action for "The Soul of Agriculture:
A New Production Ethic for the 21st Century," 1997 conference

Juggling thirty, forty, even seventy crops is not to the taste or within the technical capabilities of every farmer. Instead, groups of farmers are choosing to associate to produce the crops for a Community Supported Agriculture (CSA) project, though few seem to be attempting to form integrated communities like Temple-Wilton Community Farm, where three farm entities merged into one. Consumer-initiated CSAs are purchasing crops from multiple farms instead of focusing on only one, and farmers' markets have attracted additional customers by combining produce from several farms or selling shares and allowing customers to select from all the farm stands. There are advantages to sharing the work and sharing the risk, but coordinating harvesting schedules and deliveries, apportioning crops fairly, and agreeing on quality standards are organizing challenges each group must overcome. Once a solid group of farms has formed, however, the possibilities for additional cooperation—joint purchases, sharing equipment, developing other markets—are limitless. Multifarm CSAs allow participants to make farming as big or small a part of their lives as they choose, and enable new farmers to learn their craft in association with more experienced farmers without taking on the full risk of doing a CSA on their own.

Farmer-Initiated Group CSAs

Lynn Byczynski and Dan Nagengast, renowned in the small-farm world as editors of *Growing for Market*, started a CSA on their own farm near Topeka, Kansas, then decided to expand it by inviting other farms to join them. In the winter of 1993–94, with seed money from a Kellogg grant, the Rolling Prairie Farmer's Alliance formed as a farmer-owned cooperative in Lawrence, Kansas. The founders no longer participate in Rolling Prairie: Dan's dedicated work at the Kansas Rural Center and the success of Lynn's flower enterprise leave them too busy, but the co-op is thriving. In its 14th season in 2006, Rolling Prairie has a membership of six small family farms supplying shares for 330 families. The CSA bills itself as a "vegetable subscription service," and reassures its customers that, since the farms span four counties, production is reliable: "The six farms in the alliance operate as a cooperative, which serves as a kind of insurance policy for the subscribers. If one farm gets frosted, hailed out, dried up, or attacked by grasshoppers, chances are the others can take up the slack." Paul Johnson, from East Stone House Creek Farm, a co-op member, reports that Dr. Rhonda Janke of Kansas State University has updated the Cornucopia Project data (a Rodale study from the eighties on the percentage of food each state imports) and found that despite the growth of farmers' markets, Kansas imports a whopping 97 percent of its fruit and vegetables. Rolling Prairie makes a contribution to lowering this percentage.

In December the farmers get together to decide on the crop mix for the coming year, based on surveys of the members. Each farmer makes a commitment to the dollar value he or she wants to sell through Rolling Prairie. Most of the farms sell at farmers' markets as well, so Rolling Prairie is not under pressure to absorb all of the members' production, and there is plenty of backup supply. Who grows what is an evolving process, according to Paul Johnson. He began by selling strawberries through Dan and Lynn's CSA, and has expanded to include many new crops, including asparagus. The farms share responsibility for the harder-to-grow and labor-demanding crops. The co-op pays one farmer to act as produce manager, which entails calling all the other farms each week, deciding what goes in the weekly share and which farms will provide it, and calling in the orders to the farms. The manager aims for a weekly bag value of as close to $15 as possible for the regular share and $12 a week for the economy bag. Several farms often split crops, such as peas and green beans. The manager is also empowered to do quality control but has had to pull out only a few items. The co-op asks its members for their best quality, and as the years go by, all the members' skills are improving. Occasionally, because of deficiencies in supply, the co-op cuts the value of the bag for the week. Paul says the co-op could handle more shares but needs to improve promotion.

In 2007, Rolling Prairie is charging $15 per week for twenty-four weekly bags of fresh produce. The farmers designed the $12 economy size to attract more senior citizens and other lower-income members. The co-op's accountant, also a farmer member, uses local retail prices as a guide to determine pay-

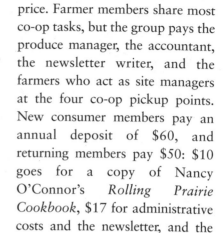

Farm Direct Co-op pays farmers the best price the members will accept, roughly 80 percent of the retail price in an upscale produce market.

ments to farmers. To help cover some of its administrative costs, the co-op also buys a few crops wholesale and resells them to consumer members at retail price. Farmer members share most co-op tasks, but the group pays the produce manager, the accountant, the newsletter writer, and the farmers who act as site managers at the four co-op pickup points. New consumer members pay an annual deposit of $60, and returning members pay $50: $10 goes for a copy of Nancy O'Connor's *Rolling Prairie Cookbook*, $17 for administrative costs and the newsletter, and the balance applies to the final month's produce bill. The co-op bills monthly; members pay after receipt of the food. Besides the weekly share, the farmers also sell organic chicken, eggs, beef, lamb, frozen raspberries, bulk quantities of produce for canning or freezing and cut flowers, and they offer an early-bird special of greenhouse crops and asparagus.

The co-op has added three distribution sites to its original pickup at the Community Mercantile, a natural foods cooperative store in Lawrence. Synergy between the two co-ops has benefited everyone; Rolling Prairie has brought new members to "the Merc," and many sharers do other shopping when they come for their bags. Merc nutritionist Nancy O'Connor supplies weekly recipes for the newsletter, and the deli cooks up samples for members to taste. A different farmer coordinates each of their four sites. The farmers are excited about the new Kansas City sites, one at a natural foods supermarket and the other at a community college with a famous culinary school. Anything left over from the shares goes to emergency food providers—Shalom House in Kansas City or Achievement Place in Lawrence.

Rolling Prairie's goal is to make buying from local farms as easy and attractive as possible. The co-op unabashedly follows the subscription model for its CSA, asking consumer members for no participation beyond purchasing shares. The cooperative efforts of the six farmers greatly reduce the risk to sharers

Ceili and John Leahy, children of members of the Fredricksburg Area CSA, help Julia Lewis, who has been packing shares since she was two.
PHOTO BY HEIDI LEWIS.

of crop failures or shortages. Sharers pay only for what they get. For the consumers, the farmers have tried to build a sense of community and provide education through their newsletter. In the early years they offered farm, garden, and greenhouse tours; flower walks; and potluck dinners. Though they let this slide for a while, in 2006 they held a successful potluck and a tour. For the farmers, the close working relationships with one another have value beyond the business connection, enabling them to share friendship and information; equipment, such as a tub and grinder for chicken feed; cooler space; and even fields. (Land-poor farmers are growing crops on the larger farms.)

In other parts of the country, groups of farmers have emulated the Rolling Prairie model, some establishing farmer co-ops, others associating less formally. In Weare, New Hampshire, David Trumble was showing the strains of life as a stay-at-home dad/farmer, trying to keep Good Earth CSA running while his wife worked full-time. In 2002 he read about Rolling Prairie in *Growing for Market*,

and called Lynn, who encouraged him to set up a cooperative CSA and gave him helpful advice about how to do it. Together with the Northeast Organic Farming Association of New Hampshire (NOFA-NH), David was able to get a SARE grant to start Local Harvest CSA. David's idea attracted seven other farmers and a baker, all certified organic, who spent a full year in preparation, getting to know one another, building trust, and agreeing on the many details of operation, including incorporating as an agricultural marketing cooperative. David stresses the value of this legal form, which sets up the rules that govern the relationships among the farmers. Much of their internal process resembles Rolling Prairie, with the co-op paying members to do administration. One of Local Harvest's goals is to spread crop commitments among the farms so that no one farm dominates. There is an annual bidding process for crops, with the farmers' bidding rights based on their sales for the previous year; the co-op rewards reliabilty. Two of the members have written a guide to forming a multi-farm CSA.

Local Harvest CSA has leveled off at 250 shares, the upper limit for the church hall they rent in Concord, with total gross sales at about $150,000. It offers two sizes of produce shares for eighteen weeks, a five-week fall extension, and a bread share, and member farms bring many other products to the pickup site, as well as offering bulk sales by special order. As with Rolling Prairie, quality control has not been a problem. The farmers set as their standard the best quality from each farm. The site manager, one of the farmers, notes any problems, and the production manager, another farmer, talks them over with the producer.

Having run his own CSA, David Trumble is sensitive to the greater distance between farmers and consumers and stresses that the co-op does its best to make the relationship personal: "The co-op allows us to develop a community of farmers. We lose out on the close contact with members that we used to have when the CSA was at our own farm, but our new group CSA has tried hard to have the farmers meet as many members as possible on pickup days and at group events we sponsor."

Corry Bregendahl, researcher with the North Central Regional Center for Rural Development in Iowa and for seven years an organizer of the Magic Beanstalk, has been investigating collaborative CSAs in her state.[1] She found that there can be serious drawbacks to cooperative efforts: insufficient income for each farm, lack of control over the entire process and linking your reputation to the performance of others. On the positive side, Corry confirms what David says about Local Harvest CSA, and in particular, his assertion that cooperative CSAs can give a boost to struggling new farms and contribute to the expansion of local agriculture. As David puts it:

> One of the stated goals of our SARE grant was to help beginning farmers. We have had some beginning farmers and also some small commercial growers who were in the early stages of their career when they joined the co-op. We are no longer the same actual group of farmers. We have lost two of our larger growers and yet we con-

Members pick up their shares from the Farm Direct Co-op in Marblehead, MA. PHOTO BY R.C. HARRISON.

tinue to keep our co-op going. At least three farmers in the co-op have increased their total farm size dramatically since they joined the co-op and I think it is fair to say the co-op has been instrumental in that growth. In NH, there is a dearth of commercial farms . . . the co-op helps farmers go from that beginning stage to the viable commercial stage. Starting a CSA with little farming experience is a hard row to hoe. Joining a group of experienced farmers gives an outline and a path to follow.

Several groups of very small farms have clustered together to offer shares in their communities. Held together by a core group of growers and members, the Fredericksburg Area CSA Project in Virginia has combined food from three or four small farms for a full decade. The number of shares fluctuates from year to year, ranging from thirty-two to seventy-five, depending on the energies of the organizers.

The Common Wealth CSA near Greenfield, Massachusetts, includes two farms with less than half an acre in cultivation. Two larger farms supply the bulk of the produce and could certainly do without their tiny colleagues, but the farmers became close friends while selling for many years at the Greenfield Farmers' Market. As tax resisters, Wally and Juanita Nelson of The Beanpatch lived on the $5,000 worth of produce they sold from their market garden, so the four or five shares' worth of vegetables for the collaborative CSA was a significant contribution to their subsistence. Wally continued running the tiller until only a few months before his death—at the age of ninety-two. In 2006, Juanita became an emeritus member of the co-op, though she still grows her own food. The younger farmers in Common Wealth worked out a payment formula that favored the Nelsons, a sort of community social security for two people who are deeply loved and revered throughout the Pioneer Valley. Ryan Voiland, at twenty-eight years old and a veteran of fifteen years of growing food, says it is worth his while to provide over half the produce for Common Wealth because he likes to keep in touch with the many supporters in the Greenfield area, where he got his start. Ryan's Red Fire Farm grosses over $300,000 in sales, quite a contrast with The Beanpatch, though Ryan learned a lot from Wally and Juanita about how to live on very little and to farm with integrity. The Common Wealth brochure sums up its philosophy with a quote from a 1649 poem from the Diggers at St. George's Hill:

> "We come in peace" they said "to dig and sow
> We come to work the land in common
> And to make the waste ground grow
> This earth divided we will make whole
> So it will be a common treasury for all."

Simpler teams of just two farms are running CSAs on a variety of scales. Mariquita Farm and High Ground Organics have united in Two Small Farms CSA in Watsonville, California, to supply over nine hundred shares with thirty drop-off points in the Bay Area. (The concept of smallness is clearly relative; with thirty-five acres in vegetables, Mariquita Farm would be one of the larger organic vegetable farms in the Northeast!) To their north, Live Power Farm fills CSA baskets during the summer, while Good Humus Farm takes care of the winter and also provides fruit shares. Full Plate Farm Collective, made up of three farms, Stick and Stone Farm, Three Swallows Farm and Remembrance Farm, offers members pickup at

Farm Direct Co-op 2005 Budget Summary		
Revenue	$118,241	
Cost of Goods Sold	$68,305	58%
Gross Margin	$49,937	42%
Labor	$23,644	20%
Depot	$1,065	
Trucking	$7,169	6%
Insurance	$2,699	2%
Printing	$979	
Taxes	$979	
Telephone	$1,036	
Other	$3,412	
Expenses	$40,980	35%
Net Ordinary Income	$8,957	7.6%
Scholarships	$1,388	
Donations	$1,187	
Interest income	$557	
Net Income	$9,313	7.9%

the farms or home or workplace deliveries in the Ithaca, New York, area.

Grow Alabama is a multifarm CSA of a different kind, the creation of Jerry Spencer, a chiropractor turned farmer and entrepreneur. Driven by his determination to raise, from a lowly 5 percent, the amount of Alabama-grown food eaten within the state, Spencer is trying to build a CSA with statewide production and distribution. So far, he has recruited twenty-five widely separated farms and delivers 250 shares to Montgomery, Birmingham, and Huntsville. Busy with growing some of the food at his Mt. Laurel Organic Gardens, and doing all of the organizational work and much of the distribution, he has not had time to create a support group among either the other farmers or the members of his CSA. By his own admission, suffering from stress and overwork, Spencer dreams of finding a businessperson to take over CSA management so that he can form a nonprofit that would allow him to devote his time to easing the transition to organic for more farmers.

Consumer-Initiated CSAs

"What do you do if you live in a suburb where there are no farms?" asked Sarah Lincoln-Harrison and her husband, Richard "Pic" Harrison. Their answer: They organized the Marblehead Eco-Farm in Massachusetts as a charitable corporation and started a CSA. (See p. 31 for the story of its birth.) Farmers hungry for markets in other parts of the country will shake their heads at the irony of how hard the Harrisons had to hunt to find a suitable farmer. Over the years, Marblehead Eco-Farm has evolved into the Farm Direct Co-op, "a not-for-profit member organization with the mission of promoting environmentally sustainable lifestyle options through the distribution of locally grown organic food."

In 2006, its fourteenth year, Farm Direct had 355 members and three drop-off points in suburbs near Boston. A steering committee of seven, including the Harrisons and a farmer representative, oversees the project. Farm Direct pays farmers the best price the members will accept, roughly 80 percent of the retail price in an upscale produce market. Members make five payments from December through August, but farmers receive monthly paychecks year-round, with amounts based on the prior year's sales. The total budget for 2006 was $120,000, of which 62.5 percent went to the farmers and the rest covered co-op expenses. Members have a choice of share sizes—a single vegetable share for $150, a family share for $250, a single fruit share for $120, a family fruit share for $175, an herb share for $40, and a cheese share for $60. Another $50 gets them an extended season share. Home delivery costs another $75, and there is a $40 membership fee to cover administrative costs.

Sarah describes the members as "representing a broad spectrum of understanding and commitment, from those who simply want fresh, local produce to those who desire above all to obtain organic food and to support the viability of regional farms." Therefore, the group's policy is "not that purist." Not all of the twenty-one farms supplying the co-op are certified organic, though their primary vegetable farmer, De Witt Thompson of Full Bloom Farm in Amherst, is. Cider Hill Farm, in Amesbury, which uses integrated pest management (IPM) practices, provides the bulk of the fruit. Until 1998 the Harrisons and treasurer Don Morgan did all of the administrative work for Eco-Farm, Pic putting in fifteen hours a week on office tasks and Sarah working full time on the CSA and the group's other projects. The bookkeeping system Pic uses is Microsoft Access, and he is willing to share his pricing and payment formulas.

Sarah says the CSA has become less intimate as it has grown to 355 families in 2006, and members are having a harder time grasping the notion of sharing the risk with the farmers.

Farm Direct asks all members to pitch in and spend at least two hours helping out at the three depots. Many do much more. "Once you participate on the fringe of sustainable agriculture," Sarah explains, "it opens you up for deeper involvement." Sarah hopes that her work will contribute to making life on this planet more sustainable. At the very least, Sarah says, the Farm Direct Co-op, "has brought balance to a community that is heavily involved in soccer and sailing and fosters an appreciation of the land that continues to disappear with indiscriminate development."

———————————————•———————————————

Like the old argument among actors about whether you achieve the most genuine emotion on stage by reliving the feelings from the inside or by imitating the outward gestures (a great performer can get there either way), CSA can arise from the collective efforts of farmers reaching out to consumers or consumers reaching out to farmers. Some projects use financial incentives to get people to work; others rely on volunteer energy. The revival of local food systems through agricultural development economics is the goal. As with genuine emotion, we will recognize it when we get there.

CHAPTER 19 NOTE

1. Bregendahl, Corry. "The Role of Collaborative CSA: Lessons from Iowa." Paper delivered at SARE National conference, Aug. 14-15, 2006. Available on www.ncrcrd.iastate.edu/ projects/csa/index.html along with other papers by Bregendahl.

MATCHING BIODIVERSITY WITH SOCIAL DIVERSITY

Everyone does better when everyone does better.
—JIM HIGHTOWER, *There's Nothing in the Middle of the
Road but Yellow Stripes and Dead Armadillos*

A powerful stream of concern for social justice runs through the movement for a sustainable food system. In "Peace with the Land, Justice among Ourselves," Marty Strange suggests that "the growing income inequality that divides our world" is the greatest threat to ever achieving a truly sustainable agriculture. He asks, "Would we be comfortable with a dual food system in which the rich paid a premium for food produced by agronomically wholesome means, while the poor ate cheap food produced by making war on the land?"[1] Recoiling in distaste from the notion that organically grown food is exclusively for "yuppies," many Community Supported Agriculture (CSA) projects are seeking creative ways to balance financial support for their farmers with including members who have little money to spend on food. Community organizers eager to empower the poor to take charge of their own food supply see the potential of CSAs to involve and train people as active members and staff.

But money is not the only obstacle to diversifying CSA membership. At a workshop titled "What Accessibility Means for CSA" at the Just Food CSA in NYC Conference in April 2006, core group activists brainstormed a list of the many issues involved: addressing barriers, nutrition education, language, comfort, inclusivity, the familiarity of the vegetables and how to prepare them, affordability, location, cultural food needs, fitting people's lifestyles, understanding CSA, health, applicability to all diets, breaking down stereotypes, and building communication and community. Trina Semorile, a core member of the Hell's Kitchen CSA, spoke about the danger of putting the poor in "stocks" by lack of sensitivity to the language we use in CSA outreach. People from different cultural groups or people with disabilities may assume that they are not welcome if we do not reach out to them in welcoming ways. Redesign of CSA distribution, share sizes, and payment schedules may be in order.

Connecting to the Emergency Food Supply

The simplest way to make sure that at least some of the CSA food reaches low-income people has been to donate leftover shares to food pantries or soup kitchens. Likely candidates are local churches, meals-on-wheels programs, or caring societies that supply groceries or meals to the sick or infirm elderly. In some areas, county gleaning programs or groups of gleaners organized by food banks or churches will come to farms to pick or pick up food. CSAs that do their distribution in churches often find a ready connection with existing food programs. In the past, the Genesee Valley Organic Community Supported Agriculture (GVOCSA) has used two churches as distribution centers. One gave the unclaimed shares to

Gleaners harvest kale and mustard greens in the snow at Rose Valley Farm, Rose, NY. PHOTO BY ELIZABETH HENDERSON.

senior citizens who came to the church for meals; the other had a program for distributing packets to people in the parish. They decided to allocate the CSA leftovers to elderly people, who appreciated the fresh vegetables and knew how to cook them. After we moved our pickup site to the Abundance Cooperative Market, the Southeast Ecumenical Ministry took over the distribution of leftovers to senior citizens in the neighborhood. When boxes of food remain at the end of the agreed upon hours, site hosts for Angelic Organics call other members who have volunteered to take the boxes to food pantries.

If sharers have to miss a week, many CSAs give the food to the hungry. Winter Green Farm in Noti, Oregon, makes a special effort to contribute to the struggle against hunger in its community. First, the farm's CSA, the Winter Green Community Farm, gives a discount of up to 50 percent for members who cannot afford the full price; $6 of regular share prices supports these low-income members, seventeen households in 2006. The CSA's contract with members solicits contributions to the Food for Lane County Program, which supplies area food pantries and other emergency food programs. In 2006 members donated over $1,500, which purchased five

shares for participants in Womenspace, a transitional home for women escaping abusive family situations. "We have developed a close link with our local food bank, and through the combined effort from our labor and donations from the community we provide thousands of pounds of produce to those in need. Some of this food is distributed as memberships in our Community Farm for women and their children who are in the process of healing from abusive partner relationships." Wintergreen also participates in Back My Farmer, an annual event, organized by John Pitney, bringing all local CSA farms together with congregations to raise money for low-income shares.

One of the recipients expressed her gratitude in these terms: "I felt special that I could have fresh food. When I was down on my luck, it helped me through the hard times." The farm also gives unclaimed shares and excess produce in bulk to Food for Lane County, delivered by farm staff or CSA members. Full Belly Farm in Guinda, California, donates five CSA boxes a week to the Charlotte Maxwell Complementary Clinic, which treats low-income women with cancer. Member donations pay for these contributions.

Gardeners know that some pretty ugly-looking food can be perfectly good to eat. When you get full from eating the factory rejects yourself, you need to find other hungry mouths to feed. While chefs at restaurants or institutional cooks may not want this food, people who run soup kitchens are often willing to take the time required to cut off bad spots. The county Extension office or local council of churches will usually be able to give you a list of soup kitchens and free meal providers in your area. We invite several groups to come to the farm to glean vegetables that are edible though less than market quality.

Taking a much more ambitious approach to getting fresh produce into the emergency food supply, a few food banks have established their own farms. The Food Bank Farm of Western Massachusetts (I tell its birth story on p. 31), under the energetic leadership of Michael Docter, serves a CSA of eight hundred families with 660 shares while supplying an equal amount of food to its parent food bank. In 2005, the farm produced 251,000 pounds of food for agencies supplied by the Food Bank. Only a tiny percentage of that passes through the Food Bank warehouse; the agencies pick up directly from the farm three days a week. Since the farm has no refrigeration, Michael and his crew don't grow many tender greens for the agencies, concentrating instead on "hardware" such as carrots, cabbages, and summer and winter squash. They also grow produce for the Food Bank's brown bag program, supplying the right number of units directly to the various distribution sites. The farm gets some volunteer labor from Food Bank supporters but does not make training recipients of the food in farming skills a priority. Until 1997, Michael and Linda Hildebrand, his partner in the early years, had to balance dual roles as farmers and employees of the Food Bank. Tensions resulting from missed staff meetings when farmwork took precedence led them to negotiate a new structure with the Food Bank: The farm formed an S corporation, which has a contract to run the farm autonomously and pay for the lease with 40 to 50 percent of production plus some cash.

In selling its shares, the Food Bank Farm stresses its social mission. Its brochure carries this message:

Your share of the harvest provides nutritious foods for:
- the unemployed family at a Springfield soup kitchen
- the battered wife in a shelter in Greenfield
- the mother feeding her family from the Northampton Survival Center
- the hilltown widow trying to get by on a social security check

Michael insists, though, that it is the economically produced, high-quality food that retains members once they join.

Other food bank farms have not yet managed to be as impressively productive, and a few have been total flops. A food bank in upstate New York tried to benefit from a piece of muck land that a well-wisher allowed it to use free of charge. Unfortunately, the free land did not come equipped with a free farmer, farm implements, or any infrastructure. Not too many of the potatoes that grew made it out of the ground and into hungry mouths.

For a few years, the Greater Pittsburgh Community Food Bank sponsored the Green Harvest program, which had an ambitious list of interrelated goals: to build local support for local farms and urban gardens, to eliminate hunger, and to encourage self-reliance. Green Harvest sponsored several farm stands in inner-city neighborhoods, recruited experienced gardeners to train new participants in the urban gardens it had established, and coordinated volunteers to do gleaning and farmwork on the farm it tried to consolidate. Green Harvest took the Food Bank Farm of Western Massachusetts as its model—and failed to replicate its success. The Green Harvest Farm never achieved its goal of self-funding through sales of CSA shares. From 2001 to 2005, the Pittsburgh Food Bank had a contract with The Food Farm, providing as many as fifteen hundred volunteers a year to the farm in exchange for most of its production. In 2005, The Food Farm donated one hundred thousand pounds of fresh produce to the Food Bank and also sold some of its vegetables to the Food Bank's farm stands. This farm too failed to get on its feet financially.

229

Meanwhile, the Food Bank has been successful at working with a dozen community organizations around Pittsburgh in running food stands in low-income neighborhoods. The Food Bank contracts with area farms for the food, delivers the food to the stands, and sells the food to them at cost. Without this support, the stands would not be viable on their own and low-income people around Pittsburgh would be unable to make use of the coupons from the federally funded Farmers' Market Nutrition Program.

Most community food security projects are sponsored by nonprofit organizations, which are seeking innovative ways to solve the complex and interrelated problems of hunger and poverty in the current food system.

Including Low-Income Members

In addition to giving away food, CSAs have sought ways to include people of diverse income levels and ethnic backgrounds among their members. One reason the GVOCSA decided to ask all members to work was to keep the share price low. We also offer a sliding scale for payments, emphasizing that those who pay on the high end are subsidizing those who pay on the low end. Peacework Farm accepts food stamps, and the GVOCSA treasurers are willing to adjust payment schedules to accommodate people with low incomes or to satisfy food stamp rules. In its third year the GVOCSA set up a scholarship fund to support membership for people who could not even pay the lowest rate of the sliding scale. That year we gave three scholarships to residents of Wilson Commencement Park, a two-year program for single mothers. To this day we give eight to ten scholarships a year to families that suffer from financial hardships of different kinds—temporary unemployment, low pay, extended illness. Rochester area churches have generously provided grant money and tithing contributions for the fund, which is also fed by members' donations and sales of the *FoodBook* and T-shirts. Despite all these measures, and strong efforts by Alison Clarke, the staff person for Politics of Food, only a few inner-city, low-income people have stayed on as members for more

than a year or two. Each person has a different story; either the food did not suit family tastes or a combination of family, health, and job disasters made a steady weekly commitment impossible. Middle-class people with a job or a career and health insurance find it hard to imagine the upheavals that poor people suffer in their daily lives. The most effective way we have found to support diverse members is to buddy them up with other members who have reliable automobiles and more stable lives. Genuine friendships are the strongest basis for retaining members.

In 2004 and 2005 the Hunger Action Network of New York State (HANNYS) undertook a program to stimulate CSAs to increase low-income participation. HANNYS is a nonprofit antihunger coalition that "combines grassroots organizing at the local level with state level research, education and advocacy to address the root causes of hunger, including poverty." Its initial survey of existing New York State CSAs revealed that 28 of the 41 CSAs that responded have some way to accommodate lower income members and together reach 690 low-income families. The farms reported using these approaches: "flexible payment plans (20), scholarships/sliding scale (14), working shares (10), and acceptance of food stamps/EBT (5). Additional diverse payment options included student shares (1), senior shares (2), paycheck deduction (1), 1 free share per 40 paid (1), bartering arrangements (1), and low share price (1)." The most common barrier to low-income outreach was the low incomes of the farmers themselves. HANNYS then offered competitive grants with one year of funding to CSAs with innovative approaches. Our CSA provided shares to the families of the children in the Rochester Roots school gardens club. The children took some of the food home and cooked some of it for city soup kitchens. A 2005 Community Food Projects grant to Rochester Roots allows the expansion of this program to three inner-city schools. (HANNYS has a write-up on this project: "Model

CSA Projects in New York State: Profiles Exploring How New York's Farmers Are Providing Low-Income Families with Healthy, Fresh and Nutritious Fruits and Vegetables.")

Regional support organizations, such as the Madison Area Community Supported Agriculture Coalition (MACSAC), Northeast Organic Farming Association of Vermont (NOFA-VT), and Just Food, are supplementing the efforts of individual CSAs. MACSAC's Partner Shares Program in Wisconsin is subsidizing shares for low-income and special needs households and community groups. As with the GVOCSA, everyone pays something, if only a few dollars per week. MACSAC coordinator Laura Brown describes their system this way: "Low-income households can choose any CSA that has been a member of the Coalition for at least one year." Replicating Just Food's revolving loan fund concept, MACSAC pays the farm up front for the cost of the share at the beginning of the season. Households then pay MACSAC back for 50 percent of the share price in five to six monthly installments. Eligibility is based on federal poverty guidelines. Word of mouth attracts many of the families. Although MACSAC does some outreach, Laura thinks it works better now that each CSA promotes the program through its own brochure.

Partner Shares works with one community organization, Porchlight, Inc, which serves free meals to people who are transitioning from homelessness into homes and jobs. Instead of shares, Porchlight orders what it needs directly from Vermont Valley Community Farm. Abby Bachhuber, Partner Shares coordinator, reports that it has been difficult to recruit other organizations "primarily because the transition from ready-made frozen/canned foods to fresh foods is perceived to be too much of a challenge for current kitchen staff to take on." In 2006 the sixteen residents in a senior living facility split four regular shares to meet their needs as one-person households, and the staff of a YWCA formed long-term residents into a "buying club," providing a new drop-site location for a farm. The retention rate for Partner Shares participants has been about two-thirds; however, the rate improved in 2006. Over the years, money for Partner Shares has come from grants and from "Empty

Bowl" events. At these benefit dinners, potters and members of Chefs Collaborative 2000 combine crafts to sell bowls full of gourmet food, with the proceeds going to Partner Shares.

MACSAC has found that farm-a-thons provide the steadiest support. Similar to crop walks or marathons except that instead of walking or running for money volunteers from church congregations, student groups, and other organizations enlisted by MACSAC get sponsors to pledge money for doing farmwork.

Since 1994, NOFA-VT has been coordinating Vermont Farm Share, a program to enable low-income people to eat more fresh produce, learn about growing and preparing food, and connect with local farms by joining CSAs. Twenty-six of the forty-five CSAs in Vermont participate, providing shares to 180 families a year. Initially, in partnership with the Vermont Anti-Hunger Corps, local social services, and the University of Vermont's Cooperative Extension, Farm Share identified appropriate families or centers and linked them with a local CSA farm. Once the program was established, Farm Share reduced its involvement, leaving outreach to the twenty-six participating CSAs. Enid Wonnacott, executive director of NOFA-VT, admits that this program does not cover people who suffer dire emergencies. It is designed for families on the edge of food insecurity; the eligibility level is 185 percent of the federal poverty line, with a limit of $30,000 a year for a family of four. To raise money to subsidize the share fees, NOFA-VT solicits donations from individuals and holds an annual Share Our Harvest event, in which people dine out all over Vermont and the restaurants contribute a percentage of their proceeds from the meals. A hundred restaurants and stores participate, generating annual proceeds of $10,000 to $12,000.

During the first two years, Farm Share paid the entire fee for each farm resulting in a program without much participant ownership or involvement. In 1997 the program asked the families to pay on a sliding scale and grouped them so they could divide both large-size shares and the responsibility for pickups among two or three families. Starting with smaller quantities of food made membership easier. Even with a subsidized fee, lump-sum payments were beyond the means of most of the families, so the program asked the farmers to arrange more flexible payment schedules. Experience has shown that most families can pay 50 percent of the CSA share price. NOFA-VT asks the farms to split the remaining 50 percent with 25 percent raised by the farm through "supported share" donations from other CSA members who can send their contributions to NOFA-VT to get a tax deduction. NOFA-VT confirms the eligibility of the participants and then matches the donations with funds from the Share Our Harvest fundraiser. Food and farm education is one of Farm Share's goals, but since low-income people must so often jump through hoops to receive services, the program does not require formal classes. Farm Share offers member families free participation in NOFA Vermont's summer workshop series and requires each farm to host a field day. A lot of the learning happens naturally through the CSA process with recipe exchanges, newsletters, and farm visits.

In 2004, NOFA-VT received a Community Food Project grant to expand the program to congregate meal sites, such as summer feeding programs, day cares facilities, and senior meal sites. Each site receives shares to use for snacks and lunches. Enid reports that there are a lot of similarities with efforts to integrate local foods into school meals. Supporting the professional development of food service providers, she adds, is an important strategy. The administrative costs of Vermont Farm Share are low, and Enid is pleased that a small amount of funding leverages a lot more money for CSA shares.

The year Kate Larson ran Vermont Farm Share she did extensive telephone interviews with participants at the end of the season. Of twenty-one interviewed, nineteen wanted to continue for another year if they could pay the same amount. These are some of the comments:

"I really appreciated the program. I feel healthier, and no one in my family got sick this winter. I think it is because they ate so many vegetables this summer."

"Having this food for our teenage clients meant that they were not going to McDonalds or the Kerry Quick Stop as much."

"I am unable to purchase fresh produce now because it costs too much in the winter. My finances are tight. I just saved $40 using coupons, but there aren't any coupons for fresh food. I sometimes purchase dented canned vegetables at a discount store."

Since 1997, Just Food has been persuading New York City community organizations that already work with low-income groups to add CSA to their missions. The Sixth Street Community Center, located in a neighborhood with many working poor, was the first organization to adopt CSA. Executive Director Annette Averette met Kathy Lawrence, the founding director of Just Food, and "grasped her vision." Annette is a cancer survivor; given two months to live ten years ago, she took her diet in her own hands and is very much alive today and very determined to help other low-income city dwellers gain access to organic food they can afford. She introduced me to Citlalic Jeffers, an impressively well-informed sixteen-year old intern in the center's Seeds to Supper program who teaches sessions for other teens on how to change their diets. Catalpa Ridge Farm in New Jersey provided weekly deliveries to the center for five years until the number of members outgrew its capacity. These days, their food comes from Amy Hepworth's farm in the Hudson Valley.

Paula Lukats, Just Food CSA coordinator, reported at the 2006 CSA in NYC Conference that fourteen hundred of the ten thousand city CSA members have lower incomes. Whether they include community organizations or not, each core group makes its own decisions about how to involve lower-income members. Of the forty-two city CSAs in 2006, twenty-nine are mixed-income projects that accept food stamps or have payment options that make membership more accessible. Just Food–initiated CSAs use all of the methods mentioned so far—fund-raising for subsidized shares, sliding scale payments, flexible payment schedules including biweekly payments for food stamp recipients, donations to emergency food suppliers—and some creative innovations of their own. A few of the CSAs have revolving loan funds. At the beginning of the season, the loan fund pays the farmers for the low-income shares so that the farmers benefit from the start-up money. Over the course of the season, the lower-income members make their payments back into the loan fund, replenishing it to start again the next year. If members suffer financial difficulties, they can postpone payments without hurting the farm. The South Bronx CSA in the Bronx raised the money for its revolving loan fund from the City Council and the Department of Youth and Community Services. With a revolving loan fund of $4,000, 30 percent of the Chelsea CSA are low-income families.

The Upper West Side CSA on 86th Street in Manhattan charges members $75 if they fail to do their three work shifts. Roxbury Farm allows this CSA to use that money as matching funds for member contributions for subsidized shares. Tinamarie Panyard was so delighted by membership in this CSA that she decided to start another one at the Stanton Street Settlement House on the Lower East Side, where she works with children from low-income families. At this new CSA, supplied by Windflower Farm, there are enough members who can pay the entire fee up front to allow some members to pay every two weeks in cash and food stamps. Gradually, families whose children attend programs at the Settlement House are joining the CSA.

The W. Rogowski Farm works closely with the East New York Project, where low-income teenagers learn to grow food at a community garden and sell it at a farm market. The Rogowskis supply thirty-five family shares and whatever the community garden cannot grow. In 2005 the John D. and Catherine T. MacArthur Foundation declared Cheryl Rogowski a "genius." My first thought on hearing this was Why Cheryl? She is doing no more nor less than hundreds of organic farmers around the country. But then I thought, wow! Cheryl *is* a genius, and so are many, many other farmers whose stories have not reached the MacArthurs. Perhaps Cheryl is only the first of a string of farmer geniuses.

In any case, she is definitely a worthy candidate. A hardworking yet gentle and compassionate woman, Cheryl helped save her family's farm, transforming it from a conventional onion operation into a diversified, ecological farm growing over two hundred varieties for 150 CSA shares, eight farmers' markets, a farm store, and several restaurants. Cheryl has sought opportunities to support struggling immigrants and battered women and to preserve local farmland. Sixty shares from her farm go to two groups in New York City—the East New York site and the Bushwick Sister to Sister CSA. Located in a low-income area where gentrification is creeping in, the East New York CSA sells shares on a sliding scale to Jamaican and Caribbean families. The farm designed the share contents to suit the culinary tastes of the members for basic vegetables: collards, onions, potatoes, cabbage, bicolored corn, turnips, only red tomatoes, cilantro, and hot peppers. Cheryl's brother Mike makes the city deliveries and works with the community gardeners. "It's the people who make it so amazing," Cheryl reports, "and the synergy between the garden and the farm." (See chapter 3 for more on Just Food.)

Senior Farm Shares

When the United States Department of Agriculture (USDA) began offering Farmers' Market Nutrition Program funding for seniors, the Maine Department of Agriculture jumped right in with a well-defined program and snatched up $759,000 of the $15 million available for the whole country. Instead of giving each eligible senior $20 to $30 worth of coupons redeemable only at farmers' markets, as other states had been doing, Maine offered $100 of fresh produce per senior direct from farms. According to Deanne Herman, the Maine Department of Agriculture marketing manager who runs the program, since the coupons are like money, processing them is expensive, 8 cents for each $2 coupon. By paying $100 directly to farms, the Maine approach puts more of the money into food for the seniors. An advisory committee, which includes experienced

CSA farmer Jan Goranson and senior advocates, works with Deanne to set policy and screen the farms. The committee insists on farms with established production experience and set a maximum of two hundred shares per farm. In 2005, 165 farms provided senior shares for seventy-five hundred low-income seniors, 10 percent of the eligible seniors in the state. Each farm designs its own system and recruits the seniors. Fifty percent of the farms have the seniors collect their vegetables at farmers' market stands, while others have a mixture of on-farm pickups and deliveries. The program includes careful accountability and record keeping to be sure that the food really goes to seniors. If a senior drops out, the farm has to find another person to use the rest of the funding the same year. In Vermont, NOFA-VT administers a similar Senior Farm Share program distributing $75,000 a year to 750 senior citizens, working through twenty-six senior housing sites. A fight is brewing over the USDA regulations for this program, which propose a $50 limit per senior, hardly worth the effort, according to Deanne. Taking the Maine program as a model and dramatically increasing the funding for Senior Farm Shares so that all states could participate would give a big boost to senior nutrition and local farms alike.

Community Food Security

As the government safety net weakens, and awareness of the meaning of community food security spreads, projects have been springing to life around the country to help people meet their own food needs. Linking with Community Supported Agriculture has seemed a natural idea. Some of these combinations have really taken hold; others have crumbled from lack of ongoing resources or when well-meaning but overly complex projects with multiple and conflicting demands have burned out inexperienced farmers.

The Coalition for Community Food Security defines food security as "all persons in a community having access to culturally acceptable, nutritionally adequate food through local non-emergency sources at all times." The Coalition encourages the establishment of local and regional food system councils to do long-range planning, and the creation of community-based networks and coalitions to strategize and implement multifaceted programs. Most community food security projects are sponsored by nonprofit organizations, which are seeking innovative ways to solve the complex and interrelated problems of hunger and poverty in the current food system. Only a few of these projects are more than fifteen years old, and many are just getting under way as I write. Although a few of them involve independent farmers as advisers or as sources for some of the food, most, like the Food Bank of Western Massachusetts, have established new farms. Like Community Supported Agriculture, community food security projects are experiments with new social and organizational forms, working with the slimmest of resources and combining inherently fragile human materials.

One of the earliest of these projects is the Homeless Garden in Santa Cruz, California. I was fortunate to have had the chance to tour the garden in 1992, when it was in its second year at its original site, in the middle of a neighborhood of fairly expensive homes. (Read Jered Lawson's vivid account of the Homeless Garden, written for *Rain* magazine in 1993, in the sidebar on pp. 236–37.)

When Jered wrote his article on the Garden, its situation was precarious: The city owned the land and was planning to subdivide it to sell as building lots. By 1998 the Garden had finally been forced to move to a two-acre piece of land on loan from Barry Spenson Builder but was negotiating with the city for twelve acres in the Pogonip, a greenbelt planned for the edge of town. According to Rick Gladstone, director at that time, its fate depended on the outcome of a debate raging between "deep" ecologists, who wanted to keep the land as a nature preserve, and "social" ecologists who favored the Garden. The social ecologists won; however, many years have passed and the Garden still has not been moved because the city of Santa Cruz lacks the funding to improve the road to the Pogonip and put in waterlines. As I write in 2006, the Homeless Garden persists at Natural Bridges Farm

A lunch of greens and early spring vegetables from Holcomb Farm at Peter's Retreat, a residence for homeless men and women with HIV/AIDS in Hartford, Connecticut. PHOTO BY JASON HOUSTON. (FROM *ORION* ARTICLE)

on the donated land. When I visited in 2004, the CSA was not functioning, but homeless women in the Women's Organic Flower Enterprise were growing flowers there and making bouquets and candles to sell at their newly opened Santa Cruz garden shop. By 2005 the CSA was up and running again with sixty shares in a partnership with Maria Inés Catalán's Laughing Onion Farm. Maria herself is a successful graduate of ALBA that assists motivated farmworkers to become farm owners. Maria grows the heat-loving crops, while the Homeless Garden grows crops that do better on their ocean-cooled site.

Director Dawn Coppin told me that most of what Jered Lawson described in his 1993 article is still valid except that the salary for the homeless workers has risen to between $7.50 and $8.50 an hour and all participants are working 20 hours a week. Jered's conclusion is as true today as it was in 1993: "The Garden demonstrates that ecologically sound, socially just, and economically viable projects are possible. What's needed now is the motivation, determination, and commitment of individuals who recognize the potential in people, and the land, to heal, take root, and grow. Michael Walla of the Garden says, 'The Garden is showing we're people

with pride, people willing to struggle. . . . We don't need someone who will carry us. We need someone who's willing to help get us on our feet.'"

The Hartford Food System's Holcomb Farm CSA

The Hartford Food System (HFS) in Hartford, Connecticut, is one of the oldest, most innovative and productive of the community food security nonprofit organizations in the country, providing a model of the systemic approaches advocated by the Coalition for Food Security, of which it is a founding member. Since 1978 the HFS has worked to plan, develop, and operate local solutions for the City of Hartford's food problems. To help save area farmland by improving the earnings of local farmers while increasing the supply of fresh, nutritious food for city residents, the HFS rallied Hartford city agencies and community organizations to establish the Downtown Farmers' Market, the first of forty-eight farmers' markets in the state. To enable more low-income people to shop at these markets and increase sales for local farms, the HFS worked with the Women, Infants and Children

The Homeless Garden

By Jered Lawson

A year and a half ago, Bill Tracey stood on the corner of Chestnut and Mission clutching a piece of cardboard that read "Homeless and Hungry: Will Work for Food." Now Bill works with a group of homeless people who take these words literally, growing food not only for themselves, but for the surrounding community as well. In just two years, over forty homeless people, a committed staff, and countless volunteers have turned a 2 1/2-acre vacant urban lot in Santa Cruz, California, into a thriving organic garden.

After the gardeners take their portion of the harvest, much of the food goes to community members, or "shareholders," who support the garden financially. A percentage of the produce is sold to local stores, restaurants, and folks at the farmers' market. The rest is donated to homeless shelters and free-meal programs. Bill says "Other homeless projects can give you files, reports, and statistics, but we can give you a flat of strawberries."

The garden offers diverse flora with a mixed crew of gardeners. There's Peter, a homeless trainee; Darrie, a mother of two; Paddy, a volunteer handyman and gift-giver to the garden; and Phyllis, a vivacious 82-year-old who asserts, "I don't have to die to get to heaven . . . this place is heaven on Earth." Both Mac, a humorous and stately homeless man, and Mike a "practical idealist" university intern, work with groups of children in the garden. According to Lynne Basehore, the project director, "The garden has been useful to those who simply need to witness life's abundance. Most of all, it has been a renewal for long-term jobless and homeless citizens of the community."

With over two thousand homeless people in Santa Cruz County, it's no wonder there's a waiting list for the fifteen paid positions available at the Garden. When a position does open, prospective employees volunteer a short while to see if they are truly interested in the work. If so, they begin at a minimum of twelve hours a week and attend the weekly meeting. Workers are familiarized with procedures of the garden, and then choose an area for in-depth training. For Skooter it was compost, for Octaciano, the greenhouse. Jane Freedman, the garden director, trains the gardeners in bed preparation, composting, cultivating, planting, harvesting, and selling produce in the farmers' markets.

The weekly meetings provide group members with an opportunity to air concerns, make collective decisions, and work through any pressing problems. A rules committee—made up of five of the homeless workers and two of the staff—compiles and presents a list of rules that are then agreed upon by the larger group. Developing and enforcing their own rules gives the workers a voice in decision-making that they were generally denied elsewhere. Some of the rules: "When scheduled for work, do not come high, drunk, or hungover: If you do, you will be sent away immediately and the consequence is suspended paid work until nine hours of volunteer work are completed." "No sleeping at the Garden. Anyone caught camping is kicked off the project."

The Garden's pay, $5–6 an hour for twelve hours a week, may not be a living wage. But Darrie Ganzhorn, a garden employee, feels the money is "only one piece of the puzzle. It's part of the network needed to get one's life together." Another gardener says, "The work has grounded me. It's stabilized me to where I can actually go out and enroll in school. Otherwise I'd be too scattered. You know, hustling to get this or that. Since I've worked here, I've moved to a safe place to sleep at night."

History

Paul Lee, an internationally renowned herbalist, former UCSC Professor of Philosophy, and longtime advocate for the homeless, inspired Lynne Basehore and Adam Silverstein in May of 1990 "to transform the vacant lot into a healing, productive garden." Paul, after receiving a donation of herb plants from a store in Carpinteria, California, knew that "if we had a couple thousand plants on hand, we would have to get them in the ground; hence, the Homeless Garden Project!"

Lynne began recruiting homeless workers from the shelter to come and work for a few hours here and there, getting the herbs in the ground. Since the herbs needed watering, and since Adam had experience in irrigation systems, he too became a part of the crew. Jane Freedman became director in November 1991. In reference to the "horticultural therapy" aspect of the garden, Jane once joked, "We may not have any couches, but we certainly have a lot of beds."

The Garden uses Alan Chadwick's French Intensive/Biodynamic, raised-bed method of gardening. Local restaurants, cafés, horse stables, landscapers, and neighbors give [organic materials] to an innovative composting system.

Money for salaries and wages comes from a variety of sources. One-third of the budget is covered by the CSA, as well as through the sales of produce and flowers at the local farmers' markets, restaurants, and natural foods stores. Funds are also raised through special events, grant and letter writing, awards, and direct campaigning. The project was selected by Visa cardholders of the local Santa Cruz Community Credit Union to receive 5 percent of the money generated from the use of their cards. The New Leaf Community Market began a unique system of fundraising, issuing 5¢ "enviro-tokens" to shoppers returning paper bags. The tokens are given to the nonprofit organization of their choice. So far the Garden has been the community's favorite, generating more than five thousand tokens in three months. And finally, the Garden receives subsidies from the local Job Training Partnership Act (JTPA), the American Association of Retired People and the Veterans Affairs Job Training Program.

The gardeners also receive benefits from the community. One vegetarian restaurant gives the project free meal tickets in exchange for produce. A local laundromat, "Ultra-mat," provides the gardeners with a monthly allotment of "Ultra-bucks" to use for washing their clothes. Some of the gardeners have buddied up with volunteers who assist with basic needs; from a bedroll for the night, to a job or housing opportunity. The Garden is also a magnet for contributions: clothes, a computer, and even a couple of trucks.

In the fall of 1991, the Garden adopted the CSA model. Shareholder Steven Beedle says "The CSA has meant guaranteed access to the freshest organic produce at a great price, supporting the much-needed assistance to homeless people, and having a say in all issues that are confronting the Garden. There's a sense of involvement, with the people doing great work and benefitting in the process."

(Source: Rain vol. 14, no. 3 (Spring 1993): 2–7)

(WIC) program, the Connecticut Department of Agriculture, and other agencies to pilot the Connecticut Farmers' Market Nutrition Program, which annually provides over fifty thousand WIC recipients with $396,000 in special vouchers that they can spend only to purchase fresh produce from area farmers' markets. The HFS has also worked with agencies for the elderly to fund $90,000 worth of these coupons for six thousand senior citizens.

Since he helped initiate this program in 1987 in Connecticut, Mark Winne, the founding director of the HFS, has facilitated the spread of the Farmers' Market Nutrition Program all over the country with joint funding from the USDA and the states. Throughout the 1990s, HFS program director Elizabeth Wheeler worked with the public schools in Hartford on Project Farm Fresh Start to find ways to incorporate locally grown organic foods into the school lunch program. The HFS has also helped bring new supermarkets to inner-city neighborhoods and has initiated city- and statewide food policy councils to plan for greater food security. Mark Winne was a guiding member of the Community Food Security leadership, which achieved passage of the Community Food Project Program as part of the 1996 Farm Bill, providing $2.5 million a year in competitive grant money to projects such as NOFA-VT and Rochester Roots. Remarkably, the HFS has survived the departure of both Winne and Wheeler and continues to do innovative work, such as improving healthy food choices in bodegas and small market stores in inner-city neighborhoods.

In its efforts to address the critical need for sources of fresh, nutritious produce in Hartford, the HFS has worked to create links between growers and low-income urban consumers and for many years sought the opportunity to establish a CSA. In 1993 the Friends of the Holcomb Farm, a nonprofit organization, invited HFS to participate in its plans for the Holcomb Farm estate, which was bequeathed to the Town of Granby by Tudor and Laura Holcomb. In public hearings, and not without some dissenting voices, the people of Granby approved two main goals for the farm: to maintain the farm for agricultural use and to make the farm available to the wider community, with special outreach to Hartford residents. The plan aimed to create a CSA that would cultivate sixteen acres of fruits and vegetables, half of which would go to low-income Hartford residents and community organizations. The lease signed by the HFS also stipulates that for every five acres cultivated, the farm will provide two shares to Granby social services. Instead of trying to sell produce to low-income individuals one at a time, the HFS devised a strategy based on creating partnerships with community organizations that already were working with low-income people, allowing each organization to distribute the food in a way that suits its mission. When Just Food decided to concentrate on extending CSAs in New York City to low-income people, it took a clue from the HFS.

Elizabeth Wheeler related some of Holcomb Farm's history at the Northeast CSA Conference in November 1997:

> There was some emotion broiled up among the citizens of Granby by this new approach. To do a subscription farm in a town that has an existing farm community was quite a radical idea. I did get a few phone calls from people wanting to know when the buses with "those people" were coming and whether their social security numbers would be written down, and a few calls from irate farmers claiming we were unfairly competing with them because we were going to steal customers.
>
> In Hartford, it was my job to find the folks we were trying to reach. Hartford is the eighth-poorest city of its size in the U.S. To give you an indicator, 80 percent of the twenty-three thousand children who go to school in Hartford are on free or reduced-price lunch. We have sections of the city where there is 40 to 50 percent unemployment and some fairly dire circumstances. Obviously, for a low-income individual in Hartford who has not set foot outside the city in some time and doesn't have a car, the CSA was not going to work. The best approach was to go to community organizations that were already working in the city on economic development, housing, and the like, and recruit them to take on responsibility for distributing the food.
>
> In terms of connecting diverse people, it's a real slow process. Folks who aren't used to being with one another are not necessarily going to want to be with one another, and it's taken a little bit of social engineering to bring them together. We have arranged some special events. When we bought our new tractor, we invited the black clergy from the city and some of the white suburban clergy to bless it, a wonderful symbolic event. Many little connections and bonds have been made. It's exciting to see this happening through the leveling connection of food.

For most of a decade, the HFS staff ran the CSA operation, handled all the money, and did all the hiring and firing of staff for the farm. HFS fundraising covered most of the expenses of purchasing equipment and improving the land. The organizing and support work of Elizabeth Wheeler, which earned her the nickname "the queen of donations," was critical to setting up Holcomb Farm as a well-equipped and functioning enterprise. In 2003, Sam Hammer took over the job of farm manager and filled the leadership vacuum created when Mark and Elizabeth departed shortly after hiring him. Since then, Sam has succeeded in expanding the acreage under cultivation to twenty-five. Family shares that pay the full value of the produce will reach 325 in 2006, further supplemented by increased sales to restaurants and a farmers' market. The farm operating budget has allowed for significant investments: a new tractor with a loader, a cultivating tractor, irrigation equipment, a new greenhouse, and four thousand feet of woven wire deer fencing around nine acres of the land. Eleven community groups pay about 25 percent of the food value of a bulk share ($1,200). The HFS makes up the difference with grants, donations, and federal senior nutrition funding and expects to continue so that the CSA need not bear that burden. Waltham Fields Community Farm in Massachusetts modeled its low-income group shares after the HFS program but finances the donation of 30 percent of its production entirely by the sale of 250 regular shares. Farm manager Amanda Cather attests to the strain that places on the farm crew.

The Holcomb Farm staff consists of a farm manager, an assistant manager, five interns, and a bookkeeper; an outreach assistant facilitates the participation of the Hartford groups. The farm offers working shares—all you can eat in exchange for 60 hours of work—and endless opportunities for volunteers from the CSA membership and community service groups. Like most other farms run by nonfarming organizations, the Holcomb Farm had a steady turnover of farm managers until Sam settled into the job. For a young farmer, the chance of earning a steady salary with benefits while running an established farm is attractive. He tells me he loves the farm and intends to stay as long as he can keep control of farm management.

As the project has matured, the HFS and Holcomb Farm have been able to sort out more clearly their food production and social service goals. Until 2004 inner-city youth from Grow Hartford came out to the suburban farm to experience farmwork, causing friction with the demands of production. "Going to Granby was like going to the moon for these kids," in Sam's words. The HFS has since moved Grow Hartford to a lot in the city, where the youngsters have more control over what they grow, and sell their produce at the Laurel Street Farmers' Market together with food from Holcomb Farm. Sam made an interesting observation about this market, "Currently, it isn't a big money maker, but it's important to our mission, since it is providing a mainstream food outlet in a low-income urban neighborhood. While providing food to social service agencies is an important food justice idea, I feel that helping to establish market-based outlets in underserved neighborhoods goes further in developing an equitable food system."

All together, the eleven community groups distribute bulk shares to twelve hundred low-income Hartford residents. Staff members of Family Life Education, which serves over two hundred families on welfare, give bags of food to the families who come to the office and sometimes deliver to their homes. Year-round, families inquire about the vegetables, which have become very popular. For ethnic groups not familiar with some of the selections, the staff offer cooking lessons. A few times, they have been able to bring families to the farm to help. The Migration and Refugee Services at Catholic Charities make up food starter packets for newly arrived refugee families, including fresh produce from Holcomb Farm. Volunteers from Catholic Charities give cooking lessons for the refugees and bring groups of Somalis and Bosnians to the farm for tours and the chance to pick their own beans or strawberries. At Shepherd Park Senior Housing, home to 450 lower-income senior citizens, Diana Maldonado, program manager, with help from residents, sorts their bulk shares into small bags for the seniors. "It's hard for them to get to a food store," she reports, "and most of them rely on already prepared food from the Meals on Wheels program. The farm's produce not only helps them stretch their social security checks, it encourages them to cook."

O.N.E. C.H.A.N.E., Inc., a nonprofit devoted to rebuilding Hartford's North End through community organizing, ownership housing development, job training, and youth programs, gave out shares through its block clubs for several years until it ran into serious funding problems. Frederick Smith, the new executive director, hopes to revive the program as soon as he gets the organization back on its feet. Peter's Retreat, a residence for twenty-six homeless men and women with HIV/AIDS, prepares their meals from its share of the produce. Once a week, Janet Candela, the chef, brings groups of the residents out to Holcomb Farm to pick peas or tomatoes or just hang out. "My clients are poor and isolated,"

Candela explains. "They have burned themselves and they have been burned by society. For them, the farm eliminates the class differences. They mingle freely there with many kinds of people and don't feel out of place. It's a funny thing to say about tough guys from the street, but they feel safe there."[2]

Turning Farmworkers into Farmers

One of the most unusual and exciting CSA-related training programs is taking place at the Agriculture and Land-Based Training Association (ALBA) in the Salinas Valley in California. ALBA's mission is "to advance economic viability, social equity and ecological land management among limited resource and aspiring farmers." Building on the work of the Rural Development Center (RDC), founded by José Montenegro in 1985, ALBA provides agricultural training, education, English as a second language instruction, and marketing experience to farmworkers to give them the skills they need to run their own farms. The project has assisted over five hundred families; ALBA staff estimate that 10 percent of them now run their own farms. On completion of PEPA (Programa Educativo para Pequeños Agricultores), a six-month course in agricultural production and farm management, participants can use up to 5 acres of land plus water, equipment, and continuing technical support for up to three years at the 196-acre Triple M Ranch at the Farmer Training and Resource Center. In 2006 there were twenty-seven farmers using ALBA land. A 1997 Community Food Project grant enabled the RDC to expand its training program for farmers, community gardeners, and schoolchildren; to set up self-supporting food distribution and marketing; and to establish a Public Education and Policy Council to promote food security efforts.

Staff member Luis Sierra told me that before 1991 the 112 acres of RDC land was managed using conventional farming methods. That year, one of the participants grew an acre of zucchini organically. The productivity and earnings were so impressive that by 1998 eighty-six of the acres had been converted to organic methods. Since that time, ALBA has expanded

to two pieces of land totaling over 300 acres, all in organic management. The project has developed its own well-equipped packing facility and brokers produce from its model farm as well as that of student farmers through ALBA Organics and its own CSA.

The ALBA CSA is small—thirty-six members in 2005. Selling shares in the economically distressed towns near the center, where farmworkers make up 30 percent of the population, is difficult, even with a lower-priced share for families earning less than $30,000. Communication Director Gary Peterson says they are exploring a strategy like HFS's to partner with social service agencies and fund-raise to support shares for low-income families. At the RDC farm, the CSA occupies a central two-acre plot so that all the farmers in training have the chance to observe the unusual crop mix and organic techniques. ALBA Organics makes up the shares from all of the farmers and explains who grew what in the CSA newsletter.

Maria Inés Catalán, one of the first participants to try organic methods, marketed her organic vegetables for three years through brokers. In 1996, with the help of Luis, who had interned at Full Belly and Live Power farms, she started a CSA, and then moved to land in Hollister. Gary Peterson describes Maria Inés as very outgoing and a good organizer. She grows about forty-five different crops, with twelve available throughout the season. An ad in a free local newspaper and recruiting by members of the RDC advisory board helped sign up her first sharers. The members of her CSA are mainly white, middle-class residents of Salinas, Carmel, and Monterey, where Maria has drop-off centers.

Maria does not speak English, which is an obstacle to organizing consumers, though some members are attracted by the idea of supporting a farmworker turned farmer. Sometimes she trades shares for help with communications or relies on her four children to translate. (My son Andy interviewed her for me, since he speaks fluent Spanish.) Maria has produced a CSA brochure in Spanish in the hope of recruiting in the Hispanic community. Finding Latino members has proved difficult because the CSA is in the middle of a major vegetable crop region. Many Latinos obtain free food through connections to the vegetable industry, or grow their own. Maria has found that she can attract Latinos by growing varieties, such as chiles, that are not available commercially in the area. Her crop list—potatoes, onions, garlic, Mexican corn, jicama, cilantro, tomatoes, and sweet and hot peppers—would be helpful for CSAs elsewhere that seek to serve Hispanic communities. (For the full list, see p. 203.) Besides running her own CSA, Maria Inés is partnering with the Homeless Garden Project, supplying the tomatoes and peppers that are harder to grow along the coast. Another PEPA graduate, Tomatilla Martinez, initiated a CSA in 2006. She is determined to serve her home community, the impoverished farmworkers from the Oaxaca area who live in the town of Greenfield.

Community Food Projects

The availability of grant funding from the Sustainable Agriculture Research and Education (SARE) and Community Food Projects programs has stimulated the expansion of a number of projects that combine training in food production for low-income people, the use of university land and student energies, and CSAs. To receive Community Food grant money, projects must: meet the food needs of low-income people; increase the self-reliance of communities in providing for their own food needs; and promote comprehensive responses to local food, farm, and nutrition issues.

The outstanding qualities that unite these projects, however, are the enthusiasm, high energy, and determination of the organizers and the unusual mixture of social service, farming, church, Extension, university, and community organizations that have come together around local food. No two of these projects are the same. For the first edition of this book, I wrote profiles of five of them. In 2006, at least elements of three out of the five are thriving; two have disappeared almost without a trace. Isles, Inc., a twenty-six-year-old community development organization, tried to start a CSA and flower farm to put people on general assistance to work. Ron Friedman, the organizer, told me that it was a fun project "but

undeniable as a dog's smile

in just an instant
the belief was made manifest.
the early evening
light was aslant and full
of flying creatures,
myriad motes and swirling
insect constellations.
swallows pirouetted above the melon
vines in an ecstasy of bug catching
while a phoebe sat atop
the fork handle and watched,
flicked her tail then hovered
and nabbed a meal.
black cat stalked amidst foliage
where soldier bugs sucked away at
immature potato beetles, fat and
            orange and
softly similar to the ladybugs walking by.
in that instant,
as sudden and clear as the light
in a loved one's eyes,
the belief in our interconnectedness
was visible.
and with it, undeniable
as a dog's smile, comes the need
to honor the balances
and to nurture the creation
instead of
randomly
extinguishing its parts.

—Sherrie Mickel, 1992

not well conceived on my part, given the politics of Trenton."[3] Sea Change in Philadelphia lost its garden site, where an experienced farm manager tutored forty teenagers in food production, to a city redevelopment agency parking lot. However, there are new start-ups every year. We need to understand better what social fuel and community resources keep these hopeful beginnings chugging along.

Starting in 1997, the Field to Family Community Food Project in Iowa, with funding from the Kellogg Foundation, the Leopold Center for Sustainable Agriculture, and the Community Food Projects Program, did outreach to low-income families through church groups and social services to involve them in the Magic Beanstalk CSA, nutrition education, hands-on farming and garden work, and leadership training. Members of Magic Beanstalk initiated the Field to Family Project to foster more self-reliant local food production and to help lower-income families gain access to wholesome food. Robert Karp, codirector of Field to Family, explained its mission: "Instead of just being given a food handout, low-income families are invited to participate in a community process that supports local farmers." Robert developed the idea of using CSA as a focus for delivering social services to low-income families. Grants from national churches subsidized seventeen shares for low-income families in 1997 and twenty-five shares in 1998. Magic Beanstalk donated over thirty-five hundred pounds of food to local food pantries. (See chapter 19 to read about what has become Farm To Folk in Ames, Iowa.) In 1998, Field to Family organized well-attended community meals, started a new downtown farmers' market, held monthly cooking classes for adults and children, sponsored Iowa-grown meals with several conference centers, and began planning a food-processing microenterprise. Although the project itself is gone, as a result of Field to Family important elements of the original activities have been institutionalized. Ames, Iowa, has a popular farmers' market, and a local ecumenical group raises funds each year for healthy food vouchers that low-income families can use to join the Farm To Folk CSAs.

Smokey House Center in Danby, Vermont, employs local teenagers, who learn work skills on the job in carpentry, forestry, animal husbandry, and gardening. The primary markets for the center are a forty-member CSA, a local farm market, and area restaurants. Besides vegetable shares, Smokey House offers pork, beef, and lamb shares. Under the tutelage of Theresa Hoffman and a staff that has grown to ten full-timers year-round plus summer interns, the youngsters produce maple syrup, hardwood charcoal, Christmas trees, blueberries, pork, beef, lamb, and wool as well as the vegetables. The nonprofit's main product, however, is the training for the teenagers.

In Missoula, Montana, Josh Slotnick, his students, and volunteers started the Garden City Harvest CSA in 1997 on two acres donated by the University of Montana. A few years later, moving from that land turned out to be a gift in disguise—the new site has over six acres of tillable soils, comes with a ten-year renewable lease from the City, and is a higher-profile site. Garden City Harvest CSA continues to grow food for seventy shares and gives away as much to the local food bank (twenty thousand pounds in 2005). The students do six hours of farmwork per week for Josh's PEAS course (the Program for Ecological Agriculture and Society) and some work through the summer. Local people can "Volunteer for Veggies" and earn a whole week's share by working eight hours, or a half-share for four hours.

To make Garden City financially self-sufficient, the project sells the shares, offering one share size but three fee choices: "living lightly"—$180; "middle of the road"—$250; and "have enough to share"—$320. The Environmental Studies Program at the university pays 75 percent of Josh's salary. Josh reports they have gotten off the grant bandwagon thanks to local support for two annual fundraising events and one annual appeal. When they need a new piece of equipment, Josh does write grants and has funded a new barn and a spader. Garden City keeps its budget low by functioning without an executive director as a collective of the three program directors.

Youth Harvest is Garden City's newest program. With county funding, Tim Ballard, a trained counselor and a farmer, selects five teenagers, found guilty by the county Drug Court, to work at their farm. These at-risk youngsters from depressing, impoverished backgrounds work side by side with the optimistic and more privileged PEAS students. This combination has been working well for both sides, resulting in strong mentoring relationships. Josh says, "It is a powerful thing to give someone the chance to care for something, to belong to a project and take pride in their work." In recent years, Garden City has also formalized its connections with the local school district. In the fall, Garden City provides daily tours for fourth and fifth graders, and in 2006 it launched six sessions of a weeklong summer camp. The Garden City slogan could serve all of these projects: "Together we are learning self-reliance, and the medium is food."

Again in 2005 the Community Food Grants funded five more projects that involve CSA along with impressive lists of other social service goals. Cultivating Communities in Portland, Maine, intends to attract low-income families to finance their CSA memberships with Time Dollars, a sort of local currency based on providing services to others. Gateway Greening, an organization with a long track record of supporting urban gardens in St. Louis, Missouri, plans to distribute food grown at an urban farm through CSAs and corner stores. The Ecumenical Ministries of Oregon, in Portland, is going to provide low-income members of faith communities with subsidized shares in local CSAs. The Sunfield Education Association in Port Hadlock, Washington, is hiring a farmer to turn a fifteen-acre field of grass into a CSA farm. Rochester Roots in Rochester, New York, will be buying shares in the GVOCSA for the children in its school gardens programs and bringing the children to our farm for educational sessions and farmwork.

As someone who has farmed on a small scale in the quiet of the countryside, I find the projects I have just described incredibly complex organizationally, admirable in motivation, and perilously ambitious in scope. Every one of these projects deserves to have its full story told. I hope their daily struggles and moments of victory will be carefully documented. Their ultimate success will depend on many fragile elements coming together in just the right balance. If Community Supported Agriculture does nothing more than to help redress some of the injustices in our present food system, it will have earned its place in history.

CHAPTER 20 NOTES

1. Marty Strange, "Peace with the Land, Justice among Ourselves," CRA Newsletter (March 1997). Carlo Petrini of Slow Foods notes the irony that only the rich can afford to eat like peasants. In A Fate Worse Than Debt (Grove Press, 1990) and Ill Fares the Land: Essays on Food, Hunger and Power, Susan George analyzes the social prestige in developing countries associated with eating fast food, leaving only the poorest to eat the indigenous diet of whole grains, roots, and greens that is imitated in fancy restaurants.

2. Mark Winne, "The Food Gap," Orion (September–October 2005): 65–66.

3. When I called Isles, Inc., to follow up on the CSA Friedman started, no one in the office could tell me anything about it. By one of those flukes of serendipity, in March 2007, Ron Friedman e-mailed me to tell me he is moving to Rochester, New York, and would like to join the GVOCSA. I asked him what happened to the Isles, Inc. CSA. He responded: "The farm was located on the grounds of Mercer County Community College, about twenty minutes from Trenton. We tried our best to make the produce and the farming education available to the Trenton population. It was received well by some; however, the idea of getting African Americans out to visit or work on a farm was almost a brick wall. One of the town politicians was quoted in the newspaper that she would steal food rather than work on someone's farm. We had support from the community college president, but he resigned and went elsewhere. I also had the support of the head of the Ag. program at the college, but she went off to another job also. It was all very difficult. A great portion of the food went to the Trenton Area Soup Kitchen. We also had a weekly farm stand present at the Downtown Farmers' Market."

AGRICULTURE SUPPORTED COMMUNITIES

"Creativity is about the extravagances of variety."
—Comment by SISTER MIRIAM McGILLIS in
workshop at NOFA Summer Conference, August 12, 2006

No one licenses or certifies Community Supported Agriculture (CSA) projects. While farmers have created most of the existing CSAs in North America, reaching out to local consumers for financial and moral support, a variety of communities are making use of the CSA concept to serve their own missions. Community Supported Agriculture—Agriculture Supported Community; as Robyn Van En noted at the beginning, the mutual relationship is what is important. As long as there is someone involved who knows how to grow good food reliably, there is no limit to the additional purposes that may be served. The creators of each project are free to use their imaginations, combining their values, resources, talents, and site-specific conditions as they will. In this chapter we will offer snapshots of CSAs that caught my attention: CSAs initiated by faith-based groups, projects that provide work for adults with disabilities, a trade union supported CSA, and The Food Project, which successfully weaves together the complex strands of community food security, education for youth, and sustainable agriculture. (See appendix for two more profiles—"The Intervale: Community Owned Farms" by Beth Holzman, and "Hampshire College CSA" by Nancy Hanson.)

Faith Community Supported Gardens

"We see the despoiling of the environment as nothing less than the degradation of God's gracious gift of creation."

("A Social Statement on: Caring for Creation: Vision, Hope, and Justice," adopted by the Evangelical Lutheran Church in America, August 28, 1993)

Tormented that hunger and poverty persist amid plenty (the same dilemma that led me to begin my own investigations of the food system), Sister Miriam McGillis started on the path that led to Genesis Farm and inspired a whole series of CSA farms on land owned by religious orders. By the mid-1970s, Sister Miriam, a member of the Dominican Sisters, a Roman Catholic order, had become disenchanted with current political realities, identifying the Judeo-Christian worldview as a root cause of militarism, injustice, and the industrialization of agriculture. As she put it, "I longed for something to make sense." Thomas Berry's New Cosmology helped her articulate her intuition that the earth is a living organism and gave her a theoretical basis for practical action.

In 1978 she proposed to her Dominican congregation that they transform a 140-acre farm they had inherited in Blairstown, New Jersey, into a place where it is safe to ask questions, "a learning place for re-inhabiting the Earth." They started Genesis Farm in 1980, retiring the development rights on most of its 226 acres through the state's farmland conservation program. A few years later, a young Swiss biodynamic farmer came to help and stayed for thirteen years, training their gardeners. But it wasn't until 1987 when they saw *It's Not Just about Vegetables*, Robyn Van En's film on Indian Line Farm, that Sister Miriam and her colleagues decided to pursue CSA. In 1990 they also initiated their Learning Center for Earth Studies to delve into the more theoretical aspects of their mission and to offer classes and workshops to others. The successful intertwining of practical farming and earth literacy thinking are what make Genesis Farm such a special model. Their Web site explains, "The Community Supported Garden is not simply pre-buying vegetables; it is also helping to support itself through the running of a farm. The direct link between members and farmers puts the 'culture' back in 'agriculture.' As a result, our community begins to reconnect to the earth and to each other."

Steady development since 1988 has turned the Genesis Farm Community Supported Garden into one of the best-established CSAs in the country, with an active core group and deeply committed members. The Garden is a separate not-for profit and has a fifty-year lease on fifty-one acres from the Dominican Sisters of Caldwell, New Jersey. The sisters have a share of the vegetables, and the farm keeps them informed of its doings. Three hired farmers, several apprentices, and many volunteers do the farmwork. Membership has grown from 100 to 250 households. Contributions from members have helped fund the construction of a central garden house with a root cellar and distribution center, the establishment of fruit trees, grapes, hardy kiwi, and blueberries; and the purchase of a grinding mill and a precision seeder. In 1999 the farm installed a new well and the next year constructed a mile of deer fencing. The farm's 2005 wish list focused on equipment for cultivating grains and beans. In 2006 a neighbor left the farm corporation a challenging bequest—eighty acres with a house and barns but no money for maintenance.

Providing just wages for farmers was one of the

Solar panels provide electricity for the Learning Center at Genesis Farm, Blairstown, New Jersey. PHOTO COURTESY OF GENESIS FARM.

founding impulses for Genesis Farm. Gradually, the benefits of the three full-time farmers have improved, adding 15 percent of their salaries for a pension plan in 1996 and health insurance in 2000. According to the Web site, "The farm is working toward removing itself from the inequities, injustices and ecological devastation of the present market system. It is a social as well as an economic alternative." Despite the hefty share prices—$1,656 for a family and $861 for a single per year—head "gardener" Mike Baki says he "still feels we subsidize people's food bills a little."

The founding core group—Sister Miriam, Chan Moore, Heinz Thomet, and Judy Van Handorf—met monthly for the first twelve years. More recently, the core group adopted a committee structure and meets as a whole only four or five times a year. The working core of ten members continues to be responsible for the newsletter, payroll, and Web site, while the "gardeners" make decisions about the farm. Mike Baki drafts the annual budget and presents it to the core group for approval. He says they could ask more of the core.

Although not much involved in the day-to-day work, Sister Miriam saves seed from tomatoes and beans and sees to the small landscaping gardens. Her most vital role has been to inspire and to teach about the critical need to connect with the rhythms of the earth. Infused as her writings are with sadness about the state of the world, she finds hope through the practical activities of improving the farm and contributing to the local economy. In her letter to farm supporters in 2005 she asks three questions:

"How shall we help to sustain a viable economy in this Ridge and Valley region?"

"How shall we help our faith traditions sustain hope at the creative edge of the chaos we experience in these times?"

"How shall we bring our young into the experience of their deepest selves as an expression of the wholeness of Earth and Universe?"

While couched as an appeal for financial support for the farm's Earth Literacy Program, capital improvements for the garden, and energy-saving and recycled supplies for the center, her conclusion reveals the solid interconnection between her spiri-

Sister Miriam McGillis. PHOTO COURTESY OF GENESIS FARM.

tual searching and her leadership in practical action for a more sustainable world. In "Food as Sacrament," she writes: "When we understand that food is not a metaphor for spiritual nourishment, but is itself spiritual, then we eat food with a spiritual attitude and taste and are nourished by the Divine *directly*."[1] No wonder that she has inspired other faith communities to undertake the complex work of starting farms and CSAs. If you are seeking usable quotes for your CSA brochures you could do no better than to mine the well-thought-out and well-worded publications from Genesis Farm.

Responsibility for significant land resources combined with concerns about the environmental degradation caused by conventional agriculture has led Catholic religious orders in many parts of the country to organic farming and CSA. The National Catholic Rural Life Conference lists fifty "Sustainable Communities on the Land," including seven existing CSAs in Indiana, New Jersey, New York, Massachusetts, Michigan, and Ohio and plans for three more in Illinois, New Hampshire, and Ohio (and I know of several more that are not on this list). Having land available and the intention to grow food, of course, are only the beginnings of a farm. Religious orders face the same struggles as anyone else in developing the infrastructure for a CSA. With an aging and diminishing population, few orders can count skilled vegetable farmers among their members. Like other not-for-profits, religious orders usually have to hire someone to do the farming. Partnering with a faith community can be a tremendous opportunity for a young landless farmer.

In the fall of 2001, Sister Anne Rothmeier invited me to speak at the first harvest dinner of the Canticle Farm CSA in Allegany, New York. She told me the farm was on land owned by her order of Franciscan sisters. Although I followed her directions, I drove right past the farm on my first try. On closer investigation, I found a ten-acre open field with vegetables growing in it—no barn, not even a shed, only a small canopy over some equipment. Since that first season, contributions from the Franciscan friars at nearby St. Bonaventure University, from the Holy Name Province Benevolence Fund, from the sisters, and from the community, along with the hard work of farmer Mark Printz, have turned the field into a semblance of a farm. The founding core group was fortunate in finding Mark, a young man with gardening experience who was tired of his job in marketing. Mark has blossomed into a skilled farmer. The sisters love him, and he loves his job. With support from donations, Canticle gives 20 percent of its produce to nearby service organizations and low-income families and has created a revolving loan fund to pay the farm up front, allowing low-income members to pay back into the fund as they can.

The Sisters of Charity of Saint Vincent de Paul of New York were similarly blessed in their choice of farmer for Sisters Hill Farm in Stanfordville. David Hambleton had only had a few years of internships under his belt when he became their farmer, but he is doing an outstanding job since he helped the sisters start the CSA in 1999. David's relationship with the sisters is based on trust. The CSA is the property of the sisters, and he has no contract. To feel he is the resident farmer, David purchased land adjacent to the farm from the sisters to build a house, and meanwhile he lives with his wife and child in a cottage on the farm. He stays on, he says, because he has invested so much: "I started a good thing, and I want to see it continue to flourish."

Other faith group CSAs have had a harder time finding and keeping a farmer. Michaela Farm in Indiana and Sophia Garden at Queen of the Rosary in Amityville, New York, have both suffered from the frequent farmer turnover that plagues other nonprofit farms. Despite the lack of a steady farmer, Sophia Garden, on land that has belonged to the Sisters of St. Dominic since 1875, has kept in production since 1998. Victoria Gagliano, a former intern at our farm, reports that her parents have become addicted to the farmwork requirement of fifteen hours per season.

One summer day, while we were harvesting together, a member of my CSA, Jessica Conde, told me the remarkable story of Erda Gardens, an urban farm founded in 1994 by Marie Nord, a Franciscan nun. Marie was one of the leaders of her order and an environmental activist with an arrest record for protesting against the construction of nuclear power plants. With a vision similar to Sister Miriam's, Marie started her farm on land belonging to a bed and breakfast business that also rents to Los Poblanos Organics, sandwiched between the city of Santa Fe and the farms that stretch toward the Rio Grande River. Development pressure makes this very valuable real estate. Long periods of drought and sharp daily temperature swings make organic vegetable production particularly difficult in New Mexico. In "The Challenges of Constructing 'Shared Community': CSA Organizations in New Mexico," Lois Stanford documents the social obstacles to creating CSAs.[2] Yet, within a few years, Erda Gardens had grown into a community gathering place and education center with one hundred member families. According to Jessica, Erda Gardens was a stunningly beautiful place: "Marie brought it out of nothing and created a huge community around having a good food supply."

In the spring of 2000, with the garden only half planted, a tragic car crash caused Marie's death. At an emergency meeting that evening, the members of the farm decided to do whatever it would take to keep it going in Marie's memory. They struggled through that season, working along with inexperienced assistant farmer Dave Peacock. At a low point,

Erda Gardens. PHOTO BY CLARK CONDE.

he remarked, "I'm going to have to go to the grocery store to buy bananas to distribute." The interns moved into Jessica's two-bedroom house along with her, her husband, two small children, and a newborn baby. There wasn't a lot of food grown that year, but no one complained, and everyone signed up for the next year. The members discovered that Marie had been using her stipend as a nun and all her savings to support the farm, which was seriously in the red. The core group made the difficult decision to raise the share price and to do fund-raising to pay off the debt. Gifts from members and the Franciscan sisters plugged the hole and provided a salary for Dave. The next year, the core hired a farmer, Jimmy Petit, a

friend of Marie's who had farmed in the valley for many years. In 2005 the owners of the land they had used for eight years asked them to move. Kim Phopal, a member of the core group, reports that though they lost almost half their members, Erda Gardens found a new home and is making plans for improved marketing for 2007.

Like Genesis Farm, most of the CSAs initiated by religious orders also provide educational services to their communities. The Center for Ecology, Spirituality and Earth Education at Crystal Spring in Plainville, Massachusetts, has a forty-member CSA and offers classes in cooking for adults and children, organic gardening instruction, and bird watching walks as well as workshops on the new cosmology. At-risk youngsters from the "Weeds and Seeds" Program find paid work and ecological learning at Sophia Garden. Similarly, at Erda Gardens, Marie Nord gave workshops on biodynamics and permaculture.

Other denominations as well have grasped the importance of CSA. In 1993 the Evangelical Lutheran Church in America adopted "Caring for Creation: Vision, Hope and Justice," a statement committing the church and its members to earth keeping based on "participation, solidarity, sufficiency, and sustainability," and "personal life styles that contribute to the health of the environment," with special emphasis on the preservation of farmland. The next year, the church's Environmental Stewardship and Hunger Education Program published farmer Dan Guenthner's "To Till It and Keep It: New Models for Congregational Involvement with the Land," a call to "Congregation Supported Sustainable Agriculture." (See Dan's "What Can We Do to Respond to the Needs of the Land?" on p. 41.)

Much of the initial networking for Dan Guenthner and Margaret Pennings's Common Harvest Farm in Osceola, Wisconsin, was through churches, and until three years ago churches served as their drop-off points. Logistical problems have since led them to shift to neighborhood deliveries. Although their connection with particular churches has loosened, Dan says the core members of their CSA are progressive Christians who are integrating acts of faith with their lifesyles. Membership in Common Harvest has grown to 225 families, with over 100 who have been with the farm for more than fifteen years. Issues of peace and justice continue to be a unifying thread connecting the farmers and their members, who include non-Christians with the same sensibility. Dan says they are comfortable talking about faith issues in a political realm in their communications with members: "We are first and foremost spiritual beings. People want to bring that feeling to the dinner table."

Satisfied that the Genesee Valley Organic Community Supported Agriculture (GVOCSA) was running smoothly, organizer Alison Clarke turned her sights in 2004 on congregation supported agriculture in partnership with the Unitarian Universalists in Canandaigua, New York. Unable to find a farm in that town that was ready to start a CSA, Alison appealed to Peacework to help her train a core group by supplying them with shares for one year. The initial group of half a dozen recruited members within their congregation and used the church as the distribution site. That winter they put out an appeal to farms located closer to their church, inviting them to submit proposals. I helped explain to the five farmers who responded what would be involved in producing for a CSA. By this time, Andy Fellenz was ready. He and his wife had spent the 2000 season as members of our CSA to see what the experience was like from the customer's perspective. They purchased a house with seven acres of good-quality farmland in Phelps, New York, and Andy started growing vegetables and planting fruit trees, selling through farmers' markets with a few sales to stores. Only then did he give up his full-time job as an engineer. In 2006 the Unitarian CSA grew to fifty shares and the Fellenz Farm started a second group with members in Geneva.

Union Supported Agriculture (USA)

I wish I could announce that USA was a program of every labor council across the United States. Although the labor movement is beleaguered—rep-resenting a mere 13 percent of the workforce as compared with 80 percent in some European countries—unions still represent millions of working people. The idea of connecting farmers and urban workers has deep roots in our history. Three-quarters of a million people voted for the Farmer-Labor Party candidate for president, Parley Parker Christensen, in 1920, and in Minnesota the party put three governors and four senators in office during the 1920s and 1930s. Eventually, the Minnesota Farmer-Labor Party merged with the Democrats. Few people besides history buffs remember its heyday. The USA that does exist is a valiant effort on the part of one little farm in association with the Greater Bangor Area Central Labor Council and a nonprofit called Food AND Medicine.

Beginning in 2002, Laura Millay, a journeyperson (an advanced farm intern in the Maine Organic Farming and Gardening Association [MOFGA] farmer training program) on King Hill Farm and a staffer for Food AND Medicine, has been slowly building USA. Her work with Food AND Medicine put her in touch with union members who often had to choose between those two essentials when they lost their jobs. The Bangor area has suffered the loss of over four thousand shoe and wood product manufacturing jobs in recent years as corporations have moved production overseas. Food AND Medicine and the local unions have been doing educational work on the dislocations caused by Free Trade. A core group of union members, representatives of the Central Labor Council, and four farmers from two farms conceived of Union Supported Agriculture as a way to connect farms and unions by offering shares as a benefit for union members.

For the first season, union leaders recruited members at regular union meetings. These unionists understood why an alliance with local farms and a more just economic system made sense for them. Unfortunately, their unions were having internal troubles with participation; a union with two hundred members might attract only five or ten people to a meeting. Some of the unions donated farm shares to laid-off members. The first year, only ten people joined USA, not enough to make it worthwhile for more than one farm to do production. The members included teachers, postal workers, and mill workers. The cleaning staff at a nursing home purchased a share to redistribute to their union members. King Hill Farm donated a share to the family of a union member who had been injured in an accident. For the second year, the farmers took over recruiting, doing presentations at community functions where union members were present. By 2006 there were thirty members, with word of mouth responsible for attracting most of the new people. Laura reports that there are a lot of steps to take for people who have never purchased food directly from farms.

At King Hill Farm, Laura has become a 20 percent partner. Her former grade school teacher, Jo, owns the farm with her husband, Dennis, and his brother, Ron King. A big influence in Laura's life, Jo is a major supporter of the USA effort. Laura discovered food and farm issues while living in Thailand, where she worked with the Green Net Cooperative, an outstanding model of organic Fair Trade. At King Hill, Laura manages the 3-acre market garden. The 260-acre farm has 70 acres under cultivation with a herd of beef cows, thirty-five ewes, and a woodlot. USA sells sixteen-week shares for $340, and ten-week summer shares for $160. Members pick up at the farm or at drop-off points at a papermill and a union hall. USA members prefer the standard vegetables: lettuce, spinach, carrots, beets, peppers. Mill workers were not excited when the farm offered mesclun mix.

Besides supplying USA, King Hill has its own CSA and does thirty Senior Farm Shares.

Laura says USA has turned out differently from what they originally envisioned. Some people who are very dedicated to the union movement still do not get why buying locally grown food is important—and besides, they are not big vegetable eaters. Most of those who join have some personal connection with farming through family or past farmwork. Yet Laura and the Bangor Labor Council are determined to get other union groups interested in the USA concept. In the national and international struggles over "Free" Trade, family-scale farming

groups and labor unions have found themselves allied against big corporate interests. As was said above, the Farmer-Labor Party reached its peak almost a century ago. Is it unreasonable to dream that farmers and union members might make common cause again? Can we realistically hope to build sustainable communities without this alliance?

Including the Disabled

Although most nonfarmers think of farmwork as terribly hard physical labor, there are many jobs on a farm that can be done by people with various disabilities, both mental and physical. In the past some mental institutions put their troubled inmates to work in gardens and on farms because growing things can be very soothing. Taking care of animals has proven to be effective therapy for people who have trouble communicating with humans. No doubt, there have been cases of exploitation, but when done in a caring way, farms can provide dignified, meaningful labor. In this country and abroad, Biodynamic Camphill Villages provide farm based homes for the disabled and involve them in growing food for their own community, as does Innisfree Village in Virginia. Sankanac CSA at Camphill Village Kimberton and a few nonprofits use the CSA concept to combine food production with employment for the "differently abled."

Red Wiggler Community Farm in Clarksburg, Maryland, is an outstanding example. Initiated by Woody Woodroof in 1996, Red Wiggler's mission is "to create meaningful employment for adults with developmental disabilities. This is why we were formed and it is the core of everything we do. The jobs created by the CSA enable our activities at Red Wiggler to thrive. . . . Our vision is to create fertile ground to nourish a healthy and inclusive community. The CSA is integral to our mission." The number of shares Red Wiggler offers is modest; eighty households participated in 2007, including three group homes, which received subsidized shares. The farm also donates several thousand pounds of food a year to local food banks and low-

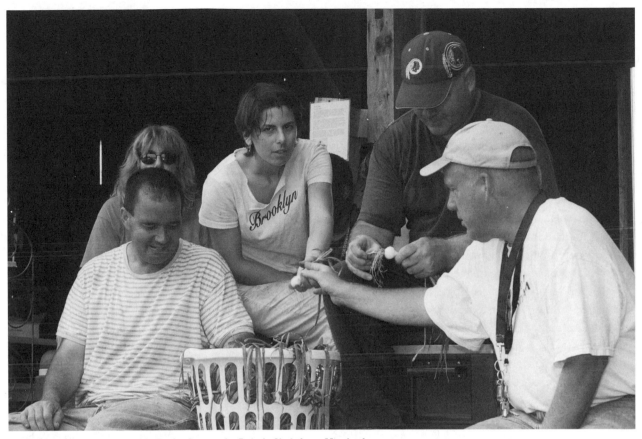

Sorting onions—trainees at Red Wiggler Community Farm in Clarksburg, Maryland. PHOTO COURTESY OF RED WIGGLER COMMUNITY FARM.

income families. The share price for twenty weeks is $475, with an additional $100 for a four-week fall extended season. Produce sales for 2006 were $43,290, a small portion of the project's total expenses of $251,000. Pushing production to the maximum that could be achieved on five acres is not one of their primary goals.

With a staff of four, Red Wiggler provides jobs for sixteen men and women with developmental disabilities. Woody's passion is to ensure that these marginalized people get the chance to "discover their true inner potential and ability in a nurturing and supportive environment." The "growers" work from 9:00 A.M. till 2:00 P.M. five days a week, a total of 2,465 hours in 2006, which earned them $16,675. The new site for the farm makes it possible for some of them to take public transportation to work, a huge step toward independence. Previous to joining Red Wiggler, some of them had worked only in sheltered workshops. A contract with the Montgomery County ARC (formerly known as Association of Retarded Citizens) brings its landscaping crew to the farm one day a week to do all the mowing. The farm also involves 289 youth and over 100 adult volunteers and gives workshops, tours, and educational programs on organic growing and environmental stewardship for schools and youth organizations in the area. During the winter the farm sets up collaborative art projects between the growers and regional artists. Bringing the able and disabled together in the farmwork is central to Red Wiggler's vision of "a vibrant, dynamic community farm where each participant has a place at the table."

With a classic organic approach, Red Wiggler focuses on building healthy soil with cover crops, natural fertilizers, and compost. Most of the work is done without heavy machinery so that there are plenty of jobs involving many hands. Shannon Varley, who worked as farm manager, describes the farmwork:

Here's an example of how we split job responsibilities: During planting season, for instance, bulbs (including garlic, shallots, onions, etc.), potatoes, and winter and summer squash are planted by the growers. They are larger and easier to handle. The farm manager is responsible for the smaller seeded crops (salad mix, chard, beets, spinach, etc.) although there are instances when a grower might be interested in helping out with those chores and we give it a whirl.

The growers' responsibilities also include laying mulch, placing and replacing hoops and Remay® fabric in the fields, hilling potatoes, weeding, weed-whacking, building deer fences, pounding tomato stakes, taking out tomato stakes, prepping the hoop houses, and seeding. Not to mention the thrice-weekly harvest, the twice-weekly CSA preparation, and the bagging and labeling of vegetables for five months a year. And I'm quite certain that I'm forgetting a thousand little things that keep this place running.[3]

Woody Woodroof has overseen the steady development of Red Wiggler. A huge step forward occurred over the winter of 2004–2005 when the farm signed a twenty-year lease for ten acres of the Ovid Hazen Wells Park owned by Montgomery County. The farm immediately put up a seven-foot-high deer fence around five acres. The new location makes the farm more accessible to its "growers" and its network of volunteers. In 2005 the University of Maryland donated an 800-square-foot solar home that was a part of the '05 Solar Decathlon sponsored by the Department of Energy. The house serves as "affordable workforce housing" for a Red Wiggler staff member. Further public support is crucial for its next project, the construction of a 2,000 square feet "green" multi-purpose building. This structure will enable the farm to run programs year-round, doubling the number of jobs for disabled people while expanding the educational activities to include food and nutrition training for all CSA members. The building will be LEED (Leadership in Energy and Environmental Design) certifiable. Funding for the $400,000 project is coming from individual farm members, local businesses, foundations, and state and local government. A Community Food Project grant will enable the CSA to expand to 120 shares, of which 25 percent will be available on a "sliding scale" to area group homes serving adults with developmental disabilities.

Red Wiggler's approach to funding and development is an instructive model for others who seek to combine CSA with educational missions. Red Wiggler does not expect its farm to pay for more than direct farm expenses. The revenue from produce sales cannot cover the cost of the programs for the disabled and local youth. Instead, support comes from grants from private foundations and government and strenuous fund-raising activities with local citizens. Having realistic expectations for what a farm can generate is critical to combining CSA with other purposes. The CSA concept is perfectly adapted for outreach into the broader community, but overloading the farm with financial obligations can lead to stressed farmers and eventual failure.

Near Lansing, Michigan, Giving Tree Farm also combines jobs and horticultural therapy for the disabled with a CSA. Susan Houghton, who has been running the farm since 1998, receives a salary from Community Based Interventions (CBI), a not-for-profit organization that works with people with all kinds of disabilities; CBI purchased the farm in 2005. Giving Tree gives on-the-job training and pays minimum wage to five people who have suffered traumatic injuries to their heads. The staffing ratio is one on one for some of them and one on two for others. Susan is in charge of production and finding the right jobs for the trainees. One of their favorite jobs is staffing the farm's stand at a farmers' market. When participants complete the program, Giving Tree helps them find employment with horticulturally related businesses in the area.

Figuring out how to keep the farm financially viable while providing services to the disabled has taken several years. Initially, Susan managed the twenty-one-acre farm for its board of directors, selling as many as one hundred CSA shares, as well as supplying restaurants, a farmers' market, and wholesale accounts. Making ends meet was difficult. The land belonged to

Caroline and Bob Bower, who had a child with Down's syndrome. To create a safe place for that child, the Bowers started a garden in 1991 and then invited other families with similar children to join them in a "Leisure Gardening Program." The group garden still exists in a separate area on the farm. Since the families wanted pesticide-free food, the group decided to hire a farmer. Together they also built a house for disabled children, but when they could not find adequate funding to keep it running, they leased it to CBI. The umbrella of the larger organization has solved Susan's financial struggles, and she is redesigning the marketing mix and scaling back CSA membership to the farm's most ardent supporters. By concentrating on a smaller number of crops, microgreens in the hoop houses and root crops in the muck soils, Susan is creating the regular routine that best suits her disabled workers.

Inspired by the experience of Camphill Villages, Community Homestead in Osceola, Wisconsin, provides homes and work for adults with special needs. Living as an intentional community in six extended-family households, the forty members of Community Homestead share a common budget supported by an impressive mix of enterprises. On 220 certified organic acres, they have a 7-acre garden, which supplies a 150-member CSA, a forty-cow organic dairy herd shipping through Organic Valley, 3 acres of orchards, a bakery, chickens, beef and pigs, and craft workshops. For eighteen to twenty weeks of produce, CSA members pay $475 for a large or $375 for a medium share and can order additional products from the farm online. In 2006, Community Homestead invited its supporters to contribute work and funds for a new $350,000 processing center and store.

The members with developmental and other disabilities live in households with the other residents, both adults and children. Christine Elmquist, a member of the farm, describes their approach:

> All our activities are created to provide a lifestyle that includes our people with special needs in a dignified, therapeutic and sustainable setting. We are all-volunteer, pooling all our resources, human and economic, to enable all of us to live with meaning. We work with a shared budget according to need and not ability and enjoy each other a lot! Our activities—dairy, woodshop, crafts, etc.—are all together about 45 percent of the budget. The CSA is approx. 15 to 20 percent. We count on our activities to sustain us. Sustainability is very important for us all! We work as a team with a point person or people for each area. Our farmer, who organizes the whole farm, has 25 years experience organic gardening and farming. His wife has 15 years experience in organic gardening and she has particular organizational responsibility for the garden. Another person has day-to-day responsibility for all gardeners (of all abilities) and more day-to-day stuff. She has three years experience. About seven people put in half day or equivalent through the high season and maybe half that in slower times. Then there are sixteen or so people with special needs (this is a tough thing to describe because some of our 'disabled people' are more capable than a young 'non-disabled' summer person, so I use these categories for clarity only!!) who help out half day or more too.

The farm Web site includes these lovely words about their garden, "Here is a safe and peaceful place to learn, to nurture, and to accomplish. Rhythm, reliability, and self-esteem are well-rooted in the soil of the garden."

Breaking Down Racial and Economic Divisions: The Food Project

When I moved to Boston with my four-year-old son to take a teaching position in 1975, the public schools in that city were so segregated by race that the courts had issued an order to force integration. My plan had been to choose a decent school for my son before renting an apartment in the neighborhood served by that school. Traveling around the city, I was amazed and horrified at the devastation in some areas—parts of Roxbury looked like the rubble left after a war. I visited several excellent schools and some really poor ones. But I could not

Youth participants and staff at The Food Project, work on the farm in Lincoln, Massachusetts. PHOTO COURTESY OF THE FOOD PROJECT.

find out which street went with which school; the desegregation order forbade the schools from providing this information to parents. My personal solution was to settle in Cambridge. The De Mau Mau, a group of African American Vietnam veterans led by John Clinkscales, confronted the poverty and racial strife through a Black Panther–like program of free breakfasts for children and gun-carrying street patrols. They ended up with long prison terms. Ward Cheney, the founder of The Food Project, took on the challenges of healing the earth and the community by inviting young people—black, white, and Hispanic—to grow food for the hungry together.

From a modest beginning in 1991 with three unpaid staff people and eighteen youth, The Food Project has blossomed into an established nonprofit with a budget of $2 million employing thirty full-time staff, over one hundred youth trainees, and thousands of volunteers of all ages. On four urban sites and thirty-one acres of farmland in Lincoln,

Massachusetts, the project annually produces 250,000 pounds of naturally grown food, of which half is sold and half donated to food shelters and soup kitchens. The Lincoln farmland produces the vegetables and fruit for 233 shares that pick up at the farm and 100 shares for delivery to drop-off points in Cambridge and Jamaica Plain. The youth sell much of the food from the urban gardens at a farmers' market and take some home to their families. Along with the many volunteers, the young people provide the labor for the farm and gardens. Yet, impressive as these statistics may be, they do not convey the tremendous social significance of The Food Project, which is transforming inner-city and suburban youngsters into skilled producers and articulate spokespeople for sustainable agriculture and inspiring replications around the country and abroad.

Farmer Ward Cheney had a vision of a peaceful path to change through putting teens from middle-class suburbs together with inner-city minority and low-income teens and teaching them how to grow

Food Project youth staff booth at Farmers Market in Boston, Massachusetts. PHOTO COURTESY OF THE FOOD PROJECT.

food to donate to the hungry. Since none of the young people come from farm families, the farm-work puts everyone on even ground. As The Food Project Web site states, "Our mission is to create a thoughtful and productive community of youth and adults from diverse backgrounds who work together to build a sustainable food system. . . . We consider our hallmark to be our focus on identifying and transforming a new generation of leaders by placing teens in unusually responsible roles, with deeply meaningful work." An indication of how successful they have been is the fact that Elise LeClair, the current director of agriculture at the farm in Lincoln, started as a teenager with the project as a summer youth in 1994.

The meaningful work that The Food Project offers its teenage recruits and pays them to do takes many forms: they grow food at four urban gardens as well as at the farm, sell at a farmers' market, process some of the food into salsa and holiday pies, and distribute the food directly to the needy at food shel-

ters. They also learn to supervise the work of the volunteers. The youngsters spend half of each day doing physical labor. The other half, however, they spend learning more about one another, confronting racial and economic diversity, and training to be effective trainers and leaders themselves. Through a process they call "Straight Talk," everyone at The Food Project, youth and adult staff alike, pairs up for feedback sessions in which they exchange praise and constructive criticism. Elise LeClair explains, "The power comes from the chance to listen to somebody else talk about yourself."

The youth also participate in workshops in how to manage their money, and get practice in public speaking by visiting legislators and making presentations about the project to school groups and at conferences. As they learn new skills, they have the opportunity to pass them on to others by supervising younger participants. Sixteen-year-old Will Quayle said about his first season with the Food Project, "It's more responsibility than I had ever had before."

Greg Gale, who runs the youth program says, "We don't make work for them. We don't pretend that the mission is important—we actually really feel it. They feel that, and that's where this deep commitment comes from in a lot of them. I mean kids stay with us for years to keep serving."[4]

I have observed Food Project youth conduct workshops at sustainable agriculture conferences—their poise and ability to express themselves are impressive. At meals and dances, the members of this very diverse group of youngsters seem to be genuinely enjoying one another's company. They do not divide into separate groups by sex or color the way teens often do in school cafeterias. I have asked both African American and white participants what they like about The Food Project. They all seem to agree that one of the things they value the most is the chance to make real friends by working and learning with people they would not have gotten to know otherwise.

The CSA grew out of the project's need to find a way to support its farm. When Don Zasada took the job of farmer in 1998, the main emphasis was on growing food for hunger relief. Contributions and grants were covering the work with youth and food for the hungry, but the project was having trouble raising money for the farm. Don expanded the CSA, and today it covers the farm's expenses and development.

Being the head farmer for The Food Project is rather different from running one's own farm somewhere out in the country. Elise describes her job as managing production and also managing people—a lot of people. During the summer program, four crews four days a week, forty of the youth do four and a half hours of fieldwork. They are divided into crews of ten with a slightly older person as crew leader. Elise trains the leader, who comes an hour earlier, in the work for the day, and then the leader trains the crew. People who buy shares also are welcome to volunteer, and many do. This abundance of ready workers dictates some of the production practices. For example, they hand-transplant all the sweet corn, an effective technique but one that is far more time consuming than direct seeding. Twice a week they harvest for the CSA.

Elise also has the job of developing the farm's budget, which gives her basic control over expenses. The sale of shares has to cover production costs for the CSA and for the 40 percent of the food that the farm gives away. When they discovered that the food shelters were throwing out the salad mix, they focused on crops the shelters can use, such as peppers, tomatoes, cucumbers, zucchini, and winter squash. Two of the shelters pick up the leftovers after CSA distribution. Elise, who grew up in Carlisle, one of the towns the CSA serves, plans to stay with the farm for another year. It means a lot to her to be a producer in her own community. She says that she loves the farm and loves the people but would like to have her own farm. Working for a nonprofit is a balancing act with advantages and disadvantages, even though The Food Project understands her needs as a farmer.

The Food Project makes it very easy to learn more about its work. Half of what it does is to serve as a resource center for other organizations in this country and worldwide. Through its Web site, one can access an array of helpful materials: *Growing Together: A Guide to Building Inspired, Diverse and Productive Youth Communities*; *Dirt: The Next Generation*, a fifteen-minute video; a series of manuals on running programs for the academic year, on "how to run a sustainable production farm while integrating thousands of youths and volunteers throughout the season;" three volumes on the summer youth program; and one each on farmers' markets, urban agriculture, and managing volunteers. The Food Project initiated BLAST (Building Local Agricultural Systems Today), which encourages young people around the country to undertake local projects; and the staff offers regular teleconferences on organization and funding. Together with groups like the Hartford Food System, The Food Project instigated "Eat In, Act Out," an annual series of events in many localities.

I asked Elise and Don to what they attribute The Food Project's success. Don answered that it has been very important that three of the founding members are still on the staff. They remember the organization's roots and help newer people understand their history. This institutional memory serves them well in sorting through what works and what does not. Founder Ward Cheney set them on a path of close cooperation with existing neighborhood churches and projects in Roxbury, such as the Dudley Street Neighborhood Initiative, a resident-developed community revitalization project. Elise stresses the "open culture around communication. This is a place where you can walk around and be yourself." Jen James, the associate director, told me that five of the full-timers are alumni of the youth program. "People love being here," she says, "Everyone has a voice and we pay attention to everyone. We are true to our mission of growing food and involving youth." Attention to process combined with solid grassroots organizing is a formula for sustainable development.

When we come together around good food, our creative juices start flowing. The flexibility of CSA allows for so many inventive and meaningful combinations. To build sustainable communities, we need to make alliances among as many different groups and perspectives as possible—educational institutions, religious orders, labor unions, environmentalists, health practitioners, and town and county governments. Like a strong rope or cable woven together from many individual strands, these multifarious interests can bring great strength to our future and victorious party of the local economy.

CHAPTER 21 NOTES

1. Therese McGillis, "Food as Sacrament," in *Earth and Spirit: The Spiritual Dimension of the Environmental Crisis*, ed. Fritz Hull (New York: Continuum, 1993), 163.
2. Lois Stanford, "The Challenges of Constructing 'Shared Community': Community Supported Agriculture Organizations in New Mexico" (paper presented at the session "Alternative Food Movements and Culture: New Directions in Anthropological Research" of the American Anthropological Association meetings, Chicago, November 19–23, 2003).
3. Shannon Varley, "Farming Felt Deeply, and without limits," *The New Farm* (February 16, 2006).
4. Susan Chang, "Cultivating the Soil, Cultivating Youth," *The New Farm* (September 15, 2005).

CSA AROUND THE WORLD

In many highly urbanized, industrialized countries, both consumers' longings to reconnect with the soil where their food is grown and farmers' need for loyal customers have led to the birth of Community Supported Agriculture (CSA) projects, as well as direct sales of other kinds. By contrast, in places where most of the population lives on the land and people grow their own food, CSAs are much less common. As Greg Pilley, author of the Soil Association study "A Share in the Harvest," puts it, "A walk through any African city or village reveals a thriving and highly productive local food economy." In many parts of Africa and Asia, jobs outside of agriculture are scarce; working the land is the only option for income and the hope for food security. Difficult as conditions may be, people are still connected with the land. But in Europe, North America, and Japan, a century of "development" has broken that connection, and a few decades of free trade have driven family-scale farms to the point of desperation. Small farms that make growing for the conventional wholesale market their primary strategy find themselves in a deep hole. CSA offers a return to wholeness and economic viability.

Human history abounds in examples of specific groups of nonfarmers being connected with specific farms—the medieval manor, the Soviet system of linking a farm with a factory, or the steady attachment of particular customers to the stand of a particular farm at a farmers' market. In Cuba today, all institutions are obliged to be self-sufficient in food,

so companies and schools have farms or garden plots. But none of these is like the form of organization we refer to as CSA.

The modern CSA originated in Japan. In 1971, Sawako Ariyoshi, the Japanese Rachel Carson, alerted consumers to the dangers of the chemicals used in agriculture and set off the movement for an organic agriculture. That year, concerned housewives joined with university researchers to form the Japanese Organic Agriculture Association (see the short history by Cayce Hill and Hiroko Kubota on p. 267). That same year, Yoshinori Kaneko realized that his family farm, besides providing for the subsistence of his own family, could also supply other people. He calculated that the farm produced enough rice for ten more families. To recruit local housewives, he invited them to join a reading circle, where they discussed such themes as "Oneness of Body and Environment," the value of whole foods, and the healthfulness of the traditional Japanese diet. After four years of "education and communication," in 1975, he made an agreement with ten families to provide them with rice, wheat, and vegetables in return for money and labor.[1]

Contracts between groups of highly educated consumers and farmers like Kaneko launched the *teikei* ("partnership") movement. Within a few years, farmers and consumers formed remarkably similar organizations in Switzerland, yet no one has found a link proving Japanese inspiration. There are, however, photos of Western visitors in Kaneko's book *A*

Farm with a Future (1994). Could one of them have brought the teikei idea to Europe? Or did the concept arise separately out of the principles of biodynamic and organic farming? Someday, surely, a graduate student sleuth out there will answer this question for us.

In the late 1970s several biodynamic farms in Switzerland recruited members to buy shares in the harvest and help with some of the work. Rudi Berli, one of the collective of ten farmers at Les Jardins de Cocagne near Geneva, told me that the founders were inspired by the collective farms in Chile during the Allende years and by the peasant-worker movement in Brittany in France. Reto Cadotsch and a few comrades started Les Jardins de Cocagne in 1978 with fifty members. The first year all they ate were turnips. They had poor tools, rented land, no irrigation, and no houses, but their members were very supportive. The farm still rents its land, growing fifty vegetable crops, apples, grapes, and berries on seventeen hectares in 2005. The four hundred members do a minimum of four half-days of farmwork a year. Those who do not work pay an additional $40 for each half-day they miss. Payment is on a sliding scale from $600 to over $1,000 for the eleven months of shares. The farm crew packs the bags of produce; members deliver half the shares, and a hired delivery van does the other half. The crew of ten are employees of the farm, earning the average wage for Swiss workers plus benefits. For the past twenty years, 1 percent of the farm's budget has gone to a North-South solidarity project in the Sahel region of Africa. According to Rudi Berli, there have been three CSAs in Switzerland for many years, but recently six new ones have formed, inspired by the French.

Jan Vandertuin, the man who brought the CSA concept to Robyn Van En in 1985, helped set up Topinambour, a collective farm providing vegetables to members near Zurich. In 1986, Trauger Groh brought his version of CSA from his experiences at Buschberghof in Fuhlenhagen, Germany, to the founding of Temple-Wilton Community Farm in New Hampshire. Subsequently, Buschberghof has adopted much of Temple-Wilton's approach, such as the regular membership and the annual meeting at which members pledge what they can afford to support the farm budget. According to the Soil Association study "A Share in the Harvest: A Feasibility Study for CSA" (Soil Association, 2001), there are many direct marketing initiatives in Germany where they are called Erzeuger-Verbraucher-Gemeinschaft (EVG), with twenty to one hundred members and over two hundred box schemes (see below for definition)—but not many farms organized like Buschberghof.

At meetings of the International Federation of Organic Agriculture Movements (IFOAM) in Asia in the early nineties, teikei farmers like Shinji Hashimoto made presentations about their farms and marketing, yet not many Asian farmers outside of Japan have tried to replicate the teikei model. With the help of IFOAM contacts, I was able to hunt down one CSA-like farm fifty kilometers south of Kuala Lumpur in Malaysia. GK Organic Farm was founded in 1994 by Gan, a university graduate who gave up a career as an agrochemical dealer to become an organic farmer. According to Gan, the organic farmers in Malaysia are educated people like himself, not the peasants who make up the vast majority of the people who work the land. In 1996, Kazumi joined Gan, and thus their farm name, GK Farm. They offer several sizes of vegetable and fruit baskets, a weekly "surprise," which they deliver to shops in Kuala Lumpur for their members to pick up. Most of what I know about GK Farm I have gleaned from its Web site (www.geocities.com/gankaz2000). The farm has a guest lodge and places for tents, but as of November 2005 was no longer accepting visitors. The photos and list of crops, including bananas and papayas, would entice me to visit were I ever anywhere near Malaysia. Their philosophy of farming is attractive, too: "For us, organic farming is not merely a method of food production or a way of earning a living. It is a specific lifestyle, which orientates diet choices, consumption patterns, and even ways of feeling, thinking and behaving. Organic farming is also one piece of an ever-challenging jigsaw puzzle that we ought to complete—a picture of Nature and Humans living in harmony."

A young American named Keefe Keeley, who is spending 2007 traveling around the world visiting organic farms, sent me this description of a CSA-like farm in India:

> At Vasant and Karuna Futane's farm, they market almost all their crops through CSA. Their lifestyle is profoundly influenced by Gandhi, Vinoba, and Fukuoka. It was a truly remarkable family. They live modestly, even simply, as inspired by Gandhi and Vinoba. They also work with the local villages in various constructive programs that span education (for adults and children), women's empowerment, tribal rights, anti-drunkenness campaigns, and educating farmers about the dangers of getting caught in the debt cycle of buying GM seeds and chemicals from agribusiness companies. They have a thirty-acre farm where they have been adapting Fukuoka's philosophy of natural farming to the conditions of central India. Their biodiverse farm is visually arresting; it is immediately distinguishable from the orange orchard monocultures that surround it. Around 90 percent of the food they consume (and serve to the frequent and numerous guests) is grown at the farm, and the rest is sold to friends in towns within a moderate radius of the farm. At the beginning of each year, they ask their friends how much of which crops they will want that year, and plant accordingly. They have received offers to sell their produce in megacities such as Mumbai and from countries in Europe where the affluent are willing to pay premiums for "natural" produce, but they are committed to supplying their local area with quality food. . . . Perhaps more than anything else during my year of travel, my stay at this farm with this family has inspired me to become a community-focused farmer.

During the 1990s, small organic vegetable farms all over England set up "box schemes," subscription services where the farm provides regular boxes of produce to people who sign up. These box schemes do not require much member involvement in the growing, harvesting, or distributing of the food. On its Web site, the Soil Association provides a guide, "How to Set Up a Box Scheme." The subscription approach is popular in several European countries, but it has taken off like a rocket in Denmark. Thomas Harttung, the owner of the Barritskov Farm in western Denmark, told us at the Northeast Organic Farming Association of New York (NOFA-NY) conference in 2004 that his farm started delivering shares to 100 families in 1999. Organized as Aarstiderne, a Web-based organic food delivery service, it had surged to 44,000 customers by 2004. Although still based at Harttung's farm, Aarstiderne offers six hundred organic products from over one hundred farms and employs 110 people and thirty delivery trucks. Along with the boxes come recipes from Chef Soren Ejlersen, a key player in the success of this business, and stories about the farms from which the food comes. Harttung claims that, as in CSA, there is a sharing of the risk because the customers pay in advance, though only for one month at a time. Farms in Holland sell "Green Guilders": customers pay 1,000 guilders in advance and then shop at their chosen farm.

The Soil Association feasibility study cited above came to the conclusion that CSA has many benefits for both farmers and consumers: "Consumers have access to fresh food from an accountable source with an opportunity to reconnect with the land and influence the landscape they live in. CSAs deliver environmental benefits of few food miles, less packaging and ecologically sensitive farming, and see the return of local distinctiveness and regional food production with higher employment, more local processing, local consumption and circulation of money in the community enhancing local economies."[2] As a result, the Soil Association engaged in a project it called "Cultivating Communities," to facilitate the development of more CSAs in England, and created a CSA Action Manual, a fifty-seven-page guide to setting up a CSA, which can be downloaded from the project Web site (www.cuco.org.uk). The manual defines CSA as "a partnership between farmers and consumers where the responsibilities and rewards of farming are shared. . . . CSA is a shared commitment to building a more local and equitable agricultural system, one that allows farmers to focus on good farming practices and still maintain productive and profitable farms." The manual covers all the relevant topics: how to find land, how to recruit members, sources of funding, production practices, sample operating costs, and profiles of CSAs of different kinds.

In 2005 the Soil Association identified one hundred consumer-farmer partnerships in England, which ranged from the familiar vegetable, meat, and fruit shares to a rent-an-apple-tree project, intentional communities, urban gardens, and conservation-based projects. Looking over the profiles, there are several thriving projects—EarthShare in northeast Scotland, Stroud Community Agriculture in Gloucestershire, and Tablehurst and Plaw Hatch CSA in East Essex—having some novel practices that CSAs elsewhere might consider. Both Stroud and EarthShare accept payments in LETS (Local Exchange Trading Systems, local alternative currencies). Tablehurst and Plaw Hatch gives people the choice between buying shares and simply investing in the farm to lend it financial support.

Although I found a box of organic vegetables at my doorstep in La Cadière in July 1977, CSA did not catch on in France until 2001, but then it spread like wildfire, with the number of participating farms reaching three hundred in 2006. (See p. 29 for the story of my first encounter with CSA.) Anyone who has traveled in France and experienced the wonderful farmers' markets in almost every town would share my surprise that these markets have not provided the economic support family-scale French farmers need to stay in business. Competition from cheap imports and fast food (what José Bové translates as "*mal bouf*" (bouf is slang for food) has undercut their markets and pushed them to the edge of bankruptcy. According to farmer Daniel Vuillon, "The social reality of farmers is very hard. The disappearance of small farms is in process. In Bouches-du-Rhone and the Vaucluse some 3,500 farmers are in bankruptcy. . . . In Hyeres, 600 vegetable farms are in severe economic straits."[3]

Daniel's farm, Les Olivades, in the town of Ollioules, Provence, was in similar trouble. The

Denise and Daniel Vuillon created the first AMAP (Association pour le maintien d'une agriculture paysanne) in Oulioulle, France. PHOTO COURTESY OF THE VUILLON FAMILY.

Vuillon family has owned and worked the ten hectares of Les Olivades since 1789. Daniel and his wife, Denise, took over in the early 1980s, opening a stand at the farm in 1987 and selling to supermarkets. In 1999 they discovered CSA when their daughter, Edith, traveled to New York and met up with Just Food. The next year, the parents went to see for themselves and toured Roxbury Farm. When they came home, they met with consumer activists in the nearby town of Aubagne and explained the economics of their situation. By April 2001 they were distributing their first forty shares to the members of the first Association pour le Maintien d'une Agriculture Paysanne (AMAP), the name they gave to CSA. A televison reporter at the opening ceremony asked one of the members whether it wasn't a lot of trouble to shop this way, and whether she was happy with it. André Bregliano, a loyal member to this day, replied: "Oh, yes!!! I am happy to have freshly picked vegetables, I am happy to eat in a healthy way, and if, as a result, the farmer can stay on his farm, that is the most important."

Within two years, the Vuillons were selling the entire production of their farm to three seventy-household AMAP groups, two picking up at the farm and one in Aubagne. Their improved finances have allowed them to hire four full-time, year-round employees. They provide their members with a weekly newsletter—two pages with farm news, share list, and recipes. Members can also purchase bread and chicken/egg shares from neighboring enterprises. Support from their members has been critical in restraining the municipality from expropriating the farm's land for a new tramway line. I was able to visit Les Olivades in the winter of 2005. Within the dense line of trees around the perimeter, you would never know that houses and businesses crowd right up to the edge of the property. An aerial shot would look somewhat like Michael Ableman's well-known photos of Fairview Gardens in Santa Barbara. Inside this rural oasis, there are 4 hectares of vegetables, 4 hectares of fruit trees, and 1.5 hectares of greenhouses as well as a picturesque old farmhouse.

Not content to save only their own farm, Daniel and Denise have spread the word about AMAP to farmers and consumer activists all over France. In May 2001 they founded Alliance Provence, an organization dedicated to helping farms form AMAPs in their province. The regional government of Provence–Alpes Côte d'Azur soon became a partner in the venture because of the promise of viable economic development. In 2004, with close to one hundred AMAPs in the region, they restructured the organization into six geographic sectors with a team of an experienced AMAP farmer and an active consumer in charge of each. At their farm and at meetings around the country, Daniel gives workshops on AMAP production and Denise covers organization. A national network, the Alliance of Peasants-Ecologists-Consumers (Alliance Paysans-écologistes-consommateurs) supports the development of AMAP. Like the Soil Association, its Web site provides detailed information on establishing an AMAP and also guides consumers on finding the one nearest where they live (www.alliancepec@free.fr). The text suggests making membership possible for lower-income members through weekly payments

and work exchanges and provides an outline for the workings of a core committee to support the farm. The Web site sets out these principles:

> AMAP participants seek healthy food, produced with respect for Human Beings, biodiversity and the rhythm of Nature.
>
> AMAPs participate in the struggle against pollution and the risks of industrialized agriculture, and stand for responsible, shared management of the common goods. Through this local partnership between farmers and consumers, AMAPs promote social dialogue between the city and the countryside, and facilitate coexistence between recreation and productive activities, and the multiple use of agricultural spaces.

According to the Alliance, these are the terms that distinguish an AMAP:

For the consumer:
1. Payment in advance
2. Economic and moral solidarity in sharing the risk
3. Commitment to participate in the life of the group

For the farmer:
1. Providing products of high quality
2. Commitment to involvement in the life of the group
3. Commitment to assure transparency on the farm in regard to economics, method of production, and origin of products

The infectious energy and enthusiasm of Denise and Daniel Vuillon and Alliance Provence have also helped bring about the establishment of an international CSA network, *Urgenci*. So far, the network has sponsored two international conferences, in France in 2003 and in Portugal in 2005. Edith Vuillon organized the first and a Frenchman, Samuel Thirion, who lives in Portugal, organized the second. The centerpiece of the programs was sharing experiences with CSA/AMAP/ASC/Reciproco/Voedselteams/Teikei.

URGENCI: The International CSA Network

MISSION OF THE NETWORK

1. The mission of the URGENCI network is to promote, on international scale, local solidarity based partnerships between farmers and the citizens that are fed by them. We define the solidarity based partnerships such as:
    - mutual commitment of supply (by farmers) and sales (by consumers) of foodstuffs produced within each agricultural land,
    - fair remuneration, advance and sufficient payment to insure worthy life for farmers and their families,
    - sharing risks and advantages of a healthy production, that is adapted to seasons and respectful of environment, natural and cultural heritage and health.

2. In more general terms, the objective is to promote various forms of solidarity based partnership between producers of goods and services and local consumers in areas other than agriculture, by participating in an economy that is local, respectful of the environment and meeting the needs of well-being and healthy lifestyles (this extension to other areas of economy is still being discussed within the network), and to promote citizenship between urban and rural inhabitants.

3. Besides the promotion the URGENCI network aims:
    - to allow exchanges between these partnerships, as well as with other forms of solidarity based economies;
    - to develop tools to reinforce their viability;
    - to coordinate the actions on international level (and facilitate consistency);
    - to create dialogue with public institutions;
    - to reinforce the mobilisation of local networks;
    - to reinforce the principles of urban-rural partnership.

(these latest points are taken from the common letter of intention that had been the conclusion of the first international colloquium)

VALUES OF THE NETWORK

The fundamental values of the URGENCI network and its members are:
    - equity, solidarity and reciprocity in the economic relations in particular by mutual commitment, fair and stable prices and sharing risks;
    - autonomy of each partnership of producers and consumers, particularly in decision making process and rules they establish for themselves;
    - quest for global consistency on different scales (local, regional, national and worldwide) regarding the objectives of equity and preservation of common goods, democratically debated and shared in a consenting way.

OBJECTIVES OF THE NETWORK

The fundamental objectives of the URGENCI network and its members are:
    - support and blossoming of a small scale agriculture (family run farms), that is capable to insure at the same time a local, healthy production that does not depend on fossil or imported energy, and the preservation of the environment, biodiversity, landscapes and natural and cultural heritage;
    - food sovereignty of each region and each community of the planet;
    - health by nutrition and fight against hunger and various forms of malnutrition;
    - development of citizenship within the economy and of social and interdependent links between the producers and consumers and urban and rural population;
    - environment and citizen education, in particular for future generations;
    - fight against exclusion and poverty by solidarity links, whether in rural or urban areas.

A Reciproco member farm in Odemira, Portugal. PHOTO BY ISABELLE JONCAS.

Most of the participants at both events came from the United Kingdom and Western Europe, but there was also a scattering of representation from North and South America, Australia, Japan, and Africa. The second conference concluded with the selection of a board charged with writing bylaws and finding a home for the network. The board completed both of these tasks. According to the bylaws, "The Mission of the Urgenci Network is to further, on the international level, local solidarity-based partnerships between farmers and consumers. We define the solidarity-based partnership as an equitable commitment between farmers and consumers where farmers receive fair remuneration and consumers share the risk and rewards of a sustainable agriculture." The primary activities will be facilitating exchanges of information and visits among partnership participants in different countries. You can read the rest of the bylaws on the urgenci.net Web site. The town of Aubagne and the province of Provence have pooled resources to provide the network with an office and funds to pay a staff person.

I had the good fortune to receive an invitation to be one of the speakers at the 2005 CSA conference in Palmela, near Lisbon, Portugal. Before the meetings, we spent three days observing the first steps of *Reciproco*, the Portuguese name for Community Supported Agriculture. Organizations dedicated to rural development are facilitating Reciproco with funding from Leader, a European Union–wide initiative that has set in motion 1,000 projects in local areas in the twenty-five member countries. The Leader method has eight characteristics: bottom-up organizing, a territorial approach, rural-urban partnerships, work by networks, decentralized financial management, interterritorial cooperation, multisector integration, and innovation. On the preconference tour, we visited ADREPES and TAIPA, two of the fifty-two rural action groups in Portugal. As pilot projects, ADREPES and TAIPA are helping farmers connect with consumers through CSA, an approach that fits the Leader model.

The economic pressures of global competition and the European Union's Common Agriculture Policy (CAP) in Portugal resemble the forces that have brought so many French farmers to adopt AMAP. Unable to compete with large-scale industrialized farms, farmers in periurban areas are selling their land to developers. In more isolated rural areas, the younger generation is abandoning their elders on their subsistence farms to seek opportunity in the cities. Reciproco provides markets for these farms and new hope that may give farm children reasons to stay in their villages.

Carlos has been growing vegetables for the wholesale market all his life in Porceirão, changing his crop mix to match market opportunities. In recent years he has seen his sales to Lisbon supermarkets shrink as lower-priced vegetables flood in from other parts of Europe. With support from ADREPES, he is cautiously trying Reciproco, supplying baskets of vegetables that consumers pick up at an attractive Lisbon store, the Portugal Rural Shop, which sells regional farm products. In the first seven weeks, the number of baskets has grown to thirty-two a week. His goal is fifty, although on his seven hectares of sandy loam soils, where he works with his wife and two hired helpers, he believes he could supply as many as one hundred. Each week, the basket includes fourteen products, some from exchanges with neighboring farms. Consumers pay by the week. Carlos is hesitant to ask for payment in advance because he fears that he might not be able to fill the baskets with everything he has promised. On our visit to the store, we saw two basket sizes filled to the brim with lettuce, turnips, cauliflower, lemons, cilantro, cabbage, potatoes, and tomatoes. The previous week, Carlos told us, he forgot the turnips and his customers complained.

Farther south in the Odemira region, TAIPA has organized a group of farmers in the tiny village of Corte Brique to provide baskets of vegetables and eggs to local consumers. The farmers make the baskets by hand and grow the vegetables using traditional methods. They bring their products to a central distribution point, combine them in baskets of three sizes of different weights, and then take turns delivering to pickup points in three neighboring towns. They learn from each consumer what vegetables they don't like, so the contents of all the baskets are not identical. Like Carlos, they do not expect payment in advance, but their customers sign a six-month contract in which they agree to pay every week. So far, customers who have not been able to pick up their baskets have followed through on their promise to pay. The baskets are less expensive than similar, high-quality fresh vegetables would be in a store. The three young women who staff the TAIPA project told us that it is hard for the farmers to change and that an important aspect of their project is to raise the farmers' self-esteem. During the years of the Salazar regime, many rural people remained illiterate. The TAIPA project provides the farmers with literacy and business training and the technical methods they would need to convert to organic production. TAIPA also promotes the Reciproco concept among consumers and organizes farm visits and olive-picking days. Over the next year, TAIPA intends to build a processing center for making jams, jellies, candies, and other products to be sold under an Odemira region label.

We visited three very small farms where most of the work is done by hand. The fields we saw were more like big gardens planted with a diversity of crops: broccoli, cauliflower, kale, lettuces, radicchio, and carrots. At this late point in the season, weeds were abundant, but that did not seem to worry the

The table is loaded with Odemira farm products for a feast at the 2005 Urgenci International CSA Conference. PHOTO BY ISABELLE JONCAS.

farmers. Each farm had fruit trees, oranges, lemons, and apples, olive and cork trees, and an assortment of poultry and hogs. One of them also had a family of wild boars in a cage. Farmer Andre Anastas told us that he puts manure in the trenches he digs with the help of a donkey. When pest pressure gets too heavy, he uses synthetic pesticides. Before Reciproco, he fed surplus crops to his hogs because he had no markets. The largest farm we saw had one hectare of land shared by three generations of one family. The men work off the farm, planting and cutting eucalyptus trees and harvesting cork. They till their large garden with a rototiller and a donkey. Most of their production feeds the family.

Anyone fortunate enough to live in the market shed of these farms could eat very well. The Odemira organizers hosted two sumptuous meals of local products for the two busloads of conference participants on the tour. They regaled us with fine wines, cheeses, smoked meats, an assortment of sausages, honey, preserves, and breads as well as salads, vegetable dishes, and fresh citrus fruit. The farmers, who had apologized for their old-fashioned methods when they showed us their farms, shared their products with us with quiet pride.

At the 2005 Palmela Conference, I heard the stories

of CSA efforts in several other countries. Robin Segrave and Kess Krabbe, a couple from Australia, came to the conference to learn more about CSA. Kess is a farmer from Holland; his Down Under wife is trying to persuade him to try CSA. They reported that shortage of water was a major hardship for the few CSAs they knew of in Tasmania. My Web searches for CSAs in Australia led me to Mimsbrook Farm CSA, launched as a not-for-profit in 2005, and Food Connect in Brisbane, which offers products from multiple farms, much like the consumer-run Farm Direct Co-op in Marblehead, Massachusetts. Patrick deBuck from Belgium told us about the Food Teams (Voedselteams) approach to CSA. Patrick works as an organizer, going from town to town forming groups of consumers. When he signs up twenty households, he connects them with the nearest farm. In 2005 there were sixteen hundred households in ninety Food Teams in Flanders.

Tidbits of information about CSA-like efforts elsewhere in the world have trickled to my attention. In Hungary, a group of people connected with the Institute of Environmental and Landscape Management at Godollo Agricultural University are running the Open Garden Foundation (Nyitott Kert Alapitvany). In Holland, Strohalm, the Social Trade Organization, has helped organized what it calls "Pergola." In Denmark there are *landbrugslauget* (agricultural guilds). Friends tell me they have heard of CSAs in both Israel and Palestine, but I have not been able to get details. Elisabeth Atangana, from Cameroon, who chairs an organization of farmers' groups in eleven states in Central Africa, attended

the 2005 CSA conference and was elected a member of the Urgenci steering committee. Although her home farm is twenty-four kilometers by poor roads from Mfou, the nearest city, she took the trouble to attend because there are some kernels from CSA that could be practical for her people. La Jimena, a farm in Granada, Spain, offers an annual box of local products and the farm's olive oil to supporters in Europe. A fisherman named Yanis El Yousi Mirmoum sells shares in his catch to forty households along the French coast of the Mediterranean.

———————●———————

Although consumers in the developed countries of the North are grasping the importance of eating locally grown food, and alternative economic projects based on solidarity, fair trade, and social and economic justice are springing up in many places around the world, the tide of multinational corporate globalization has yet to turn. The emergence of Teikei/CSA/ASC/AMAP/Reciproco/Voedselteams shows how consumers and farmers in many different localities are responding to the same global pressures. That one form of organization has so many names is an encouraging sign. Once they seize upon the basic principles, farmers and citizen-consumers in each culture are adapting CSA to their local conditions. Each local food project takes its shape from the tastes, talents, needs, and resources of its creators. The more we can learn from and support one another, the faster we will move toward sustainable and peaceful communities.

CHAPTER 22 NOTES
1. Yoshinori Kaneko, *A Farm with a Future: Living with the Blessings of Sun and Soil*, Eng. trans. (Saitama, Japan: Y. Kaneko & T. Kaneko, 1994).
2. George Pilley, "A Share in the Harvest: A Feasibility Project for CSA" (Soil Association, 2001), posted on www.cuco.org.
3. Daniel Vuillon, "Les AMAP en résistance contre la mondialisation," *La Marseillaise* (2005).

Thirty-five Years of Japanese Teikei

by Cayce Hill and Hiroko Kubota

Teikei is an alternative, direct local food distribution system that began over thirty years ago as an organic agriculture movement in response to rapid industrialization and environmental contamination in Japan. Many consumers, especially mothers bringing up small children in urban areas, were increasingly anxious about the safety of their food and organized themselves into buying groups to obtain uncontaminated eggs, milk, rice, vegetables and traditionally processed foods. Meanwhile farmers were becoming more aware of the harmful effect of agricultural chemicals on the health of humans and livestock and on soil fertility, and they began to adopt organic farming techniques. The establishment of the Japanese Organic Agriculture Association (JOAA) in 1971 united consumers and farmers through the system now known as teikei (literally, "relationship" or "partnership") and encouraged them to help one another. As a result, during the 1970s and 1980s new teikei groups shot up like bamboo shoots after a heavy rain, reaching as many as eight hundred to one thousand groups by 1990. The 1978 JOAA General Meeting established the Ten Teikei Principles based on the collective experience of fifteen leaders of pioneering direct producer-consumer organic produce groups across Japan.

Among those new farmers was Michio Uozumi, born in Yamaguchi prefecture and brought up in the town of Kawasaki City in Kanagawa prefecture. By the age of eighteen, Michio was already concerned about the consequences of *kogai*, or industrial pollution, and had made up his mind to engage in work closely aligned with nature. He entered Tokyo University of Agriculture, where he gained knowledge and inspiration from Albert Howard's book *An Agricultural Testament* and Jerome Rodale's *Pay Dirt*, considered by many Japanese organic activists to be an essential reference for organic farming. He was also fortunate enough to meet Teruo Ichiraku, the founder of JOAA, at a time when Ichiraku was just beginning to establish the organic agriculture movement in Japan, which is rooted deeply in Asian philosophy and nature.

On a visit to the seaside village of Minamata City in 1970, Michio became a member of the new Minamata Pilgrims Party, organized by victims suffering from Minamata disease, one of the most disastrous kogai diseases during the

1960s. Minamata disease is a neurological syndrome caused by mercury poisoning contracted by Japanese who ate fish contaminated by industrial wastewater from the Chisso Corporation's chemical factory from 1932–1968. Michio's encounter with Tamotsu Watanabe, a fisherman and one of the Minamata victims, greatly shaped his future.

After graduating from the university, both Michio and his wife, Michiko, also a graduate of the Tokyo University of Agriculture, worked briefly in Ibaraki prefecture for a farm that produced food for a consumer group known as Tamago no kai. The consumer group was highly organized and committed to growing, delivering, and, of course, eating food as part of a self-sufficient system. This group's experiences and profound dedication to organic farming were inspirations to the young couple in the establishment of their own farm.

In 2007 the Uozumis continue to work tirelessly, motivated by what they learned during those first years and their constant concern over the state of agriculture in Japan and the ever-increasing distance between consumers and farmers. Their family farm, in Ibaraki prefecture, an hour by train from Tokyo, consists of almost 7.5 acres (rice—0.4 acres; vegetables—5.7 acres; wheat, barley, soybeans—1.2 acres). They feed nearly one hundred teikei member families year-round.

The Uozumis grow over seventy different crops a year, twenty at any given season; this diversity enables them to harvest year-round without relying on greenhouses. During the winter months, they use row covers to protect crops from the cold. In Ibaraki Prefecture, it snows only a handful of times per year, and temperatures are typically a few degrees colder than in Tokyo. Michio, Michiko, and their son, Masataka, harvest and send their vegetables to the farm's members in a variety of shapes and sizes and usually still covered in soil, a sharp contrast to the conventional Japanese market system, which places more value on the superficial and standardized appearance of products than on food safety and flavor.

Although the white, red, and black heirloom rice the Uozumis grow is primarily for family consumption, occasionally they surprise and delight their teikei members with a small bag of rice among the vegetables in the weekly delivery. In the traditional Japanese manner, the Uozumis dry harvested rice hulls and stalks for use in compost or as a

Michio and Michiko Uozumi at their farm in Yasato-Cho, Ibaragi Prefecture, Japan. PHOTO COURTESY OF UOZUMI FARM.

A share from the Uozumi Farm. PHOTO COURTESY OF MICHIO AND MICHIKO UOZUMI.

groundcover between rows of vegetables. They also raise around six hundred chickens for eggs and slaughter. Along with large quantities of green vegetables and weeds from the farm, the chickens eat an organic mixture of crushed oyster shell, rice husks, rice bran, soybean meal, wheat, barley, and fish powder (salmon from Yamagata prefecture), to which water is added before it is left to ferment overnight.

Like most teikei farmers, the Uozumis employ three distinct delivery methods. First, on Mondays and Tuesdays they pack boxes of vegetables in bulk for members of Yuki no kai, a consumer group. Members take turns picking up the boxes and taking them back to a central warehouse near the consumer group office in Matsudo City, Chiba prefecture, about two hours by car from the Uozumis' farm. There, they divide the vegetables into thirty-five individual shares and distribute them to thirteen consumer homes, which double as drop-off points for the group and pickup points for the end members. Two to five families come to each pickup point to collect their vegetable boxes. Members can arrange to receive nearly all of their food, including tofu, fish, pork, seaweed, tea, soy sauce, oil, and fruit, through their consumer group.

On Wednesdays the Uozumis prepare around thirty boxes for transport to consumers in Tokyo via a domestic package delivery company. For the past two years, they have even delivered a large box of vegetables directly to the kitchen of a day care center. Finally, on Fridays, the Uozumis pack boxes of mixed vegetables, which they deliver to thirty families in Tsukuba City, Ibaraki prefecture, approximately one hour from the farm. They typically begin making deliveries around 5:30 in the evening and continue until around 10:00.

In keeping with the underlying principle of the teikei system, to promote mutual assistance and understanding among farmers and consumers, teikei farmers and consumer groups work to create many opportunities for face-to-face contact. Although consumer participation in actual fieldwork has declined somewhat over the years, the Uozumis organize events on their farm throughout the year, often in celebration of holidays and the changing seasons. In late autumn many of the Uozumis' consumers look forward to ochiba, a gathering devoted to collecting fallen leaves for a special compost used exclusively in seedbeds. In November the Uozumis organize shukaku sai, a social gathering in honor of the autumn harvest that is reminiscent of the Thanksgiving celebration in the United States. Children are invited to imohori, or sweet potato digging, a fun opportunity to get up to their elbows in dirt harvesting the traditional purple-skinned potatoes.

In the recent past, teikei has been much larger, more diversified, and better organized than similar movements around the world. However, as fewer and fewer young adults choose to follow careers in farming and the amount

TEN PRINCIPLES OF TEIKEI

1. *Principle of mutual assistance.* The essence of this partnership lies, not in trading itself, but in the friendly relationship between people. Therefore, both producers and consumers should help each other on the basis of mutual understanding. This relation should be established through the reflection of past experiences.

2. *Principle of intended production.* Producers should, through consultation with consumers, intend to produce the maximum amount and maximum variety of produce within the capacity of the farms.

3. *Principle of accepting the produce.* Consumers should accept all the produce that has been grown according to previous consultation between both groups, and their diet should depend as much as possible on this produce.

4. *Principle of mutual concession in the price decision.* In deciding the price of the produce, producers should take full account of savings in labor and cost, due to grading and packaging processes being curtailed, as well as of all their produce being accepted; and consumers should take into full account the benefit of getting fresh, safe, and tasty foods.

5. *Principle of deepening friendly relationships.* The continuous development of this partnership requires the deepening of friendly relationships between producers and consumers. This will be achieved only through maximizing contact between the partners.

6. *Principle of self-distribution.* On this principle, the transportation of produce should be carried out by either the producer's or consumer's groups, up to the latter's depots, without dependence on professional transporters.

7. *Principle of democratic management.* Both groups should avoid over-reliance upon limited number of leaders in their activities, and try to practice democratic management with responsibility shared by all. The particular conditions of the members' families should be taken into consideration on the principle of mutual assistance.

8. *Principle of learning among each group.* Both groups of producers and consumers should attach much importance to studying among themselves, and should try to keep their activities from ending only in the distribution of safe foods.

9. *Principle of maintaining the appropriate group scale.* The full practice of the matters written in the above articles will be difficult if the membership or the territory of these groups becomes too large. That is the reason why both of them should be kept to an appropriate size. The development of this movement in terms of membership should be promoted through increasing the number of groups and the collaboration among them.

10. *Principle of steady development.* In most cases, neither producers nor consumers will be able to enjoy such good conditions as mentioned above from the very beginning. Therefore, it is necessary for both of them to choose promising partners, even if their present situation is unsatisfactory, and to go ahead with the effort to advance in mutual cooperation.

SOURCE: JAPAN ORGANIC AGRICULTURE ASSOCIATION.

of available farmland continues to decrease, the future does not bode well for teikei. Along with other organic farmers, Michio Uozumi generously volunteers time outside the farm in several projects aimed at reversing this trend. He teaches organic agriculture techniques to laborers and children at JA Ibaraki Zenno "Pocket Farm Doki-doki" (a small farm with several animals, a vegetable field, and a restaurant) in Ibaraki prefecture just thirty minutes away from his own farm, and he is also involved in a project at Tokyo's Adachi ku Toshi Nogyo Kouen (Adachi ku City Farm Park). For the past three years, a pioneering project has been growing organic vegetables on twenty-two acres of this busy city park to introduce urban families to the philosophy and fundamentals of organic agriculture. Involving the children in farming provides a special opportunity to raise awareness about how the food they eat is produced, while giving them the fun and memorable experience of growing and harvesting organic vegetables.

Like Michio Uozumi, other farmers across Japan are working to involve consumers and educate them about farming. Yukitaka Yahiro, a farmer from Fukuoka prefecture, says, "Many young consumers say they want to eat safer food, but don't actually cook at home anymore. Therefore, as farmers we should offer not only fresh produce but knowledge about processing and cooking techniques as well. As the consumers learn about the process of making food such as soybean miso, tofu, pickles, etc., then the consumers' quality of life will gradually improve." In 1996, Yukitaka and Michiko Yahiro opened the Musubi-an house on their farm, where local consumers can purchase organic produce from their farm shop and participate in monthly cooking and traditional food-processing classes.

Farther north, in Yamagata prefecture, is Takahata machi, the area known as the heart of the Japanese organic movement. In 1973 forty-one young farmers organized the Takahata machi Organic Agriculture Association and launched the new teikei movement with consumer groups in the Tokyo area. Kanji Hoshi is a farmer and poet who volunteers his time leading various programs and activities for the Education Commissioner of Takahata machi. One such educational programs is called Takahata kyousei juku, or "rural school of living with nature," where the students experience farming and traditional rural life, which takes place in Takahata machi over the course of several days

each summer and is attended by students from universities in Tokyo. Through such programs and teikei activities, about 80 people have become farmers or residents of Takahata machi, infusing it with additional optimism and vitality.

Stories such as those about the farmers mentioned above are shared among participants at each annual JOAA National Conference. Each year the conference is held in a different region, allowing the four hundred to seven hundred participants, including both farmers and consumers, to have the opportunity to become closely acquainted with the current techniques and programs unique to that agricultural area. The 31st National Conference was held in 2004 in Imabari City, Ehime prefecture, where attendees learned about the local, organic school lunch program launched by the city of Ehime and Ehime Organic Produce Co-operatives. In 2006 the 33rd National Conference took place in Miyoshi mura, Chiba Prefecture, the site of one of the first teikei movements in 1973, where the Anzen na tabemono o tsukutte taberu kai (Organization for Growing Safe Food) continues year after year to build mutual respect among over thirty-six family farmers and eight hundred consumers. JOAA also organized a seed-saving network; publishes a monthly journal, *Soil and Health*, and leads the effort to prevent genetically modified crops (GMOs) from invading Japanese agriculture.

Nowadays there are many other groups in Japan that promote direct relationships between producers and consumers (*san-choku*). Since the 1970s, consumer cooperatives have

Harvesting rice in the city farm park, Adachiku Toshinogyo Kouen, in Tokyo, Japan. PHOTO COURTESY OF HIROKO KUBOTA.

played an important role: The Japanese Consumers' Cooperative Union unites 166 retail co-ops, including the Seikatsu-Club Co-op. Agricultural cooperatives, which include almost all of the farmers in Japan, promote san-choku with local consumer co-ops. Under the rallying cry "Chisan-Chisyo" (Grow locally, and eat locally), farmers' markets are growing rapidly. The official number is esti-mated at 2,500, and the total number, including small and private markets, is estimated at 10,000. Besides the co-ops, there are corporate organic distribution systems derived from the teikei movement, such as Daichi-wo-mamoru-kai, Polan Hiroba, and so forth, which boycott GMO foods and take part in the local food movement.

The teikei organic movement was founded to inspire in Japanese society the respect for all natural life, trust in mutual human relationships, and profound consideration for food and agriculture as the building blocks for a healthier society. Michio Uozumi considers the feeling of bare feet on freshly tilled soil to be "an astonishing and healing experi-ence," one that most people living in big cities today have long forgotten. Above all, he believes that the positive effect of such an experience on young children is fundamental to reconnecting the Japanese community to local agriculture. He and other teikei farmers believe that every possible effort should continue to be made to increase contact between farmers and consumers, and consumers and the soil, in order to spread a better understanding and appreciation of the crucial role that local, organic agriculture plays in all of our lives.

Cayce Hill holds a Master of Public Administration degree from New York University. She is originally from Austin, Texas, and currently is a resident of Tokyo and has been a teikei member since 2005.

Hiroko Kubota is a member of the Board of Directors of Japan Organic Agriculture Association and is a professor at Kokugakuin University in Tokyo.

CSAs THAT QUIT

The goal ever recedes from us. Salvation lies in the effort,
not in the attainment. Full effort is full victory.
—MAHATMA GANDHI

How many farmers and groups of members have started Community Supported Agriculture (CSA) farms and then, sooner or later, given up? No one has kept an exact count. Of the 354 CSA farms that completed the 2001 CSA Survey, 5.4 percent did not intend to continue with CSA. We have certainly not been able to track down all the CSAs that started and then quit within a five-year time frame; however, this chapter is based on interviews with farmers and sharers from short-lived CSAs, and it may shed light on some of the myriad reasons why individual projects break down and disappear. The lessons we can draw from such information can prove useful, not only for prospective CSAs to consider before they form, but also for active, vital CSAs, so that a few potential pitfalls and negative situations can be recognized and, possibly, avoided.

One common thread in many of these stories is the failure to develop a strong core group and committed member involvement. Several of the farmers who initiated these CSAs never tried to form a core group. Others did make an effort but either were unable to find sharers who would take responsibility or failed to recruit replacements when their initial supporters moved on. Finding themselves doing all of the growing, distribution, and organizing, they became discouraged. Marcy McCall, who ran Heron View in New York for five years before

giving up, explained, "It became too much of a stress for us, since we were trying to do everything ourselves. With three distribution sites for twenty-six shares, there was a huge amount of driving." Former sharers in two of these CSAs told me they were overwhelmed by the quantity of food and felt no control over the selection. More member involvement and better communications could have solved these problems.

Robert Perry in Homer, New York, blamed himself for failing to "aggressively recruit" new core people after three key members moved away. He said it was a problem "being dependent on labor for weeding, till it was overwhelming." Members assumed the CSA would function whether they showed up for their work shifts or not. Robert still believes that his five-year CSA stint was the most successful and satisfying marketing he has done. Jess and Suzanne Unger of Brook Farm in Maryland gave up after two years, blaming too much turnover, problems with recruiting, and the excessive amount of labor required to grow fifty crops. "There was a tremendous amount of running around," said Unger, who made weekly deliveries to Washington, D.C., and Leesburg and Vienna, Virginia. Too late, they realized that a core group might have helped them.[1]

A few farmers have given up on CSA because they did not like the constant pressure of supplying shares week after week. Greg Reynolds of Riverbend

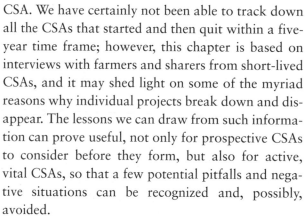

Chart from 2001 CSA survey

If you have discontinued, or are planning to discontinue, your CSA operation, why?

Total number of respondents: 32
Health reasons: 1–3.1%
Insufficient income: 11–34.4%
Lack of members/demand: 7–21.9%
Relocated: 3–9.4%
Burned out: 4–12.5%
Retired: 2–6.3%
Other: 16–50%

Farm in Minnesota told me that in a bad year, he hated worrying about filling members' expectations. By selling crops to wholesale markets and food co-ops, he can also organize his time so that he can take a summer vacation with his family.

The most frequent reason given for the dissolution of CSAs is not particular to the CSA structure but rather to the scale of the farm enterprise—the farmer finds a better-paying job. CSAs of only fifteen or twenty or even forty shares cannot offer the farmer enough money to compete with the benefits of a full-time job off the farm. At age fifty, Paul Bernacky of Way Back Farm in a rural part of Maine can earn $40 an hour off the farm but never more than $10 an hour selling summer vegetables.[2] He is proud of what they did, running a highly diversified CSA for twenty families for fifteen years. But the high turnover, the need to find as many as ten new families every year, and the low pay "got exhausting."

Jim Volkhausen of Buttermilk Farm near Ithaca, New York, also stopped doing a CSA to take another job. He had other interests, and with only forty shares could never have lived on just the farm income. Other CSAs in the area picked up his members. Despite its failure to produce enough money, Jim considered the CSA successful: "The concept of member involvement was very important, and I would encourage others to do it that way for the sense of community."

Any number of CSAs have gone under because they underpriced their shares. Pricing that is fair to the farmer is an essential ingredient of CSA sustainability. For the first year or so, farmers can postpone their own financial needs to get a project off the ground. Trying to farm in a state of continual financial need, however, is not sustainable. In his economic study of CSA in the Northeast, Dan Lass made the astonishing (to him, as an economist) discovery that some CSA farmers were not paying themselves for their work. Their budgets did not include a salary for the farmer. Anne Morgan of Lakes and Valley CSA in Minnesota writes,

> I think it needs to be stressed that lowering the share price of a CSA is ill-advised. CSAs are not sustainable if they are really GSAs—grower-subsidized agriculture. A CSA cannot grow to the point where the farmer can make a living and hire extra hands at a fair wage on $350 or $400 per share. Cheap share prices mean that the gardener is allowing him/her self to be exploited, often out of desperation to compete. And it is very difficult to raise the price of shares once members have a few seasons at a lower price. We tried that and decided to discontinue our first CSA years ago. With our second CSA, we earned no wages the first two years; the money went to the expenses of interns and equipment. Now, with 70-shares at $650 per we should have $10,000 left over for us after everything/everyone is paid. This does not fully compensate us for the time and materials we contribute, but it is getting better. As our membership levels grow, we can afford to stick with it. That's sustainable agriculture.

Marty Rice, of Country Pleasures Farm in Maryland, told a reporter for the *Washington Post* that she gave up on CSA because she had to work eighty to ninety hours a week to gross $6,300 for the year. "We did an end-of-season survey, and asked people if they would be willing to have smaller shares, pay more, or contribute some labor, and the answer was no, no, no. People didn't have an understanding of the concept, that we're in this together," Marty complained, but concluded, "It's a very important idea, but we have to figure out how to make it succeed."[3]

People who choose to live on farms but who want to do farmwork only part time invariably find that CSA is too time consuming. In different parts of New York State, Deb Teeter and Pat LaPoint both tried to keep up their farms and families, run CSAs, and work as Extension educators. Deb described her family's dilemma:

> What brought everything to a head for us was working on our business plan last fall. After three droughts in the last five years, we were facing up to the need for a reliable irrigation system. However, last year's early, continuous rain presented an additional problem: We could not get into the fields to harvest hay on time, leading to tremendous stands which took five days to dry rather than the normal three (and clogged the mower, and the balers had trouble, etc). When it looked like we had a weather window to do hay, invariably there was a conflict with the vegetable fields' needs. It was a very difficult year, producing a lot of mediocre hay and neglected vegetables. This conflict also really brought home how little was getting done on the rest of the farm. . . . Hence the buckling down with the business plan process, leading to the "what has to get done" list—and the CSA didn't make the cut.

After six years of successful operation, Pat and Mike Kane of Shamrock Hill Farm in Port Crane, New York, decided to take a year's sabbatical from the CSA. Pat's nonfarm job had become too consuming to allow her much time to help Mike on the farm, and Mike couldn't do the high-quality job he likes by himself. Their sharers were very supportive and agreed to keep the core together through the year off by focusing on bulk buying. Farmers and sharers hoped to redesign the project for the future to make it easier for the farm; however, after another year, when Pat's job continued to expand, they regretfully gave up on the CSA.

Divorce or other emotional upheaval in the farm family destroyed several former CSAs. In New York State alone, I can count four CSAs that ended when the farm family split up in a divorce. A crippling car accident forced another farmer to cut back to just a few crops that she could manage with occasional volunteer help. Silvercreek Farm CSA in Ohio was tremendously successful at building member involvement; yet, the vicissitudes of life led them to give up their CSA. In the first edition of this book, I held Molly and Ted Bartlett up as a model. In 1992, Silver Creek started a CSA with a goal of twenty-five shares. They targeted a mailing to friends and customers within one zip code and quickly recruited 47 members. Within a few years membership swelled to 165, but they cut back to 100 shares, which allowed them to get to know all the members well. Silver Creek also ran a market at the farm twice a week, attended a farmers' market, and wholesaled to store and restaurant accounts. Their farm served as a hub for other local farms as well with lamb, beef, chicken, egg, cut flower, goat's milk cheese, fiber, and "preserver" shares to fill more of the food needs of their CSA members. The Bartletts trained their members so well that when Ted had a heart attack, the crew of members was able to keep the farm going. Ted recovered, but worse trials lay ahead. This is how Molly tells their story:

> The decision to end our CSA was made after our family had some tremendous physical and personal problems that just required us to step back from our very fast-paced farming operation and take some time to heal. Ted had an aortic aneurysm burst, which was very scary and put him out of commission. That same year, our son-in-law was diagnosed with brain cancer, and he and our daughter and their infant child moved in with us. Steve died eight months later, and our daughter and whole family was devastated. I am sure you get the picture. Our plate was too full. We sold all four-legged animals (sheep, angora goats, draft horses) and told a very interested core group of CSAers that they were welcome to our farm and implements and could organize a CSA on their own and even get a farmer if they needed to. Well, they met probably about a dozen times and kept running up against self-imposed roadblocks with regard to time and money things. . . . Early that spring, they announced to us that it just was not

going to work, and we said okay. . . . Many of them were well trained here and are now growing veggies in their suburban backyards or [have] joined community gardens. Others are still buying locally by purchasing at some very popular and hip farmers' markets in the Cleveland area. I think we did a very good job at teaching the mantra of eating locally and seasonally. . . . I am so glad we did have a CSA, but I would be the first to say it was a lot of work!

Jim Rose closed down Earthcraft CSA in 2002 after his partner, Signe, left. He said he could not have done it without her and her excellent communication skills. After 9/11, Jim found that membership in the CSA was declining. Instead, he grew what he could by himself and sold at a farmers' market, making as much as he had with the CSA. "I still had the same customers," he reported, "though they no longer came to work at the farm, and I missed that relationship." At age sixty-nine, Jim has decided to retire after the 2006 season to spend more time with his children and ply his skills as a cabinetmaker. A neighboring vegetable grower will take over his equipment and three acres of asparagus.

> The most frequent reason given for the dissolution of CSAs is not particular to the CSA structure but rather to the scale of the farm enterprise—the farmer finds a better-paying job.

Farms in remote rural areas have had difficulty recruiting local sharers. Jim Gerritsen of Wood Prairie Farm in Aroostook County, Maine, was able to attract newcomers to the sparsely populated northern county as sharers, but not long-time residents. A few years back in New Mexico, Valerie Kaepler had a hard time selling the concept of CSA to her rural neighbors. Farming in areas with low population densities, Paul Bernacky in Maine and Martha Johnson in New York also found it hard to sign up members. According to Paul, half the people grow their own food; the other half are struggling to make life in the country work and have no time to drive for vegetables. He found himself arguing with people, trying to convince them that CSA was worth their effort. It was too discouraging. Farmers in other parts of the country similarly report that few of their rural neighbors care to purchase organically grown foods. In these areas, chemically sensitive people or hard-core organic advocates who lack the time to garden are the most likely candidates. Recent urbanites who are moving to rural areas and telecommuting also may want organic food and a local farm connection. Overall, it does seem easier to find potential CSA members in cities or suburbs. Adapting to the needs of local people is critical, however: There are examples of thriving projects in rural Vermont and Iowa, where farmers have been able to find the right mixture of CSA and other markets or have targeted groups such as the elderly and provided home delivery.

The pressures of development have taken their toll on CSAs as they have with other farms. In a few bitter cases, where the land belonged to several siblings or to a member of a family who was not doing the farming, CSAs have had the land sold out from under them. The redevelopment agency in Philadelphia paved over the Sea Change urban garden with a parking lot, throwing into crisis a hopeful project, which combined production for a CSA with a charter school, training for teenagers, and bicycle recycling.

High land values overwhelmed one of the outstanding early CSAs, Moore Farm in California. A hard freeze and low wholesale prices for lemons and avocados had driven Steve Moore to turn to CSA. After the Civil War, Moore's great-grandfather had homesteaded on the sixty-acre ranch in Carpinteria, California. Moore's father, a retired physician, inherited the farm in the 1960s. He hired a manager to oversee the production of lemons and avocados for the wholesale market. In 1981 he signed the farm over to his five children. That year, Steve writes, "was the worst year in history for the lemon and avocado markets. We just lost our shirts." Steve, with a PhD in civil engineering and a master's degree in psychological counseling, found himself running the farm "working one hundred hours a week and making $500 a month." Gradually, he converted the farm to biodynamic management. Then, in 1990, a

big freeze destroyed the tree crops, a loss of
$250,000. To save the farm, Steve turned to vegeta-
bles. "I knew," he later said, that "there was a community of people out there who wanted vegetables
from me." In 1997 the Moore Ranch CSA included
220 families in the San Fernando and San Gabriel
valleys, Thousand Oaks, Santa Monica, and
Malibu. Steve Moore told me that the farm was sold
in 2001. When I asked him what happened, he
replied: "I would rather not plow this field again.
Suffice it to say, it was an extremely difficult and
trying family decision that went against my own
preferences. I am still connected with ag. and BD
through a bit of consulting and working with the
Sustainable Ag. Resource Consortium at Cal Poly."

Two of the CSAs I investigated fell apart because the
grower lacked farming experience, resulting in inadequate production, disorganization, and poor quality.
One grower tried to get started in a year of a serious
drought, undercharged for his shares, and then could
not produce enough food even for that limited payment. He gave up in August. Neither he nor the two
growers for the other CSA had any previous experience in growing for market. Before allowing a new
CSA to join their network, the Madison Area CSA
Coalition (MACSAC) growers in Wisconsin ask new
farmers to fill out a questionnaire on their practices
and previous experience. If the farmers lack experience
and knowledge, MACSAC encourages them to intern
or apprentice with another farm. If their production is
too specialized, MACSAC advises developing cooperative relationships with existing CSAs.[4] Similarly, the
Senior Farm Share program in Maine requires a
couple of years of marketing experience before allocating senior shares to a farm.

The failure to create a real sense of community
participation led Harvey Harman at Sustenance
Farm near Raleigh-Durham, North Carolina, to dis-
band his CSA. For five years, the project worked
well enough; members were happy to get the vegetables and to support the farm financially, though the
forty shares generated too little income to provide a
living for Harman. With thirty to forty miles
between their homes and the farm, the members
never became active. Turnover from year to year

was high. The core group was forced to spend too
much of its energy recruiting new members. Even
social events, such as potlucks, which the farm sponsored in town, were not well attended. As Harvey
put it, the CSA was a good thing, but "the community aspect never happened."

Another important category of failed CSAs are the
nonprofit-initiated projects with overly ambitious
mixed social and farming agendas. In 1992 the
Pittsburgh Food Bank decided to replicate the
Western Massachusetts Food Bank Farm CSA, with
half the production going to CSA shares to pay for
the farm and the other half going to the Food Bank.
They imagined that the farm would increase Food
Bank control over the fresh produce supply while
providing good publicity, lots of work for volunteers, and a forum for relating sustainable agriculture and hunger issues. The Food Bank began with a
lease arrangement with an independent farmer, who
then turned over his land with its CSA to the Food
Bank. With only four years' experience in farming,
Mike Gabel took over as farm manager of the Green
Harvest CSA in 1997, the year the farm had to move
to a new piece of land. Largely owing to the new
location, CSA membership dropped from 110 to 25.
After a second season at the new site, shares covered
50 percent of the farm's funding needs. The many
other social goals of the Green Harvest project complicated the already challenging goal of increasing
food production on the new farm.

At a Pennsylvania Society for Sustainable
Agriculture workshop, Mike delivered this impassioned statement on the need for clarity in the relationship between the food bank and the farm, which
would apply with equal force to any other institutional farm arrangement:

> You need to tell the food bank realistically what kind
> of support you are going to need. . . . If you are to be
> self-sufficient, you are going to have to reach a cer-
> tain threshold of CSA members to cover the cost of
> that farm, plus a surplus, a big surplus. It's critical
> that you and the administration of this organization
> are on the same page. You also need to decide, along
> with the food bank, whether your operation is going

to have an educational component or whether it's going to be purely production in pounds. If you have all these people coming onto your farm and you are doing educational tours, who is planting stuff when it needs to be planted? Who is dealing with those groups? A lot of funders want you to do Herculean things. They want you to produce big pounds, they want you to have the most incredible, all-inclusive educational component that ever existed, reaching every inner-city kid and, by the way, why don't you put numbers on it? You've got to get your expectations straight with the administration and the farm—what is this beast going to do? Don't even put the plow in the ground until these things have been ironed out!

Unfortunately, Mike was not able to reach the level of clarity he needed with the Food Bank, and they decided to close the CSA. Several of the other CSAs funded by Community Food Projects met similar fates. In 1998, Ron Friedman attempted to initiate a farm for Isles, Inc. in Trenton, New Jersey; current staff does not remember what became of the effort. That same year, Carrie Little started the Tahoma Food System project in Tacoma, Washington, employing and training homeless people to combine food from nine urban gardens into CSA shares. Carrie worked closely with Guadelupe House, a Catholic Workers intentional community project that kept the gardens going when grant funding to pay Carrie dried up. The sixty CSA shares at $450 were not enough to sustain the project financially. Without Carrie's outgoing personality to attract members and inspire volunteers, share numbers fell and the project could afford to pay its workers only $250 a month. According to Laura Karlin, a staff person at Guadelupe House, the subsequent gardeners were hard working but more introverted. They could not maintain the administration of the project while producing the volume of food it required. The pressures

took a high personal toll, and in 2005, Guadelupe House gave up its garden project.

Kristen Markley's master's essay "Sustainable Agriculture and Hunger" documents the experience of five Northeast nonprofit organizations, four of them with CSAs.[5] Committed to both alleviating hunger and supporting small farms, Kristen provides a realistic appraisal of the pressures and obstacles facing small farmers and hunger organizations, and the benefits and difficulties of working together. She interviewed farmers employed by the nonprofits and independent farmers who sell to them, as well as office and management staff. She observed a high level of turnover among farmer-employees. She supports Mike Gabel's plea for clarifying priorities and setting realistic expectations. If a food bank expects young, inexperienced farmers to produce large enough amounts of diverse crops to make a farm financially self-sufficient while at the same time training and empowering low-income community members to produce food themselves, the frustration levels on all sides are likely to be high. Greater familiarity with farming among nonprofit staff members and better training and higher salaries for farm staff will help resolve these difficulties.

———————————— • ————————————

Some attrition of CSAs is inevitable. The same pressures that wear down other farms (and many other small, family-run businesses) do their destructive work on CSAs as well. And, of course, CSA does not suit every farmer or every consumer. In too many cases, though, farmers have given up on CSAs because they did not know how to organize the support they needed from the members, or members either did not understand or know how to give the help that would have kept their farmers going. Hopefully this book, with its outpouring of generous sharing of experiences and bright ideas, will save a few shaky CSAs from going over the edge.

CHAPTER 23 NOTES

1. *Washington Post*, September 3, 1995.
2. Paul Bernacky, letter to author, February 16, 2006.
3. *Washington Post*, September 3, 1995.

4. Marcia Ostrom, "Community Farm Coalitions," *Farms of Tomorrow Revisited* (1997): 94.
5. Kristen Markley, "Sustainable Agriculture and Hunger." Master's thesis, Pennsylvania State University, 1997.

AFTERWORD

THE FUTURE: ON ACTIVE HOPE

It's not possible to know what is possible.
—FRANCES MOORE LAPPÉ, October 21, 2006,
at "The World on Your Plate" in Buffalo, New York

When asked what he would do if the world were to end tomorrow,
Martin Luther reportedly answered, I would plant an apple tree today.
— "Caring for Creation: Vision, Hope and Justice,"
A Social Statement from the Evangelical Lutheran Church in America

Looking around us at the shopping strips and end-less parking lots, the undrinkable water, and massive fish kills; listening to the news of brutally destructive wars, of massive environmental disasters, of children murdering children, of self-serving politicians, of batterings, senseless car wrecks, and billion-dollar buyouts with downsizings in the tens of thousands—it is easy to give in to cynicism or despair. Overshadowed by the enormous forces of the global economy, each of us feels as powerless as a tiny ground beetle. Reviewing the history of the past century, however, Howard Zinn observes that, above all else, its course has been marked by unpredictability. Contrary to all odds, ordinary people have won the most unpromising and unequal of battles. And upon this unpredictability, Zinn offers hope:

> The struggle for justice should never be abandoned because of the apparent overwhelming power of those who have the guns and the money and who seem invincible in their determination to hold on to it. The apparent power has, again and again, proved vulnerable to human qualities less measurable than bombs and dollars: moral fervor, determination, unity, organization, sacrifice, wit, ingenuity, courage, patience—whether by blacks in Alabama and South Africa, peasants in El Salvador, Nicaragua and Vietnam, or workers and intellectuals in Poland, Hungary, and the Soviet Union itself. (And I would add, by people all over

Hands of peas. PHOTO BY TOM RUGGIERI.

the world to prevent a nuclear holocaust.) No cold calculation of the balance of power need deter people who are persuaded that their cause is just.[1]

We can go on eating our way to obesity, heart attacks, cancer, infertility, and total dependence on the Altrias (a.k.a. Philip Morrises) who dominate our food supply, or we can start to take charge of what we put in our mouths. Our tiny groups of farmers and aware farm members have the qualities Zinn lists, and we have a cause as just and vital as any for which people have ever struggled. Buying locally grown food is a first step toward the health of our own bodies and the health of our local communities. Joining in Community Supported Agriculture is another step toward nurturing the interdependence among humans, the soil, the plants, and other creatures that we must have if we hope for a future on this very small planet.

As Robyn Van En declared in her notes for this conclusion, "CSA is a viable contender to the reckless and unsustainable food system to which we have grown accustomed. CSA strives to be socially and ecologically responsible, to educate and empower, while providing good food, one of the basic necessities of life. . . . It is a participatory means to securing your food supply for today and for future generations."

CSA farms and gardens around the country are changing and growing. The Genesee Valley Organic CSA has doubled in size and its members have helped my farm, Peacework, to acquire secure tenure on protected land. Like Fairview Gardens, Angelic Organics has inspired the development of the CSA Learning Center, whose educational programs give inner-city youngsters and adults the chance to learn about rural life and the ecological growing of food. Solar panels on the roof at Live Power Community Farm are reducing the electric bills, while schoolchildren help the farmers put up a new post and beam, straw bale building for classes, workshops and dances. Just Food's City Harvest is bringing organic vegetables and nutrition training to the West Bronx, one of the poorest neighborhoods in New York. After years of support from the

On a CSA harvest morning, author unloads box of kale from "Lurch," the farm truck. PHOTO BY MIRIAM GALE.

Hartford Food System, Holcomb Farm has become financially self-sufficient. Grown Locally and Farm To Folk are expanding their roles as distribution centers for local products from area farms, while value added processing by both individual farms and co-ops is spreading in New York, New Hampshire, and Appalachia. A generous bequest has doubled the protected land under the management of Genesis Farm. Common Harvest is launching its second round of sales of seven-year shares to build a brick oven and a classroom for a local folk school. With its CSA members helping the farm survive a devastating flood, Featherstone Farm in Minnesota is adventuring into Local Fair Trade with certification to the standards of the Agricultural Justice Project.

In rural Iowa, coastal California, the province of Quebec, and even in the concrete canyons of New York City, regional support groups are reaching out to conventional farmers with the promise of customers who will share the risk as the farms reduce their reliance on agrochemicals. In the Northeast, New Mexico, and Oregon, farmers are increasing the production and sharing of seed. Lower-income people are no longer an afterthought and are active participants in more and more CSAs in California, New York, and Wisconsin. In New York City; Portland, Oregon; Minneapolis; and Boston, CSA food production on urban garden sites is under way and teenagers are learning to grow that food. And farmers in southern France, central England, and southeast Portugal are adopting CSA to their cultures.

Whether CSAs number ten thousand in the year 2020 is less important than whether the fifteen hundred to two thousand CSAs that already exist mature and flower. Invaluable as computer models and sophisticated indicators of sustainability may be, the real story is happening on the ground in community supported farms and gardens, a living reality for thousands of people who are learning to work together to live more sustainably. In the shadow of the global supermarket, CSAs, like one thousand farmer controlled experiment stations, are busy with research on the social and economic relations of the future.

Reflecting on future plans, Farmer Dan Guenthner says, "Once we got off the path, there was no one to tell us we couldn't do what we are doing. It's amazing the level of creativity. Members ask—what's next? They are ready to follow our leadership." Each of us may be as insignificant as a ground beetle, but together the ground beetles of the globe keep it from being buried in a thick layer of rot.

In another of his wonderful vegetable love notes to Rose Valley Farm, Josh Tenenbaum wrote:

> Thich Nhat Hanh, the Buddhist monk and peace activist, recommends that we have a great store of positive images to counterbalance the negative ones. And it is easy for me to think of the beautiful food life that you nurture, poetry in fiber and carbohydrate. . . . I know that you do the most fundamental kind of peace work, each and every day on your farm. For this I am speechless with gratitude (well, except for this letter, I am speechless) and deeply honored to know you. You live now in my images, but also in me, from your living energy that goes into your food, and from these into me. I try to keep the chain going, and pass it along to others.

I don't think Josh will mind sharing his image and allowing me to extend it to all the CSAs on this continent and around the globe, as I invite them to share from this book to help build a peaceful world of productive community farms and gardens.

AFTERWORD NOTE
1. Howard Zinn, *The Zinn Reader: Writings on Disobedience and Democracy* (New York: Seven Stories Press, 1997), 642.

APPENDIX

The Intervale: Community Owned Farms
By Beth Holtzman

Intervale Community Farm is a small farm in a "neighborhood" of small farms in a city [Burlington, Vermont]. Located in the Intervale, a 1,000-acre piece of open and agricultural land in Burlington's North End, Intervale Community Farm (ICF) is one of several farms on publicly owned land in New England, along with the Natick Community Farm, in Natick, Land's Sake in Weston, and Codman Community Farms in Lincoln, all in Massachusetts.

In 2006, the Intervale Community Farm was the largest, in terms of membership, and one of the oldest CSAs in Vermont.

Since it began in 1989, the farm has continually fine-tuned the details of share size, crop mix, member work requirements, governance, staffing, and salaries to develop an operation that is both financially stable and highly attractive to Burlington-area residents. Throughout, a spirit of inclusion and cooperation has helped this consumer-driven CSA to flourish.

"Our farm is a very family-centered farm and it's a really unique and wonderful thing about this farm," says Becky Maden, who joined the farm as co-manager in 2005. "People come down with their kids and they'll hang out for a while. They'll do the pick-your-own crops. There's a sandbox, and toys." Feedback on member surveys repeatedly shows that the high quality of the produce is just one of the attractions of membership. "People say they join to get their kids out, and to connect their families with food. The fun and community aspect of it is really important."

ICF leases its land from an umbrella nonprofit organization, the Intervale Center, which leases or owns a total of 354 acres. The Intervale Center's mission is to grow viable farms, preserve productive agricultural land, increase access to local, organic food, compost and other soil amendments, and protect water quality through manure management technology and stream bank restoration.

Since 1990, the Intervale Center has leased land, equipment, greenhouses, and irrigation and storage facilities to small, independent organic farms, many of which are in start-up or incubator status. In 2006, thirteen Intervale farms, including ICF, operated on 120 acres with over 60 full-time and seasonal workers. Additionally, the Intervale Center operates a large composting operation, a youth farm program, and a conservation nursery, which offers native, ecologically grown trees and shrubs for conservation projects statewide.

In 2005, the Intervale Center acquired 53 acres through a donation from Paul and Rita Calkin, adding to the seven acres the Calkins had donated previously. In 2006, the Center was working to acquire 199 acres it had been leasing from the Burlington Electric Department. The goal is to transfer ownership of the land to the Intervale Center with the help of a grant from the Vermont Housing and Conservation Board in cooperation with the Vermont Land Trust. The plan includes establishing easements on the property to ensure perpetual agricultural and conservation use of the land, and will allow the independent farmers to establish longer-term leases to provide more security for their farms.

The base for these efforts is the Intervale compost facility. Launched around the same time as the Intervale CSA, the compost facility is now a joint public-private venture that composts manures, bedding, leaves and food waste. Its profits from tipping fees and some compost sales support other Intervale Center initiatives. Moreover, the facility's product, ten thousand tons of compost annually, is used to restore the Intervale's sandy soils. According to Intervale Center founder Will Raap, "The compost project was the logical place to start. We were looking to renew land that was wonderful land but had been depleted over the last fifty years. Because public officials were looking for cost-effective recycling and waste management alternatives, it also provided a way to get agriculture onto the policy agenda of Burlington."

In 2006, there were seven other farms that produce mixed vegetables, small fruit and herbs, as well as farms that focus on bees, beans and grains, eggs, flowers and berries. One of the farms grows food for the Fletcher Allen Medical Center, Vermont's largest hospital. Motivated by the economics of waste disposal, the hospital first contracted with the Intervale Foundation to compost its food waste. That relationship paved the way for the hospital's food service to begin buying Intervale-grown produce, first on a purchase-by-purchase basis, and later through contracts.

The Intervale growers have formal monthly meetings, but also try to get together for lunch once a week to share experiences and provide support for each other. "It's really great to be a farmer in a community of farmers," says farm co-manager Andy Jones.

In 2006, ICF was operating on 35 acres and offering about 500 shares—in three different sizes with multiple payment plans—to Burlington-area residents. Shares include vegetables, herbs, flowers, and berries, with a weekly pickup at the farm, usually from late May through October. Nearly 100 percent of the crops the farm grows are for its CSA members.

"Conservatively you could say that two thousand people get some amount of their food from here," Jones says. In addition to the 500 growing season shares, the farm in 2005 began offering a winter share, with every-other-week pickups of storage vegetables, from November through mid-March. In addition, the farm offers an extended line of products from other local farms—goat cheese, bread, eggs, apples, and poultry—on a pre-pay basis.

"We're at the point now where we're big enough that we can provide some substantial markets for other local farms," Jones says, estimating those supplemental products represent as much as $10,000 worth of business for the other farms.

Farm management has evolved too. As a member- or consumer-owned cooperative, the farm has a seven-member board—in this case called the steering committee—elected from the membership. The board employs Jones as farm manager, and he, in turn coordinates the rest of the farm activities. This system is called *policy governance,* and it is commonly used in nonprofits as a way to help boards avoid both micro-management and detachment.

"It's very clear what my role is and what the steering committee's role is, so consequently we work together very effectively," Jones says. In the Intervale's case, the board has set limits on the scope of Jones and Maden's authority, but within that, the co-managers are responsible for all administrative and functional decisions on the farm. "It's separated farming functions from management functions," says Jones. "Steering committee members aren't doing the work of the farm except as it relates to linking with membership to find out what membership interests and concerns are."

Approximately 60 percent of ICF's members are Burlington

residents. The rest live within about 15 miles of the farm in the "bedroom" communities that surround Burlington. Someone from most member households works in Burlington.

Over the years, the CSA has always balanced improving wages and working conditions for its farmer employees with remaining accessible to the broader Burlington community. In 2006, the farm had a standing crew of about 10 people for the growing season, including Jones and Maden, who are year-round, salaried employees with benefits. The rest are seasonal, hourly employees, and the farm tries to hire people who express a serious interest in exploring a career in agriculture. In addition to the hands-on learning they obtain, the farm offers a weekly workshop series for its apprentices and workers.

Jones anticipates that a third year-round position is not too far off, in part due to improved production practices, and in part to the addition of the winter CSA. Basic production practices, particularly weed control, irrigation, and greenhouse management, have improved, as has yield per acre. "It's a function of having the right equipment," Jones says, "but also knowing how to use it and when to use it."

The farm's three share sizes are the result of "historical accident," and are something Jones says he would never recommend. Jones' focus is on providing a diverse, high-quality assortment of produce each week. But he also factors in efficiency of production in its selection. "We don't do a ton of the more specialty things, like fingerling potatoes, or purple tomatoes," Jones explains. "Since everything is in a CSA share, we don't have the cost recovery on the specialty varieties where the seed is more expensive or production is more time-consuming."

The farm has between 50 and 70 senior farm shares – a USDA program that the Northeast Organic Farming Association of Vermont organizes in Vermont (see Chapter 20, p. 231 for more details)— providing these lower-income seniors with fresh, high-quality food at no cost to them. Volunteers drop the food off at Burlington's public housing development for seniors.

The Intervale Community Farm also operates a supported share program, through which about 10 percent of the membership pays half the cost of the normal share price. ICF also accepts Food stamps and offers a "pay-as-you-go" program for these shares, so that these families can pay monthly instead of all up-front. The farm's retention rate—that is, the number of members who sign up again for another year—averages around 75 percent. And every year, there's a waiting list.

The farm has a small number of working members, whose field contributions most commonly involve harvest. Among them were Bonnie Acker and her daughter Dia Davis, who helped out as much as three days a week when Dia was in elementary school. Acker, a member of the board, summed up her family's participation in the farm this way:

"It's been the greatest adventure of our lives because it's all cooperative. Every day there's a crisis or problems that need to be dealt with, and there's always success. . . . And when we leave, we always have the sense that we've been truly helpful and truly appreciated. There aren't many places in this culture where people can participate so fully in a cooperative venture, especially children."

———————————•———————————

The Hampshire College CSA: Experiential Learning

by Nancy Hanson, farm manager

College based farms are nothing new. Land grant universities have long had research and teaching farms attached to their campuses to train new farmers and to conduct agricultural research. Starting in the early 1990s, however, a new type of college farm began to emerge on many campuses. These initiatives are often student driven and focused on sustainable agriculture. Many are built around CSAs. A recent poll by the online magazine *New Farm* identified over forty student farms from across the country. What is striking about the *New Farm* list is that many of these new college farms are based at liberal arts institutions. Driving this movement is a deep concern about our current food production system and a yearning for a new paradigm with a closer connection between producer and consumer. CSAs are uniquely suited to fulfill this need.

In existence since 1992, the Hampshire College CSA is one of the older and more established programs in the country. Hampshire College is a small, private liberal arts school situated in the fertile farming area of the Connecticut River Valley of western Massachusetts. What began as a student project with the simple goal of organic food production for thirty shareholders has grown into a fifteen-acre, two-hundred-share CSA with a professional full-time manager and a firm standing in the culture of college life. Not unlike nonacademic farms, the CSA plays a vital role in creating a sense of connectedness and community at Hampshire. Shareholders are drawn from a wide cross-section of the college community,

including student, staff, and faculty households. On pickup afternoons, it is not uncommon to see a student swapping kale recipes with a professor or picking beans with the children of a college administrator.

In addition to basic food production and community building, the college farm now reflects the larger institutional goals of a liberal arts education. The farm is integrated into academic life, first, as a "living laboratory." Science is perhaps the most daunting discipline for typical liberal arts students, and working and studying at the farm is one way to reach out to these students. As they harvest lettuce or enjoy eating fresh Sungold tomatoes right off the vine, opportunities arise to discuss a wide range of topics, from basic soil science to the unique ecology of an organic farm. As a college based farmer, I am often asked the question What does a liberal arts education have to do with sustainable agriculture? I like to respond with a quote from Daniel Webster: "Where tillage begins, the other arts follow. Therefore, farmers are the founders of civilization." There is hardly a liberal arts subject that cannot be seen through the lens of agriculture. Students are incredibly creative in weaving farm based learning into their formal education. When students leave the college, a small number do go on to become professional farmers. Most, however, carry this experience into other careers ranging from creative writing to public policy to entrepreneurial endeavors involving sustainable technology.

Operating a college CSA is largely similar to running any other CSA, but with a few significant differences. Staffing a farm and filling CSA shareholder ranks with college students is no easy matter. Approximately half of Hampshire shareholders are faculty and staff, and the other half are student households. College students are by definition a transient lot. By the time they have really learned the ropes, they graduate and move on. But that is the beauty of a college CSA. The hope is that these students will go out into the world and start or support other CSAs. As a liberal arts institution, our mission is not to train farmers. I like to think that we are training eaters. By exposing students to the difficulties and rewards of farming, the hope is that they take away an understanding of the importance of supporting sustainable farms when they make their food choices for the rest of their lives.

College CSAs are also faced with unique labor challenges. The growing season of a normal CSA does not correspond

Hampshire College students ride the harvest wagon. PHOTO BY NANCY HANSON.

with the school calendar. Much of the work of production is done while students are on summer vacation. While the Hampshire CSA does hire interns to work during the summer months, they do not start work until after classes end in late May. Starting a normal distribution season the first week of June would be nearly impossible under such conditions. Because of these constraints, the Hampshire CSA has gone to a fall-only distribution season. Regular pickups start the first week of September and continue until Thanksgiving. With some adjustments, the fall-only model has worked quite well. The shortened harvest season relieves some of the usual springtime pressure to get early crops planted and allows for a less harried summer work schedule. This has proved invaluable in maintaining the educational mission of the farm. Summer interns are given the time to learn new tasks without the pressures of summer harvesting. This schedule also allows for the flexibility to conduct on-farm research during the summer. In partnership with Hampshire faculty and University of Massachusetts Extension personnel, recent projects have included studies of potential controls for potato leafhoppers as well as organic trap cropping systems in winter squash.

During the fall harvest season, our goal is to involve as many students as possible on the farm. To that end, any student who shows up at 7:00 A.M. on harvest mornings is put to work. This means that students who have never worked on a farm before do the majority of harvesting. Because the learning curve can be steep, we have relied on returning students to help train new harvesters. Older students become mentors not only in the context of farmwork but

also in academic work and life skills as well. This builds community across the student body and also has the added benefit of fostering a sense of accomplishment and competency for these older students. On any given morning, amazing conversations about classes, living situations, and personal relationships occur across washtubs filled with cabbage or carrots.

The college CSA differs not only because we work with students but also because we operate the farm in the context of a larger institution, a situation that presents both benefits and challenges. Because education is our primary goal, our college farm is not profit driven. As the CSA manager, my salary is not exclusively dependent on the production of food. This allows me the ability to take the time to work with students in an academic context. In addition to collaborating on grants and research projects with both faculty and students, I often help to teach agriculturally related classes in the School of Natural Science. I currently co-teach a cross-disciplinary class called Agriculture, Food and Human Health. This class combines an introduction to agricultural studies, soil and plant science, epidemiology, and nutrition with hands-on experience harvesting vegetables at the farm and washing and packing twenty-five shares that are donated to local families enrolled in the federal Women, Infants, and Children (WIC) program. This partnership with the WIC program pushes students to look beyond personal health issues to the socioeconomic, financial, and political issues surrounding nutrition and agriculture.

While the CSA is an integral part of the academic program and the campus culture, we do have to constantly defend our place in the college budget. Because of our educational focus and because we operate with unskilled and highly inefficient labor, CSA income does not cover all of the expenses of the operation. Number crunchers in the college business office often do not understand the tight margins experienced by any farm, let alone one with missions beyond the bottom line. The effort to defend the farm budget, however, is made easier with the help of CSA members from across the college. Students and faculty alike feel a sense of ownership of and protectiveness toward the farm.

Hampshire College is only one example of a successful college based CSA. Rutgers University started the Cook College Student Farm in 1993. The Cook College farm is unique in that it does not have a hired manager. The farm is completely student run. They have ingeniously solved the problem of lack of student continuity by creating a guidebook written by and updated by current student farmers and passed down to incoming groups.

What does the future hold? At Hampshire there is a desire to expand the CSA season to match the full academic calendar. While we are not yet at that point, we have made significant steps in that direction. The summer of 2002 saw the construction of a new barn with a root cellar for the storage of fall-grown crops. In the summer of 2003, student interns were involved in the construction of a thirty-by-ninety-six-foot hoop house for the production of winter greens. The hope is to eventually offer a spring semester CSA share of stored crops and fresh greens.

One program that is offering a year-round CSA is based at Michigan State University. MSU now offers a formal program of study leading to an organic farming certificate. This program combines hands-on farmwork and classroom studies with the goal of preparing graduates for careers in organic farming and other sustainable agriculture fields. Students take classes such as Compost Production and Use, Organic Transplant Production, and Organic Produce Direct Marketing. Students also operate a four-season, forty-eight-week fifty-member CSA on ten acres of campus land. According to the MSU Web site, graduates can expect to "develop critical thinking skills around food systems and organic agriculture" and to "become an empowered change agent in the current food system and to have the necessary confidence to go out into the world and start their own farms, businesses and other projects"

With programs such as these springing up on many college campuses, the future of Community Supported Agriculture has never looked brighter.

CSA RESOURCES

The technical resources available to CSAs are increasing rapidly. Organizations around the country are taking on the task of maintaining updated lists of existing CSAs and distributing or circulating printed materials, slides, and videos. Rather than try to list all of the materials and risk being out of date before the ink dries, this section will list the resource centers to which people interested in Community Supported Agriculture can turn for the latest information. Please contact sources directly for current lists and prices of materials.

National Organizations

Alternative Farming Systems Information Center
National Agriculture Library Research Service
U.S. Department of Agriculture
10301 Baltimore Avenue, Room 304
Beltsville, MD 20705-2351
(301) 504-6559; fax: (301) 504-6409
e-mail: afsic@nal.usda.gov
Web site: www.nal.usda.gov/afsic/AFSIC_pubs/
Bibliographies on CSA and organic farming, lists of colleges with student farms and CSAs, and related topics.

Appropriate Technology Transfer for Rural Areas (ATTRA)
Box 3657
Fayetteville, AR 72702
(800) 346-9140
Web site: www.attra.org
Packet of basic information on CSA. Also provides packets on such topics as apprenticeships, organic farming, marketing. Staff will either send you a preassembled packet or design a packet of information to answer your specific questions. Also has tapes and videos. Great service!

The Biodynamic Farming and Gardening Association
25844 Butler Road
Junction City, OR 97448
(888) 516-7797; fax: (541) 998-0106
e-mail: biodynamic@aol.com
Web site: www.biodynamics.com
The Biodynamic Association maintains a database listing 600 CSAs throughout the country. Their list is only as good as the information we supply them, so if you are starting a CSA or stopping, please let them know. As of November 2006, many of their listings were out of date.

Publications:
It's Not Just about Vegetables, 25-minute VHS video introducing CSA, with Robyn Van En, 1988.
CSA 1, *Community Supported Agriculture*, audiotape by grower panel: Decater, Moore, Geiger, Altemueller.
CSA 2, *The Future of Community Supported Agriculture*, audiotape by Hartmut von Jeetze.

CSA 3, *Farms of the Future—Food of the Future*, audiotape by Fred Kirschenmann.

Farms of Tomorrow Revisited, by Trauger Groh and Steven McFadden, 1997.
Basic Formula to Create CSA, by Robyn Van En. 32 pp. Predecessor to this book.
Louise's Leaves, by Louise Frazier, 1994. Guide to year-round nutrition with seasonal foods and home storage.
"Introduction to Community Supported Farms/Gardens, Farm/Garden Supported Communities," 10 pp. Free booklet.
Stella Natura: Working with Cosmic Rhythms. Inspiration and Practical Advice for Home Gardeners and Professional Growers. Kimberton Hills Biodynamic Agricultural Planting Guide and Calendar.

E. F. Schumacher Society
140 Jug End Road
Box 76, RD 3
Great Barrington, MA 01230
(413) 528-1737
e-mail: efssociety@smallisbeautiful.org
Web site: www.smallisbeautiful.org
Susan Witt and the rest of the staff provide advice and information on all the ways small can be beautiful: land trusts, local currencies, sample leases, excellent decentrist library. Many useful booklets on land trusts and local economics, such as "A New Lease on Farmland," "Land: Challenge and Opportunity," "Local Currencies: Catalysts for Sustainable Local Economies."

Equity Trust
177 Avenue A
Box 746
Turners Falls, MA 01376
(413) 863-9038; fax: (413) 863-9082
Web site: www.equitytrust.org
Fund for CSA: financing through revolving loan fund for CSAs.

Ellie Kastanopolous—with years of experience in land tenure issues and land trusts—can provide information, a range of flexible solutions, and contacts with sources of funding or local land trusts.

CSA Land Tenure Initiative: land tenure counseling for CSAs, excellent source of information on land-related issues, model land lease, conservation easement and purchase option with guidelines to their use.

Localharvest.org

A Web based service for farm businesses, providing a way to advertise farm products and CSAs. For a fee, LocalHarvest will sign members up for a CSA via their Web site. Up-to-date listing of over a thousand CSAs by state with links to their Web sites.

New England Small Farm Institute (NESFI)
Box 937
Belchertown, MA 01007
(413) 323-4531
Web site: www.smallfarm.org
Courses, workshops, publications, resource referrals and individual consultation for prospective, beginning, and established small-scale farmers. Excellent library with over 5,000 books, periodicals, and videos.
Source of many useful publications:
DACUM Occupational Profile for On-Farm Mentor, 2001.
DACUM Occupational Profile for Northeast Small-Scale Sustainable Farmer, 2000.
On-Farm Mentor's Guide—practical approaches to teaching on the farm, 2005.
Cultivating a New Crop of Farmers: Is On-Farm Mentoring Right for You and Your Farm?
Exploring the Small Farm: Is Starting an Agricultural Business Right for You? A Decision-Making Workbook
Holding Ground: A Guide to Northeast Farmland Tenure and Stewardship, 2004.
Transferring the Farm Series: Access to Land, 2001.

Northeast Regional Agricultural Engineering Service (NRAES)
152 Riley-Robb Hall
Cooperative Extension
Ithaca, NY 14853-5701
Cooperative Extension publishers: excellent technical manuals on irrigation, produce handling for direct marketing, on-farm composting, refrigeration and controlled atmosphere for horticulture crops, greenhouse engineering, and more.

Robyn Van En Center for CSA Resources
The Center for Sustainable Living
Matt Steiman, Director
Wilson College
1015 Philadelphia Avenue
Chambersburg, PA 17201
(717) 264-4141, ext. 3247; fax: (717) 264-1578
e-mail: info@csacenter.org
Web site: www.wilson.edu/csacenter.org
The Center for Sustainable Living provides a home to the Robyn Van En Center for CSA Resources (RVEC), which

maintains a database of CSAs in North America. You can access this list through the Center's Web site to find the CSA nearest you. Like the Biodynamic CSA list, this list is only as good as the information we provide, so please keep them updated about your CSA.

RVEC can supply a CSA brochure, which serves as a general introduction to the concept; a CSA information packet; an 80-page "The Community Supported Agriculture Handbook: A Guide to Starting, Operating or Joining, a Successful CSA" by Randy Treicher, 1998; and a collection of clip art.

Publications; and other relevant books. "CSA: Making a Difference," 15-minute video. Interviews and footage from four CSA farms. Well-made video; provides good introduction to concept.

"CSA: Be Part of the Solution," Slide show in two versions: text slides only—you can add your own farm's slides as illustrations—or text and photo slides.

Rebirth of the Small Family Farm: A Handbook for the Starting of a Successful Organic Farm based on the Community Supported Agriculture Concept, by Bob and Bonnie Gregson.

Rocky Mountain Institute: A Think and Do Tank
1739 Snowmass Creek Road
Snowmass, CO 81654-9199
(970) 927-3851
Web site: www.rmi.org
Provides research, training, and resource materials on sustainable energy, water, agriculture, transportation, and community economic development.

Sustainable Agriculture Network (SAN)
USDA-CREES
Stop 2223
1400 Independence Avenue S.W.
Washington, DC 20250-2223
(202) 720-6071
Web site: www.sare.org
National outreach of the USDA Sustainable Agriculture Research and Education (SARE) Program. Provides publications on sustainable agriculture, information on research projects funded by SARE, videos, chat room, Web site.

Woods End Research Laboratory
Box 297
Mount Vernon, ME 04352
(207) 293-2457; fax: (207) 293-2488
e-mail: info@woodsend.org
Web site: www.woodsend.org
Compost testing and consulting; source of Community Supported Composting bags.

Regional Organizations

Agriculture and Land-Based Training Association (ALBA)
Box 6264
Salinas, CA 93912
(831) 758-1469
Web site: www.albafarmers.org
Training program in farm management for Spanish-speaking
  farmworkers.

California Certified Organic Farmers
1115 Mission Street
Santa Cruz, CA 95060
(831) 423-2263
e-mail: ccof@ccof.org
Web site: www.ccof.org
Organic farm directory; certification services.

Community Alliance with Family Farms
Box 363
Davis, CA 95617
(530) 756-8518
Web site: www.caff.org
Supports sustainable farming; directory of farms and CSAs.

CSA Learning Center at Angelic Organics
1547 Rockton Road
Caledonia, IL 61011
815-389-8455
Web site: www.AngelicOrganics.com
Distributes *Farmer John's Cookbook: The Real Dirt on
  Vegetables* by John Peterson and Angelic Organics (Gibbs
  Smith, 2006). Coordinates the Midwest CRAFT farmer
  training program; offers workshops in homesteading
  skills; supports urban gardens in Chicago area.

Future Harvest—CASA
Box 1544
Eldersbury, MD 21784
(410) 549-7878
e-mail: fhcasa@verizon.net
Web site: www.futureharvestcasa.org
Directory of CSAs in Maryland, Delaware, Virginia and
  West Virginia.

Great Lakes Area CSA Coalition (GLACSAC)
c/o Peter Seely
W7065 Silver Spring Lane
Plymouth, WI 53073
(414) 437-5971
Member organization providing networking for eastern
  Wisconsin and northeastern Illinois CSAs.

Iowa Network for Community Agriculture (INCA)
Steve Smith, Coordinator
2934 250 Street
Marshalltown, Iowa 50158
(641) 751-2851
e-mail: info@growinca.org

Web site: www.growinca.org
List of Iowa CSAs.

Iowa State University Extension Publications
119 Printing and Publications
Iowa State University
Ames, IA 50011-3171
(515) 294-7836; fax: (515) 294-2945
Web site: www.extension.iastate.edu
Statewide list of CSA farms and organizers; map of CSAs by
  county.
Publications: "Iowa CSA Resource Guide for Producers and
  Organizers," PM 1694—excellent resource booklet with
  application far beyond the borders of Iowa, with sections
  on legal issues, seed, supplies, livestock, flowers, and
  other products.
"CSA: Local Food Systems for Iowa," PM 1692

Just Food
Box 20444
Greeley Square Station
New York, NY 10001
(212) 645-9880
e-mail: info@justfood.org
Web site: www.justfood.org
CSA in New York; support project for CSAs in New York
  City area; helps connect farmers and potential members;
  guides formation of core groups for CSAs. Good mate-
  rials on core and farm CSA budgeting and responsibilities.
Publications:
CSA in New York Toolkit, comprehensive guide to setting
  up a CSA in NYC.
City Farms Toolkit, provides guidance to NYC community
  gardeners on how to grow, market, and distribute more
  food for their communities.
Veggie Tip Toolkit, provides easy-to-use information on
  more than sixty vegetables grown in the northeastern
  United States.

Land Stewardship Project
2200 4th Street
White Bear Lake, MN 55110
(612) 653-0618
e-mail: lspwbl@landstewardshipproject.org
Web site: www.landstewardshipproject.org
Newsletter; workshops; videos; on-farm training.

Madison Area Community Supported Agricultural Coalition
  (MACSAC)
Box 7814
Madison, WI 53707-7814
(608) 226-0300
Web site: www.macsac.org
Publishes annual CSA directory; provides support and net-
  working for CSA farms.
Distributes *From Asparagus to Zucchini: A Guide to Farm-
  Fresh, Seasonal Produce*, a big group effort, MACSAC,
  1996, 2003, 2004, Madison, WI.

Maine Organic Farming and Gardening Association
Box 170
Unity, ME 04988
(207) 568-4142
e-mail: mofga@mofga.org
Web site: www.mofga.org
Farmer training through on-farm internships, project to
develop more CSAs in the state; list of Maine CSAs. Great
Common Ground Fair and excellent newsletter.

Michael Fields Agricultural Institute
Box 990
East Troy, WI 53102
(262) 642-3303
Web site: www.michaelfieldsaginst.org
Offers training in biodynamic and organic farming, with the
opportunity for experience in running the Stella Gardens
CSA.

Midwest Organic and Sustainable Education Services
(MOSES)
Box 339
Spring Valley, WI 54767
(715) 772-3153
e-mail: info@mosesorganic.org
Web site: www.mosesorganic.org
Publishes Upper Mid-West Organic Resources Directory,
including CSAs; runs Upper Mid-West Organic Farming
Conference.

Northeast Organic Farming Association: Seven state chapters (Connecticut, Massachusetts, New Hampshire, New
Jersey, New York, Rhode Island, and Vermont)
Web site: www.nofa.org
Sells videotapes of conference workshops on a wide range of
topics related to organic farming, gardening, marketing,
and lifestyle; distributes *The Real Dirt: Farmers Tell
about Organic and Low-Input Practices in the Northeast*;
the five Northeast Farmer to Farmer Information
Exchange booklets on organic apples, sweet corn, straw-
berries, greenhouse, and livestock; and a series of man-
uals: Organic Weed Management; Organic Soil Fertility
Management; Vegetable Crop Health; Soil Resiliency and
Health; Compost, Vermicompost and Compost Tea; The
Organic Farmer's Guide to Marketing and Community
Relations; Organic Growers Guide: Humane and Healthy
Production of Eggs and Poultry; The Wisdom of Plant
Heritage: Organic Seed Production and Saving; Making
Milk and Dairy Products Organically.
Excellent summer conference.

Connecticut NOFA
Box 164
Stevenson, CT 06491
(203) 888-5146
e-mail: ctnofa@ctnofa.org
Web site: www.ctnofa.org
List of Connecticut CSAs.

NOFA-NY
Box 880
Cobleskill, NY 12043-0880
e-mail: Office@nofany.org
Web site: www.nofany.org
Web site includes list of CSAs in New York State, and the
Farmer's Pledge.
Distributes Tracy Frisch booklet "How to Keep Fresh Fruits
and Vegetables Longer with Less Spoilage: A Storage
Guide from Farm to Table" (NOFA-NY, 1986).

NOFA-VT
Box 697
Richmond, VT 05477
(802) 434-4122
e-mail: info@nofavt.org
Web site: www.nofavt.org
Support for low-income memberships in Vermont CSAs and
intern training.

Northeast Region Sustainable Agriculture Research and
Education Program (SARE)
10 Hills Building, 105 Carrigan Drive
University of Vermont
Burlington, VT 05405-0082
(802) 656-0471; fax: (802) 656-4656
e-mail: nesare@uvm.edu
Web site: www.uvm.edu/-nesare/
Northeast SARE offers two kinds of grants to producer-initi-
ated and -managed projects: Farmer/Grower Grants to
conduct farm based experiments to answer production and
marketing questions; and SEED (Special Evaluation,
Education and Demonstration) grants to farm-test selected,
alternative practices. Grants will be awarded on a competitive basis to farmers in the twelve-state region
(Connecticut, Delaware, Maine, Maryland, Massachusetts,
New Hampshire, New Jersey, New York, Pennsylvania,
Rhode Island, Vermont, Washington, D.C.). In the past,
grants have ranged from $300 to about $8,000. Grant
applications are due in early December; call, write, or visit
the Northeast Region SARE Web site for information.

Pennsylvania Association for Sustainable Agriculture
114 West Main Street
Millheim, PA 16856
(814) 349-9856
e-mail: info@pasafarming.org
Web site: www.pasafarming.org
Great winter conference, summer workshops, newsletter.

Southern Sustainable Agriculture Working Group
c/o The Amma Center
4203 Canal Street
New Orleans, LA 70119
e-mail: info@ssawg.org
Web site: www.ssawg.org
Sustainable farming in the southern states; quarterly
newsletter; list of CSAs.

Publication: "Making It on the Farm: Increasing Sustainability through Value-Added Processing and Marketing." 40 pp.

Periodicals and Publications

Biodynamics: A Bimonthly Magazine Centered on Health and Wholeness
Biodynamic Farming and Gardening Association
25844 Butler Road
Junction City, OR 97448
6 issues/year.

The Community Farm: A Voice for CSA
3480 Potter Rd.
Bear Lake, MI 49614
(616) 889-3216
e-mail: csafarm@jackpine.com
Web site: www.tcf.itgo.com
Quarterly newsletter.
Editors Jo Meller and Jim Sluyter also organize biannual CSA conferences in Michigan.

Dollars and Sense: The Magazine of Economic Justice
29 Winter Street
Boston, MA 02108
(617) 447-2177
e-mail: dollars@dollarsandsense.org
Web site: www.dollarsandsense.org
Bimonthly. Analyses of current economic goings-on in the United States and around the world in language normal people can understand.

A FoodBook for a Sustainable Harvest
by Elizabeth Henderson and David Stern, 1994
To order copies, write to Elizabeth Henderson at Peacework Organic Farm, 2218 Welcher Road, Newark, NY 14513.

GEO (Grassroot Economic Organizing Newsletter)
Box 115
Riverdale, MD 20738-0115
(800) 240-9721
e-mail: editors@geo.coop
Web site: www.geo.coop
Bimonthly. Covers worker-owned enterprises, co-ops, community-based businesses, community-labor-environmental coalitions.

Growing for Market: News and Ideas for Market Gardeners
Fairplain Publications
Box 3747
Lawrence, KS 66046
(785) 748-0605; (800) 307-8949
e-mail: growing4markct@earthlink.net
Web site: www.growingformarket.com
Monthly. Carries regular articles on CSAs and related small-farm topics. Also sells "Marketing Your Produce: Ideas for Small-Scale Farmers," collection of best marketing articles, 1992–1995.

The Natural Farmer
NOFA
411 Sheldon Road
Barre, MA 01005
(978) 355-2853
Quarterly, free with NOFA membership; otherwise $10/year.

The New Farm
Web site: www.newfarm.org
Online magazine of organic and sustainable agriculture published by the Rodale Institute.

Other Organizations

Accokeek Foundation
3400 Bryan Point Road
Accokeek, MD 20607
Web site: www.accokeek.org
Offers training program in CSA management.

Alan Savory Center for Holistic Management
1010 Tijeras NW
Albuquerque, NM 87102
(505) 842-5252
e-mail: savorycenter@holisticmanagement.org
Web site: www.holisticmanagement.org
Provides training and resource materials on Holistic Management. Distributes "Holistic Resource Management: A New Framework for Decision Making."

American Farmland Trust
1200 18th Street N.W. Suite 800
Washington, DC 20036
(202) 321-7300
e-mail: info@farmland.org
Web site: www.farmland.org
Leading organization for preservation of farmland. Information on land trusts, purchase of development rights, Farm Bill.

Center for Rural Affairs (CRA)
146 Main Street
Box 406
Lyons, NE 68038-0316
(402) 687-2100
e-mail: info@cra.org
Web site: www.cra.org
The CRA newsletter is the best source in this country for understanding what is going on in agriculture. The Center has projects in rural economic development, sustainable agriculture, and agricultural policy on local, state, and national levels.

Certified Naturally Grown
205 Huguenot Street
New Paltz, New York 12561
(877) 211-0308
e-mail: info@naturallygrown.org
Web site: www.naturallygrown.org
Alternative certification program for direct sales farms.

International Forum on Globalization
1009 General Kennedy Avenue #2
San Francisco, CA 94129
(415) 561-7650
e-mail: ifg@ifg.org
Web site: www.ifg.org
Newsletter, books, reports, tapes on corporate rule and how
to change it.

Organic Farming Research Foundation
Box 440
Santa Cruz, CA 95061-9984
(831) 426-6606; fax: (408) 426-6670
e-mail: research@ofrf.org
Web site: www.ofrf.org
OFRF has two funding cycles per year. Grant application
deadlines are July 15 and January 15. Projects may be
farmer initiated or should involve farmers in both design
and execution and take place on working organic farms
whenever possible. Modest proposals of $3,000 to $5,000
are encouraged. Contact OFRF office for procedures and
grant applications. A complete list of OFRF funded projects also is available on request.

Rural Advancement Foundation International-USA (RAFI-USA)
Box 640
Pittsboro, NC 27312
(919) 542-1396
Web site: www.rafiusa.org
RAFI's Just Foods program supports the integrity of the
USDA organic label and works to increase classic breeding
of seeds; the Agricultural Justice Project works on Domestic
Fair Trade. Posted on Web site: standards for social justice
on farms ("Toward Social Justice and Economic Equity in
the Food System: A Call for Social Stewardship Standards
in Sustainable and Organic Agriculture)."

Seed Savers Exchange
3094 North Winn Road
Decorah, IA 52101563-382-5990
Web site: www.seedsavers.org
Annual membership fee includes Yearbook through which
participating members can offer nonhybrid vegetable seeds
and order from other members. Other publications on seed
saving; also sell selected heirloom seeds in small quantities.

A Whole New Approach
Ed Martsolf
1039 Winrock Drive
Morrilton, AR 72110
Phone/fax: (501) 727-5659

e-mail: ed.martsolf@mev.net
Provides training in Whole Farm Planning.

International Organizations

Alliance paysans-ecologistes-consommateurs
Web site: www.Alliancepec.free.fr
French support network for Association pour le maintien
d'une agriculture paysanne (AMAP).
Web site lists AMAPs by region in France and provides
information in French on setting up an AMAP.

Équiterre
2177, rue Masson #317
Montreal, Quebec H2H 1B1
Canada
(514) 522-2000
e-mail: infoasc@equiterre.qc.ca
Web site: www.equiterre.org
Support network for *Agriculture soutenue par la commu-
naute*, the Quebec version of CSA.
Publications:
"*Je cultive, tu manges, nous partageons*," (no translation) by
Elizabeth Hunter, 2000.
A Guide for the Management of CSA Farms, by Frédéric
Paré, 2002.
"4 modèlés economiques viables et enviables d"ASC," 2005.

International Federation of Organic Agriculture Movements
(IFOAM)
Charles-de-Gaulle Str. 5
53113 Bonn
Germany
+49-228-92650-10
e-mail: headoffice@ifoam.org
Web site: www.ifoam.org
Guidelines for organic certification standards; Principles of
Organic Agriculture; Participatory Guarantee Systems;
conferences; trade shows.

The Soil Association
CSA Web site: www.cuco.org.uk.
British organic agriculture organization.
Publications:
"A Share in the Harvest: Feasibility Project for CSA," by
George Pilley, 2001.
"CSA Action Manual"
"How to Set up a Vegetable Box Scheme."

Urgenci
International Network URGENCI
Maison de la Vie Associative
13400 Aubagne, France
Web site: www.urgenci.net
The idea of Urgenci is to create a network on worldwide level
that takes into account all the initiatives aiming at citizen
involvement between inhabitants of cities and countryside.
International CSA network. Online newsletter; biannual
conferences.

ADDITIONAL READING

Ableman, Michael. *On Good Land: The Autobiography of an Urban Farm.* San Francisco: Chronicle Books, 1998.

Alaimo, Katherine. "Food Insufficiency Exists in the United States: Results from the Third National Health and Nutrition Examination Survey (NHANES III)." *American Journal of Public Health* 88, no. 3 (March 1998): 419–26.

"Alternative Certification: Workshop Reader." IFOAM, April 2004. Documents relating to Participatory Guarantee Systems can be found at: www.ifoam.org

American Farmland: The Magazine of American Farmland Trust (Summer 1998): 4 (Letters: editor's note).

Amin, Samir. "World Poverty, Pauperization and Capital Accumulation." *Monthly Review* (October 2003).

Ashman, Linda, et al., "Seeds of Change: Strategies for Food Security in the Inner City." Master's thesis, University of California, Los Angeles, 1993. Available from the Community Food Security Coalition, Box 209, Venice, CA 90294; (310) 822-5410.

Berry, Wendell. "Conserving Communities." In *The Case against the Global Economy and for a Turn toward the Local.* Edited by Jerry Mander and Edward Goldsmith. San Francisco: Sierra Books, 1996.

Bregendahl, Corry. "The Role of Collaborative CSA: Lessons from Iowa." Paper delivered at SARE National Conference, August 14–15, 2006. Available at: www.ncrcrd.iastate.edu/ projects/csa/index.html along with other papers by Bregendahl.

Brown, Lester R. *State of the World 1998.* New York: W.W. Norton, Worldwatch Institute, 1998.

"Cash Receipts and Farm Income." *New York Agricultural Statistics* (September 1995).

Chang, Susan, "Cultivating the Soil, Cultivating Youth." *The New Farm* (September 15, 2005).

Chaskey, Scott. *This Common Ground: Seasons on an Organic Farm.* New York: Viking, 2005.

Cohn, Gerry. "Community Supported Agriculture: Survey and Analysis of Consumer Motivations." Research paper, University of California at Davis, 1996. Available from Gerry Cohn, 127-B Hillside Street, Asheville, NC 28801.

Colborn, Theo, Dianne Dumanoski, and John Peterson Meyers. *Our Stolen Future: Are We Threatening Our Fertility, Intelligence, and Survival?* New York: Dutton, 1996.

The Community Land Trust Handbook, Institute for Community Economics. Emmaus, PA: Rodale, 1982.

Community Supported Agriculture: Making the Connection: A 1995 Handbook for Producers. University of California at Davis and Placer County Extension, 1995.

Daly, Herman E. "Sustainable Growth? No Thank You." In *The Case against the Global Economy.* Edited by Jerry Mander and Edward Goldsmith. San Francisco: Sierra Books, 1996.

Decater, Gloria. "Ripening into Wholeness: The Story of Live Power Community Farm." Printed in *Stella Natura: Working with Cosmic Rhythms.* Kimberton Hills Biodynamic Agricultural Planting Guide and Calendar, 2007, opposite July.

Diehl, Janet. *The Conservation Easement Handbook.* Boston: Land Trust Exchange, 1988.

DiGiacomo, Gigi, Robert King, and Dale Nordquist. *Building a Sustainable Business: A Guide to Developing a Business Plan for Farms and Rural Businesses.* SAN, Washington, D.C., 2003.

Douthwaite, Richard. *Short Circuit: Strengthening Local Economies for Security in an Unstable World.* Resurgence, Dublin, 1996.

Duesing, Bill. "Holy Day Connection." Broadcast on *Living on Earth,* National Public Radio, April 10, 1998.

Dunaway, Vicki. "The CSA Connection." *FARM: Food Alternatives with Relationship Marketing* (Summer/Fall 1996): 1–6.

E. F. Schumacher Society. *A New Lease on Farmland.* Great Barrington, MA: E. F. Schumacher Society, 1990.

Feder, Barnaby J. "Getting Biotechnology Set to Hatch." *New York Times,* May 2, 1998, D1, D15.

Forbes, Christin. "Strategies Employed by Community Supported Agriculture Farms for the Inclusion of Low-Income Consumers." Paper for course in Food and Nutrition, Montana State University, April 20, 2006.

Franzblau, Scott, and Jill Perry. *Local Harvest: A Multifarm CSA Handbook.* Concord, NH: Town and Country Reprographics. October, 2007.

Freegood, Julia. Presentation to Farmland Preservation workshop, SARE Tenth Anniversary, Austin, Texas, March 5–6, 1998.

Gaul, Gilbert M., Dan Morgan, and Sarah Cohen. "Farmer Insurance Firms Reap Billions in Profits." *Washington Post,* October 16, 2006.

Gerber, Michael E. *The E-Myth Revisited: Why Most Small Businesses Don't Work and What to Do about It.* New York: HarperCollins, 1995.

Gilman, Steve. "Our Stories." In *CSA Farm Network,* vol. II, pp. 81–83. Stillwater, NY: CSA Farm Network, 1997.

Goldberg, Ray. "New International Linkages Shaping the U.S. Food System." In *Food and Agricultural Markets: The Quiet Revolution.* Edited by Lyle P. Schertz and Lynn M. Daft, p. 165. NPA Report #270. Washington, DC: Economic Research Service, USDA, Food and Agriculture Committee and National Planning Association, 1994.

"Green Greens? The Truth about Organic Food." *Consumer Reports* (January 1998): 12–17.

Greenspan, Alan. "Greenspan: U.S Rich-Poor Gap May Threaten Stability." *USA Today,* June 13, 2005.

Groh, Trauger and Steven McFadden, *Farms of Tomorrow Revisited: Community Supported Farms—Farm Supported Communities.* Kimberton, PA: BD Association, 1997.

Growing a Healthy New York: Innovative Food Projects that End Hunger and Strengthen Communities. Albany, N.Y. and New York, N.Y., Hunger Action Network of New York State (HANNYS), 2004.

Grubinger, Vern. *Sustainable Vegetable Production from Start Up to Market.* Ithaca, NY: NRAES, 1998.

Guebert, Alan. "NAFTA Is Proving to Be a Disaster." *Agri News,* October 9, 1997.

Guthman, Julie. *Agrarian Dreams: The Paradox of Organic Farming in California.* Berkeley, CA: University of California Press, 2004.

Hamilton, Neil. *The Legal Guide for Direct Farm Marketing.* Des Moines, IA, Drake University, 1999.

Heffernan, William D. "Domination of World Agriculture by Transnational Corporations (TNCs)." In *For ALL Generations: Making World Agriculture More Sustainable.* Edited by J. Patrick Madden and Scott G. Chaplowe, pp. 173–81. Glendale, CA: WSAA, 1997.

Hendrickson, John. *Grower to Grower: Creating a Livelihood on a Fresh Market Vegetable Farm,* 2004, Center for Integrated Agricultural Systems, University of Wisconsin, Madison, WI, 2005.

Hightower, Jim. Keynote speech, SARE Tenth Anniversary, Austin, Texas, March 5, 1998.

Hightower, Jim. *There's Nothing in the Middle of the Road but Yellow Stripes and Dead Armadillos.* New York: Harper Collins, 1997.

Hoffman, Judith. "CSA: A Two-Part Discussion." *Small Farmer's Journal* 19, no. 4 (Fall): 28–29.

Howard, Phil. "Consolidation in Food and Agriculture." *The Natural Farmer* (Spring 2006).

Hunter, Elizabeth. "*Je cultive, tu manges, nous partageons.*" Montreal, Quebec, Équiterre, 2000.

Jones, Doug. *Internships in Sustainable Farming: A Handbook for Farmers.* NOFA-NY, 1999.

Kane, Deborah J., and Luanne Lohr. "Maximizing Shareholder Retention in Southeastern CSAs: A Step toward Long-term Stability." University of Georgia, 1997.

Kaneko, Yoshinori. *A Farm with a Future: Living with the Blessings of Sun and Soil.* Eng. trans. Saitama, Japan: Y. Kaneko & T. Kaneko, 1994.

Kelvin, Rochelle. *Community Supported Agriculture on the Urban Fringe: Case Study and Survey.* Kutztown, PA.: Rodale Institute Research Center, 1994.

Kittredge, Jack. "Community Supported Agriculture: Rediscovering Community." In *Rooted in the Land: Essays on Community and Place.* Edited by William Vitek and Wes Jackson. New Haven, CT: Yale University Press, 1996.

Kittredge, Jack. "CSAs in the Northeast: The Farmers Speak." *The Natural Farmer* (Summer 1993): 12.

Kolodinsky, Jane. "An Economic Analysis of Community Supported Agriculture Consumers." SARE grant #5-24544, 1996.

Kolodinsky, Jane, and Leslie Pelch. "Factors Influencing the Decision to Join a Community Supported Agriculture (CSA) Farm." *Journal of Sustainable Agriculture* 10, no. 2/3 (1997): 129–41.

Korten, David. *When Corporations Rule the World.* West Hartford, CT: Kumarian Press, 1995.

Laird, Timothy J. "Community Supported Agriculture: A Study of an Emerging Agricultural Alternative." Master's thesis, University of Vermont, 1995.

Lang, K. Brandon. "Expanding Our Understanding of CSA: An Examination of Member Satisfaction," *Journal of Sustainable Agriculture,* 26, no. 2 (2005): 61-80.

Lass, Daniel, and Jack Colley. "What's Your Share Worth?" In *CSA Farm Network,* vol. II, pp. 16–19. Stillwater, NY: CSA Farm Network, 1997.

Lass, Daniel, Ashley Bevis, G. W. Stevenson, John Hendrickson, and Kathy Ruhf. "CSA Entering the 21st Century: Results from the 2001 National Survey." University of Massachusetts

Lass, Daniel, Sumeet Rattan, and Njundu Sanneh, "The Economic Viability of CSA in the Northeast." Paper, University of Massachusetts, 1998.

Lass, Daniel, G. W. Stevenson, John Hendrickson, and Kathy Ruhf.et *CSA Across the Nation: Findings from the 1999 CSA Survey.* Center for Integrated Agricultural Systems (CIAS), University of Wisconsin–Madison, October 2003.

Lehman, Karen, and Al Krebs. "Control of the World's Food Supply." In *The Case against the Global Economy.* Edited by Jerry Mander and Edward Goldsmith. San Francisco: Sierra Books, 1996.

Lewis, W. J. et al. "A Total System Approach to Sustainable Pest Management." *Proceedings of the National Academy of Sciences* 94 (November 11, 1997): 12243–48.

Looker, Dan. *Farmers for the Future.* Ames: Iowa State University Press, 1996.

Lyson, Tom. "The House That Tobacco Built." *Food, Farm and Consumer Forum* 6 (January 1988): 2.

McFadden, Steven. "The History of CSA, Part I: Community Farms in the 21st Century: Poised for Another Wave of Growth? and "Part II: CSA's World of Possibilities." *New Farm* (January 2004).

MacGillis, Therese. "Food as Sacrament." In *Earth and Spirit: The Spiritual Dimension of the Environmental Crisis.* Edited by Fritz Hull. New York: Continuum, 1993.

Markley, Kristen. "Sustainable Agriculture and Hunger." Master's thesis, Pennsylvania State University, 1997.

Miller, Ethan. "Other Economies Are Possible," *Dollars and Sense* (July/August 2006): 11.

Mouillesseaux-Kunzman, Heidi. "Civic and Capitalist Food System Paradigms: A Framework for Understanding Community Supported Agriculture Impediments and Strategies for Success." Master's thesis, Cornell University, 2005.

O'Brien, Patrick. "Implications for Policy." In *Food and Agricultural Markets: The Quiet Revolution.* Edited by Lyle P. Schertz and Lynn M. Daft. NPA Report #270. Washington, DC: Economic Research Service, USDA, Food and Agriculture Committee and National Planning Association, 1994.

Osterholm, Michael. Interview on *Fresh Air*, National Public Radio, May 5, 1998.

Pennsylvania State Extension. "Guidelines for Renting Farm Real Estate in the Northeastern United States." Pesticide Action Network of North America. Sustainable Agriculture Network list-serve, PANNA Updates Service, panna@panna.org (May 1988).

Pilley, George. "A Share in the Harvest: A Feasibility Project for CSA." The Soil Association, 2001. Posted on www.cuco.org

"The Rich." *New York Times Magazine* (November 19, 1995): 1.

Ritchie, Mark. *The Loss of Our Family Farms*. Minneapolis: League of Rural Voters, 1979. Reprinted by Center for Rural Studies, San Francisco, 1979.

Ruhf, Kathy, Andrea Woloschuk, and Annette Higby. *Holding Ground: A Guide to Northeast Farmland Tenure and Stewardship*. Belchertown, MA, New England Small Farm Institute (NESFI), 2004.

Runyan, Jack L. "Summary of Federal Laws and Regulations Affecting Agricultural Employers, 2000." *Agricultural Handbook #719*. USDA Economic Research Service, 2000.

Scher, Les, and Carol Scher. *Finding and Buying Your Dream Home in the Country*. Chicago: Dearborn Financial Publishing, 1996.

Shute, Benjamin. *Model CSA Projects in New York State: Profiles Exploring How New York's Farmers Are Providing Low-Income Families with Healthy, Fresh and Nutritious Fruits and Vegetables*. Albany, N.Y. and New York, N.Y., HANNYS, 2004.

Smith, Miranda. *On-Farm Mentor's Guide—Practical Approaches to Teaching on the Farm*. Belchertown, MA: NESFI, 2005.

Smith, Stewart N. "Farming Activities and Family Farms: Getting the Concepts Right." Presentation at the Joint Economic Committee Symposium, Washington, DC, October 21, 1992.

Stanford, Lois. "The Challenges of Constructing 'Shared Community': Community Supported Agriculture Organizations in New Mexico." Paper presented at the session "Alternative Food Movements and Culture: New Directions in Anthropological Research" of the American Anthropological Association meetings, Chicago, November 19–23, 2003.

Strange, Marty. *Family Farming: A New Economic Vision*. Lincoln: University of Nebraska Press, 1988.

Strange, Marty. "Peace with the Land, Justice among Ourselves." *CRA Newsletter* (March 1997).

Suppan, Steve, and Karen Lehman. "Food Security and Agricultural Trade under NAFTA." Minneapolis: Institute for Agriculture and Trade Policy (IATP), July 1997. Available from the Institute for Agriculture and Trade Policy, 2105 First Ave. S., Minneapolis, MN 55404; (612) 870-0453.

Suput, Dorothy. "Community Supported Agriculture in Massachusetts: Status, Benefits, and Barriers." Master's thesis, Tufts University, 1992.

USDA. *Economic Research Statistical Bulletin #849*, December 1992.

USDA. *A Time to Act*. Washington, DC: USDA, January 1998.

USDA. *A Time to Choose: Summary Report on the Structure of Agriculture*. Washington, DC: USDA, 1981.

USDA. "Your Farm Lease Checklist." *USDA Farmers' Bulletin #2163*.

USDA Research Service. *Red Meat Yearbook*. Beltsville, MD: USDA Research Service, 1997.

Van En, Robyn. "Community Supported Agriculture (CSA) in Perspective." In *For ALL Generations: Making World Agriculture More Sustainable*, Edited by J. Patrick Madden and Scott G. Chaplowe. Glendale, CA: WSAA, 1997.

Vandertuin, Jan. "Vegetables for All." In *Basic Formula to Create Community Supported Agriculture*. Edited by Robyn Van En. Great Barrington, MA 1988.

Varley, Shannon, "Farming Felt Deeply, and without Limits." *The New Farm* (February 16, 2006).

Wilde, Parke, and Mark Nord. "Effect of Food Stamps on Food Security: A Panel Data Approach." *Review of Agricultural Economics*, 27, no. 3 (September 2005).

Winne, Mark. "The Food Gap." *Orion* (September–October 2005), 65–66.

Worden, Eva Cuadrado. "Community Supported Agriculture: Land Tenure, Social Context, Production Systems and Grower Perspectives." PhD diss., Yale University, 2000.

Zinn, Howard. *The Zinn Reader: Writings on Disobedience and Democracy*. New York: Seven Stories Press, 1997.

INDEX

the politics and practice of sustainable living

CHELSEA GREEN PUBLISHING

Chelsea Green Publishing sees books as a tool for effecting cultural change and seeks to empower citizens to participate in reclaiming our global commons and become its impassioned stewards. If you enjoyed *Sharing the Harvest,* please consider these other great books related to food, farming, and the environment.

Perennial Vegetables
ERIC TOENSMEIER
ISBN: 978-1-931498-40-1
$35 (PB)

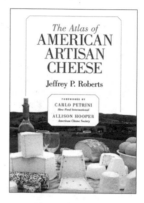

The Atlas of American Artisan Cheese
JEFFREY P. ROBERTS
ISBN: 978-1-933392-34-9
$35 (PB)

The Revolution Will Not Be Microwaved
SANDOR ELLIX KATZ
ISBN: 978-1-933392-11-0
$20 (PB)

Food Not Lawns
HEATHER C. FLORES
ISBN: 978-1-933392-07-3
$25 (PB)

the politics and practice of sustainable living

For more information or to request a catalog,
visit **www.chelseagreen.com** or
call toll-free **(800) 639-4099**.